In Search of the Causes of Evolution

In Search of the Causes of Evolution:

From Field Observations to Mechanisms

Edited by

Peter R. Grant & B. Rosemary Grant

PRINCETON UNIVERSITY PRESS

PRINCETON AND OXFORD

ISBN: 978-0-691-14681-2

ISBN (pbk.): 978-0-691-14695-9

Library of Congress Control Number: 2010931117

British Library Cataloging-in-Publication Data is available

This book has been composed in ITC Caslon 224
Printed on acid-free paper ∞
press.princeton.edu
Printed in the United States of America

1 3 5 7 9 10 8 6 4 2

Contents

INTRODUCTION TO SECTION III

INTRODUCTION TO SECTION IV

Contributors

Myra Awodey
  Research Assistant of Evolution and Ecology–University of Chicago
Christopher N. Balakrishnan
  Research Scientist of Genomic Biology–University of Illinois at Urbana-Champaign
Rowan D. H. Barrett
  Ph.D. Student of Zoology–University of British Columbia
May R. Berenbaum
  Professor of Entomology–University of Illinois at Urbana-Champaign
Paul M. Brakefield
  Professor of Evolutionary Biology–University of Leiden and Sheffield University
Philip J. Currie
  Professor of Biological Sciences–University of Alberta
Scott V. Edwards
  Professor of Organismic and Evolutionary Biology– Harvard University
Douglas J. Emlen
  Professor of Biological Sciences–University of Montana, Missoula
Joshua B. Gross
  Assistant Professor of Biological Sciences–University of Cincinnati
Hopi E. Hoekstra
  Harvard University
Richard Hudson
  Professor of Ecology and Evolution–University of Chicago
David Jablonski
  Professor of the Geophysical Sciences–University of Chicago
David T. Johnston
  Assistant Professor of Earth and Planetary Science– Harvard University

Mathieu Joron
    *Chargé de Recherche of Centre National de la Recherche
    Scientifique–Muséum National d'Histoire Naturelle, Paris*
David Kingsley
    Professor of Developmental Biology–Stanford University
Andrew H. Knoll
    *Professor of Natural History and Earth and Planetary Science–
    Harvard University*
Mimi A. R. Koehl
    *Professor of Ecological and Evolutionary Biomechanics–
    University of California, Berkeley*
June Y. Lee
    *Ph.D. Student of Organismic and Evolutionary Biology–
    Harvard University*
Jonathan B. Losos
    *Professor of Organismic and Evolutionary Biology–
    Harvard University*
Isabel Santos Magalhaes
    *Swiss National Science Foundation Fellow of Biological
    Sciences–University of Hull*
Albert B. Phillimore
    Junior Research Fellow–Imperial College London
Trevor Price
    Professor of Ecology and Evolution–University of Chicago
Dolph Schluter
    Professor of Zoology–University of British Columbia
Ole Seehausen
    *Professor of Fish Ecology and Evolution–Eawag: Swiss Federal
    Institute of Aquatic Science and Technology*
Clifford J. Tabin
    Professor of Genetics–Harvard University
John N. Thompson
    *Professor of Ecology and Evolutionary Biology–University
    of California, Santa Cruz*
David B. Wake
    *Professor of Integrative Biology–University of California,
    Berkeley*

Preface

This book is an outgrowth of a symposium held on September 5, 2008 at Princeton University to mark the occasion of our formal retirement from teaching. Dan Rubenstein, chairman of our Department of Ecology and Evolutionary Biology, generously gave us carte blanche to choose the theme and the contributors. The theme we chose is the title of this book. It exemplifies our research. The theme is not a discipline, such as ecology, and not a subject, such as insect migration, but a process of scientific discovery, a search for the causes of phenomena that cannot be fully understood when first observed. The symposium participants approach their research much as we do, and attempt to answer evolutionary questions with whatever tools are available.

Evolution touches all aspects of biology. We wanted to convey the excitement of current research and understanding across the breadth of the subject, without attempting to be comprehensive. This meant picking topics selectively, and ignoring others. The chapters are organized in four, interconnected, groups, with brief introductions to each of them. The first touches on some highlights of evolution in the past, from the origin of life to recent evolutionary history. Once the sole domain of paleontology, this field now encompasses biogeochemistry and molecular biology. The second group displays the power of molecular genetics to identify the mechanisms responsible for the development of structural diversity. The third group has a focus on behavior, and the fourth can be loosely described as ecological. We have been fortunate to add one extra chapter to each of the four sections of this book, written by people who would have been part of the symposium if more time had been available.

We thank many people for making this volume possible, starting with the contributors. David Wake introduced the symposium, and his remarks are reproduced in the final chapter. Reviewers have helped to improve the quality of the chapters. Alison Kalett of Princeton University Press gave us essential guidance and support, like a ship's navigator.

In Search of the Causes of Evolution

SECTION I

THE ORIGINS OF BIOLOGICAL DIVERSITY

Evolutionary biologists have much to explain. Our planet is occupied by a few million species of organisms. Impressive as the numbers are, they represent a few percent at most of all species that have ever lived. Where have the contemporary ones come from, why have all the rest gone, and why have the few, and no others, remained? How did a single species, *Homo sapiens*, come to dominate a planet that started out 4.567 billion years ago without any living organisms whatever? Answering questions such as these is part of the exciting task of accounting for the origin of life, the origin of species, their multiplication, diversification, and elaboration, and their extinction. This enormous challenge is being met by combinations of specialists in a broad range of disciplines, from microbiology and biochemistry to paleontology. What once looked like a linear array of experts at the forefront of science now appears to be a circle: molecular biologists and paleontologists have a common purpose, and together with ecologists, physiologists, geneticists, systematists, and others, jointly attempt to understand the origin and proliferation of life. This book is a reflection of that joint effort in the early part of the twenty-first century.

The first section is devoted to evolutionary history. It begins (chapter 1) with a broad view of the first 3 billion years of life, and presents the problem of trying to understand how the environment shaped the timing and pattern of evolution from the simplest organisms to the most complex. Years ago G. Evelyn Hutchinson wrote a book entitled *The Ecological Theatre and the Evolutionary Play*. The ecological theater changes, from scene to scene, and change largely explains both why evolution occurs and the directions it takes. In chapter 1, Andy Knoll and David Johnston extend the framework by making a distinction between an environmental (abiotic) and an ecological (biotic) context of evolution in order to focus on the physico-chemical component of the theater. They survey the biochemical evidence of conditions in the marine environment from about 2.5 to less than a billion years ago, and conclude that the deep history of life can only be understood by appreciating the dynamic nature of the physical environment. Organisms alter their environment, by releasing oxygen and depositing carbon for example; therefore, it is best to think of the whole drama of the history of life as being a tale of mutual interactions. These points are made in explaining why long periods of relatively little change such as the gradual rise in atmospheric oxygen are followed by relatively rapid transitions to new states. In more ways than one the chapter sets the stage for the rest of the book.

A time-honored method of identifying probable causes of evolution is to find trends and to search for features of the environment that are associated or correlated with them. In chapter 2, David Jablonski first gives guidance on how to interpret trends in body size and the accumulation of species in clades. This is an important overview because trends can arise in different ways. Working backwards in time to reconstruct a phylogeny, and then forward to narrate the history of a clade, he is able to identify where and how it evolved and diversified along two environmental axes: onshore-offshore gradients and tropical-temperate zone contrasts. With regard to the first, he finds that new taxa of marine invertebrates arise mostly in onshore habitats: clades that start in shallow water stay there or go deep, whereas those that start in deep water stay there. With regard to the second, he finds a strong tendency for clades to originate in the tropics and persist there for a long time. A net flow out of the tropics is caused not by species spreading from one climatic zone to another but by staying in one and giving rise to another species in a neighboring zone. This pattern raises questions about environmental determinants and the geography of speciation that are addressed later in the book (chapters 13 and 14). A major conclusion of this chapter is that the unfolding patterns of species diversity cannot be explained solely by microevolution. Clades have unequal histories and fates, and this leaves a macroevolutionary imprint on the product of microevolution.

Some faunas have had a disproportionate influence on our thinking about evolution, because they are so unusual. The Burgess Shale fauna from the Cambrian are a case in point, because they proliferated extravagantly and rapidly from poorly known, apparently impoverished, beginnings. Some taxonomic groups have the same iconic status because they too offer tantalizing paleontological puzzles. Everyone's favorite is the dinosaurs. This clade of theropod reptiles had an extraordinary history and fate. It captures our attention because of the striking diversity in form, the unequaled size of the largest members, and its rapid disappearance. These features demand an explanation. At one time the biggest question was why, having prospered, did they become extinct so rapidly? Answering this question, by invoking a major perturbation to the Earth's climate caused by a meteorite collision, removed a veil from another, perhaps more fundamental, question: who, exactly, are dinosaurs? In chapter 3, Phil Currie offers a narrative account of how modern phylogenetic and anatomical research has cast these animals in a fundamentally new light, and concludes they are still alive, represented solely by a group we call birds. Feathers, which once seemed to be unique to birds, are now known to have been possessed by their theropod ancestors. A little more than a decade of research on fossils from China has revealed numerous examples of feathers in various stages of evolutionary elaboration up to the form in *Archaeopteryx* that permitted flight. It will take time for everyone to feel comfortable with the idea of having a dinosaur in a bird cage in their living room.

Fossils are a luxury. Most evolutionary biologists don't have them, but increasingly they do have molecular tools for reaching deeply into the past to reconstruct the phylogenetic and demographic history of living organisms. Inferring phylogenetic history has become standard practice for many ecologists and behaviorists (chapters 12–16), so it is fitting to have a chapter that serves as an exemplar of what can be learned about the past without having a single fossil. It conveys the excitement of a field of inquiry undergoing rapid change as new and more powerful genomic and statistical tools become available. Scott Edwards's group (Balakrishnan et al.; see chapter 4) review the progress made on the question of how to use information on genetic variation (polymorphisms) to reach a reliable estimation of genetic relationships between populations, that is, to determine affinities in time and space. Using birds as an example, they consider why populations share polymorphisms for so long, the population-genetic context of their evolution, the way in which they can be used to construct species trees as opposed to gene trees, and the implications all of these have for efforts to identify and characterize species (see chapter 13). For example, they argue against the idea that nuclear genes are less useful for delimiting species than are mitochondrial genes. The authors conclude with expectations of future developments and recommendations for those who work with birds.

Chapter One

The Big Picture: A Tripartite View of Life and Environments through Time

Andrew H. Knoll and David T. Johnston

James Hutton, the late eighteenth-century father of Geology, recognized the goodness of fit between organisms and the environments they inhabit. To Hutton, this presented a conundrum, because inspection showed that the Earth is constantly changing. Trees and shrubs grow on mountainsides, but erosion strips sediments from highlands and carries them to the sea. Clams and seaweeds live in bays and estuaries, but the sediments eroded from mountains are continually delivered to the ocean, where they fill in those embayments. How can species be maintained in the face of such flux? Hutton's solution, presented in his seminal "Theory of the Earth" (Hutton 1788), was elegant in its simplicity: environmental constancy is maintained dynamically. Through time, the uplift of sedimentary basins approximately balances erosion, perpetuating the mountains and bays for which species are manifestly well designed. Half a century later, Charles Darwin (1859) stood this logic on its head, arguing that species are no more constant than the environments they inhabit. Darwin, of course, postulated that populations adapt via natural selection to the changing biological and physical circumstances of their environment.

Life and the physical Earth are inextricably linked at several levels. On the finest scales of space and time, evolutionary biologists can quantify the genetic and phenotypic responses of local populations to annual or decadal fluctuations in rainfall or temperature (e.g., Grant et al.

This chapter was written to celebrate Peter and Rosemary Grant. Research supported in part by NASA grant NNX07AV51G and the NASA Astrobiology Institute.

2004). At the other extreme, paleontologists can track the coevolution of life and environment through the geological record of our planet. That big picture—planetary in scale and billions of years long—is the subject of this chapter.

A Tripartite Division of Earth History

The idea that planetary surfaces not only change through time but change *directionally* owes much to the writings of an astronomer, Percival Lowell—who was describing Mars. Lowell (1908) famously, or infamously, claimed that he could discern linear features on the martian surface, and he interpreted these as canals built by technologically advanced martians to carry seasonal meltwater from the glaciated poles to parched populations at low latitudes. From this, Lowell concluded that Mars was once much wetter than it is today, a view widely accepted by present-day planetary scientists (e.g., Carr 2007), although decidedly not for the reasons advanced by Lowell. A century after its publication, however, Lowell's general point seems strikingly modern in outlook: the habitability of a planetary surface may change systematically through time. Today, we have the geochemical tools to reconstruct our own planet's past. And we have come to appreciate that Earth has witnessed both long-term state changes and transient perturbations in environments through a history more than 4 billion years long. This is the context in which biological evolution has proceeded.

In 2002, Ariel Anbar and one of us (AHK) published a diagram (fig. 1.1) in which we attempted to capture a then emerging view of Earth history. The left-hand panel, uncontroversial then and now, shows an Archean (> 2500 Ma) Earth with little or no free oxygen in the atmosphere or ocean. Sulfate concentrations were low, but ferrous iron was readily available as both nutrient and electron donor for photosynthesis. Molybdenum, a cation of central importance to the biological nitrogen cycle, was also in short supply, as were other trace elements that are less soluble in the reduced form.

The right-hand panel is equally uncontroversial. During the Phanerozoic Eon (literally, the age of visible animals—542 Ma to the present), oceans have generally been oxygenated from top to bottom. Anoxia has developed episodically in subsurface water masses of Phanerozoic

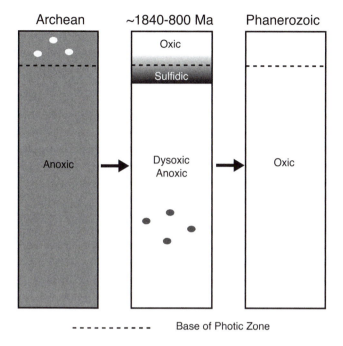

Figure 1.1. Hypothesized tripartite division of Earth's environmental history, modified from Anbar and Knoll (2002). Archean (and earliest Proterozoic) oceans contained at best low and transient amounts of oxygen gas (white ovals within photic zone). In contrast, Phanerozoic oceans have been oxic from top to bottom at most times and places. Between lies Earth's middle age, first proposed by Canfield (1998), in which the atmosphere and surface ocean were mildly oxic, while deep water masses remained dysoxic or anoxic. The figure shows sulfidic conditions in the oxygen minimum zone and, perhaps regionally or transiently, in deeper parts of Earth's middle age ocean (gray ovals).

oceans; however, in the face of higher oxygen levels, such occurrences have become increasingly uncommon through time. As a direct consequence of higher O_2 levels, iron concentrations also dropped, whereas the abundance of other essential nutrients and electrons acceptors, such as molybdenum, copper, and sulfate, increased. What was new and controversial in 2002 was the middle panel. Following the lead of Donald Canfield (1998), we depicted a long-lasting intermediate state of Earth's biosphere, with limited oxygen in the atmosphere and surface ocean, but anoxic and sulfidic waters at depth. Such an ocean was predicted to have low abundances of sulfate, molybdenum, *and* bioavailable iron (Anbar and Knoll 2002).

What evidence supported such a tripartite view of planetary history and how does the picture look now? First, as recognized decades ago (see Holland 2006), the widespread distribution of iron formation in Archean sedimentary basins records early oceans quite different from those of the present day. Transport of iron through the ocean requires oxygen-free water masses capable of carrying Fe(II) in solution. Thus, iron formation cannot form on continental shelves of well-oxygenated oceans like those that exist today—early oceans were different from those we know from direct experience. More recent evidence from oxygen-sensitive minerals such as siderite ($FeCO_3$) and uraninite (UO_2; Rasmussen and Buick 1999), the chemical profiles of ancient soil horizons (Rye and Holland 1998), and an unusual pattern of mass independence in sulfur isotopic fractionation (Farquhar and Wing 2003) make it clear that, at best, oxygen was a rare and transient feature of air and water in Earth's youth. In fact, all lines of geological and geochemical evidence suggest the initial rise of O_2 as a persistent component of the atmosphere and surface ocean approximately 2450–2320 Ma (called the Great Oxidation Event by Holland 2006). Anoxic, ferruginous water masses characterized much of the oceans' volume before this time.

For many years, majority opinion held that the Great Oxidation Event resulted in a vertical structure in ocean chemistry somewhat like today's, with iron in deep waters depleted by reaction with oxygen. In 1998, however, Donald Canfield proposed a dramatically different scenario. The earliest Proterozoic oxygenation, he hypothesized, generated only a modest amount of O_2, perhaps a few percent of modern levels. Although enough to oxygenate the atmosphere and surface ocean, the biologically derived flux of O_2 was too low to overwhelm the deep ocean reservoir of reduced species. In this circumstance, the intermediate state of Earth's oceans took form. Environments in Earth's middle age are closely linked to a second prominent product of Earth's initial oxygenation, the sulfate ion (SO_4^{2-}). Rather than oxygen, suggested Canfield, it was an increased sulfate flux to the ocean that was key to iron removal from the oceans.

It is not sulfate per se that removes ferrous iron from seawater, but rather sulfate's metabolic complement, sulfide. In Canfield's view, export of organic matter from the photic zone would have supported anaerobic respiration in anoxic water masses below the surficial mixed layer. Presuming small standing pools of electron acceptors like nitrate, microbial sulfate reduction would have provided the principal means of carbon remineralization. The metabolic by-product of sulfate reduction is H_2S, and it was this sulfide that sequestered Fe^{2+} through the formation of pyrite (Fe_2S) that accumulated in sediments. In consequence, by about 1800 Ma the Proterozoic Eon came to have sulfidic waters beneath the mixed layer (Poulton et al. 2004), introducing a long-lasting intermediate state of the biosphere distinct from both the early Earth in which life first took root and our familiar Phanerozoic world.

How do we test such hypotheses? The answer is to search in ancient sedimentary rocks for geochemical signatures that shed light on paleoenvironmental conditions. For example, iron chemistry in fine-grained sediments provides insights into the redox state of water masses directly above the accumulating sediments. Specifically, both the proportion of specific iron phases (oxides, carbonates, and sulfides) sequestered in sediments, as well as the proportion of those minerals present as pyrite and Fe-carbonate differ systematically between muds (or their lithified equivalents, shales) bathed by oxic waters and those deposited beneath deeper, anoxic waters (reviewed in Shen et al. 2002; Canfield et al. 2008).

Armed with this geochemical discriminator, we can analyze the iron in ancient shales. A 1900–1800 Ma succession in the Animike Basin, North America, records a change in iron chemistry from Fe-carbonate mineralogies in its lower parts to sulfide dominated Fe-mineralogies near the top, suggesting the establishment of sulfidic conditions in this marine basin (Poulton et al. 2004). Similar techniques show that in the ca. 1500–1400 Ma Roper Group, northern Australia, shales interpreted on sedimentological grounds as coastal deposits have iron chemistry like those of the shallow oxic Black Sea today, whereas shales deposited deeper within the Roper basin carry a chemical signature like that of the anoxic deep Black Sea (Shen et al. 2003). Indeed, in northern Australia, this chemical stratification between oxygenated shallow and anoxic deep environments characterizes a mid-Proterozoic record some 300 million years long (Shen et al. 2002, 2003).

Other indicators corroborate this view. For example, both the abundance and isotopic composition of molybdenum in marine sediments are sensitive to the distribution of sulfidic water masses throughout the subsurface oceans. Mo analyses of Proterozoic shales indicate that sulfidic water masses were much more common before the terminal Proterozoic Ediacaran Period than they have been since that time (Arnold et al. 2004; Scott et al. 2008). Further, fossil pigments synthesized by purple and green photosynthetic sulfur bacteria are preserved in basinal shales from a 1640 Ma succession in northern Australia (Brocks et al. 2005). Such biomarker molecules require that sulfide existed within the photic zone at the time the sediments were deposited. In summary, then, a number of independent lines of evidence confirm Canfield's (1998) hypothesis that during Earth's middle age, the oxygen minimum zone (OMZ) tended toward euxinia (anoxic and sulfidic conditions).

The state of deeper water masses is less clear. Canfield originally proposed that all water masses beneath the mixed layer should have been sulfidic. However, such a state is not easily reconciled with hypotheses of low PO_2 in the atmosphere and surface ocean, as the net supplies of oxidants and reductants have to balance. It is perhaps more likely that the deep oceans were dysoxic or anoxic, but not sulfidic, with euxinic water mixed in dynamically like the swirls of a marble cake. (See Slack et al. 2009, for chemical evidence of mildly oxic deep waters in a 1720 Ma ocean basin.) As we shall see, however, it was the state of the OMZ that most affected mid-Proterozoic life (Johnston, Wolfe-Simon et al. 2009).

BIOLOGY OF EARTH'S MIDDLE AGE

What was the nature of biological diversity in the three stages of Earth's redox history? Most paleontological research focuses on the Phanerozoic Eon, and it is clear that since the Cambrian diversification of marine animals and subsequent radiations of plants and animals onto land, Earth's biota has maintained the broad phylogenetic pattern observable today. We know relatively little about Archean life, but have confidence that the major anaerobic metabolisms which sustain biogeochemical cycles evolved during this interval (e.g., Knoll 2003). In the anoxic early oceans of the Archean, with abundant Fe, the iron and carbon cycles were probably much more closely linked than they are today (Konhauser et al. 2002; Fischer and Knoll 2009), with reduced Fe providing electrons from photosynthesis and oxidized Fe serving as a principal electron acceptor for respiration.

If the basic biology of bacterial and archaeal metabolism took shape in the first great chapter of Earth history, and the macroscopic world of complex multicellularity characterizes its third chapter, what was life like in chapter two, the more than1 billion year–long Proterozoic Eon?

We begin with more traditional paleontological evidence, as body fossils of Proterozoic microorganisms are, in fact, preserved in a number of marine settings. Perhaps our clearest micropaleontological window on Earth's middle age is provided by a unique taphonomic setting: silica concretions formed in carbonate sediments shortly after deposition. The paleoenvironmental distribution of these nodules reflects the processes that removed silica from Proterozoic oceans, and since silica skeletons did not exist for most of this interval, SiO_2 left the oceans largely as precipitates from mildly evaporative coastal water masses (Maliva et al. 1989). Because of this, chert nodules mostly illuminate tidal flat and associated coastal environments. Cyanobacteria are prominent in the microfossil record of these cherts—unsurprising in that, even today, cyanobacteria are important constituents of tidal flats from which animals and algae are environmentally excluded. Exceptionally well-preserved microfossils show that much of current cyanobacterial diversity existed by the Paleoproterozoic Era (figures 1.2a and 1.2b; Tomitani et al. 2006; Knoll 2007). This doesn't mean, however, that cyanobacterial diversification is completely frozen in the past. For example, molecular sequence comparisons suggest the relatively recent emergence of *Prochlorococcus*; the unicells that play an important role in modern mid-gyre photosynthesis (Rocap et al. 2003). Nonetheless, the fossil record makes it clear that cyanobacteria with essentially modern morphologies (including limited cell differentiation) and inferred physiological capabilities thrived in Proterozoic oceans.

Conspicuous sedimentary features known as stromatolites provide independent testimony to the ecological importance of cyanobacteria in middle-age oceans. Laminated sedimentary structures formed by the interaction of microbial mats and physical processes (Walter 1976; Grotzinger and Knoll 1999), stromatolites occur in Paleo- and Meso-proterozoic carbonates deposited from tidal flats to the base of the

11

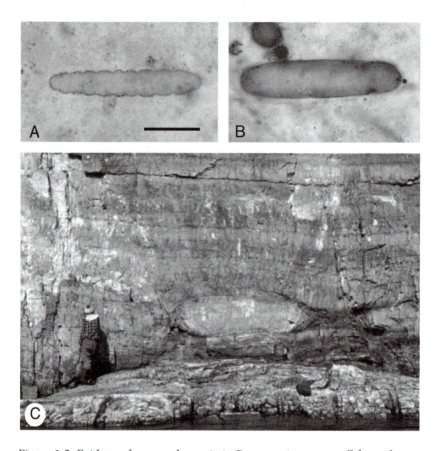

Figure 1.2. Evidence for cyanobacteria in Proterozoic oceans, all from the ca. 1500 Ma Bil'yakh Group, Siberia. (a) *Oscillatoria*-like fossil trichome. (b) Fossil akinete, a differentiated resting cell produced by nostocalean cyanobacteria. (c) Stromatolitic reefs—the m-scale structure in the lower center is a small patch reef; the geologist (2 m tall, with hat) is standing on a much larger microbial reef, and a still larger reef forms the remainder of the cliff face. Scale = 25 μm for both a and b.

photic zone (figure 1.2c). This environmental distribution would later diminish as seaweeds and sessile invertebrates evolved to compete for space on the seafloor and motile animals came to graze on benthic microbial communities.

Other microbial groups left their calling card in a different way, imparting chemical signatures to Proterozoic rocks. Biochemical fractionations of both C and S isotopes tell us that both the aerobic and anaerobic metabolisms that cycle these elements through modern oceans operated, as well, in Proterozoic seas. Molecular fossils, which are decay-resistant (mostly) lipids of known biological origin, also illuminate aspects of Proterozoic ecosystems (e.g., Knoll et al. 2007). Mid-Proterozoic shales contain molecular evidence for methanotrophic bacteria, expected in environments where sulfate levels are relatively low. Biomarker molecules also suggest that while eukaryotic organisms lived in mid-Proterozoic oceans, bacteria were the principal engines of primary production. Notably, as introduced in the preceding section, mid-Proterozoic shales from Australia preserve fossilized pigments attributable to *anoxygenic* photosynthetic bacteria (Brocks et al. 2005).

Much uncertainty attends the origins of the Eukarya, but the last common ancestor of living eukaryotes, a cell with mitochondria capable of aerobic respiration, likely postdates the Great Oxidation Event. Certainly, microfossils as old as ca. 1800 Ma can be interpreted as eukaryotic on the basis of both morphology and wall ultrastructure (fig. 1.3a; Javaux et al. 2001, 2004). A modest diversity of protistan microfossils occurs in shallow marine rocks deposited between then and 800 Ma; most cannot be assigned to specific eukaryotic clades, and they may well include stem- as well as early crown-group protists (Knoll et al. 2006). Nonetheless, beautifully preserved bangiophyte red algae in silicified tidal flat carbonates from Arctic Canada indicate that crown-group divergence, photosynthesis, and simple multicellularity had all come to characterize the domain by 1200 Ma (Butterfield 2000). As discussed more fully below, increasing eukaryotic diversity characterizes rocks ca. 800 Ma and younger.

LONG-LIVED ECOSYSTEM STATES: MAINTENANCE AND TRANSITION

Figure 1.4 summarizes current knowledge of the big picture: life and environments across our planet's full history. While Earth history can be divided into three long chapters, the nature and duration of transitions between successive biospheric states remains a subject of debate. For example, there is widespread agreement that oxygen first pervaded the atmosphere and surface ocean ca. 2450–2300 Ma and that oxygen minimum zones trended toward euxinia by about 1800 Ma, but little agreement exists on the course of environmental history between

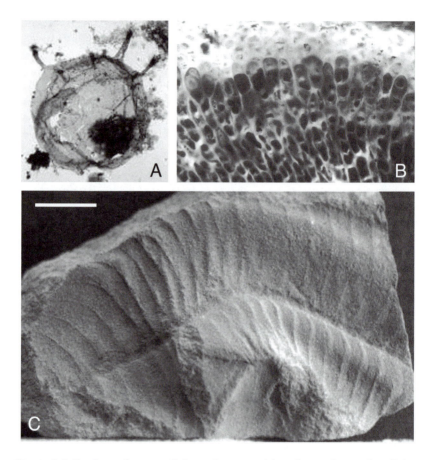

Figure 1.3. Evidence for unicellular eukaryotes (a) and complex multicellular organisms (b, c) in Proterozoic oceans. (a) Early protist from the 1500–1400 Ma Roper Group, Australia. (b) Detail of a cellularly preserved, three-dimensionally complex red alga from 580–560 Ma phosphorite of the Doushantuo Formation, China. (c) Ediacaran macrofossil in <555 Ma sandstones, Ust Pinega Formation, northern Russia. Scale bar = 100 µm for a, 40 µm for b, and 12.5 mm for c.

those dates. In particular, details of the transition from ferruginous to sulfidic oxygen minimum zones remain uncertain, and suffer from an especially incomplete rock record. In the Lake Superior region of North America, late-stage iron formations gave way to carbonaceous shales deposited beneath sulfidic waters by about 1840 Ma (Poulton et al. 2004), but whether this marks a global or regional transition, and

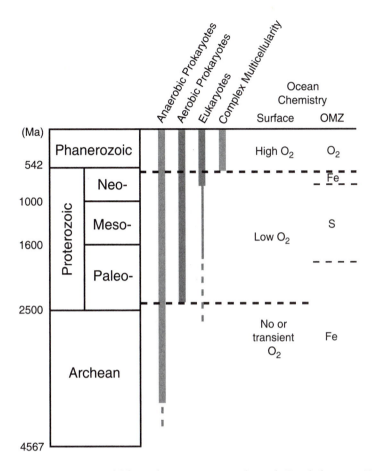

Figure 1.4. Summary of life and environments through Earth history. Early ecosystems predominantly featured anaerobic prokaryotes in anoxic oceans. Modestly oxic surface waters in Proterozoic oceans supported eukaryotic organisms, but sulfide in the oxygen minimum zone (OMZ) limited eukaryotic diversification; low availability of fixed nitrogen in oxic parts of the photic zone may also have favored prokaryotic primary producers capable of nitrogen fixation. Later Neoproterozoic sedimentary rocks show evidence for an erosion of sulfidic OMZs well before the widespread appearance of oxygen in these subsurface waters—this coincides in time with an expansion of eukaryotic diversity. Animals and complex multicellular algae enter the record as the atmosphere and oceans made the transition to a more modern state.

15

whether later Paleoproterozoic iron formations reflect the continuation or resurgence of ferruginous deep waters remains to be established.

The biogeochemical drivers of environmental transition from chapter one to two remain equally uncertain, in part because of continuing debate about the antiquity of oxygenic photosynthesis. Much opinion favors Archean origins for oxygen-producing cyanobacteria, requiring physical inhibition of oxygen accumulation until the beginning of the Proterozoic Eon (e.g., Hayes and Waldbauer 2006; Buick 2008). Kopp et al. (2005), however, have proposed that the evolution of oxygenic photosynthesis was, itself, the trigger for environmental oxygen increase. Trace metal and sulfur isotopic abundances favor at least low and transient oxygen levels in surface waters 50–180 million years before the Great Oxidation Event (Anbar et al. 2007; Kaufman et al. 2007; Wille et al. 2007), suggesting a minimum time difference between the origin of oxygenic photosynthesis and its environmental consequences. In any case, following the initial rise of oxygen, PO_2 appears to have remained low, and the suppression of ferruginous deep waters was protracted.

The late Proterozoic transition to more fully oxic oceans also appears to have been complex. Rapidly increasing geochemical data suggest that oxic conditions became widespread in the oxygen minimum zone 580–550 Ma (Fike et al. 2006; Canfield et al. 2007; McFadden et al. 2008; Scott et al. 2008). Recently, however, Canfield et al. (2008) reported an earlier loss of sulfidic OMZs, replaced not by oxic waters but by (one last time) ferruginous subsurface water masses. The proposed decoupling of sulfide loss and oxygen gain opens a new window into the interpretation of paleontological and biogeochemical records, as well as links to environmental coevolution.

Before evaluating the evolutionary consequences of the late Proterozoic environmental transition, we must ask how the intermediate state of Earth's biosphere could have been maintained for so long. One possibility is that primary producers populated a feedback system that kept PO_2 moderate and the oxygen minimum zone sulfidic. Johnston, Wolfe-Simon et al. (2009) have proposed that primary production by anoxygenic phototrophs, using sulfide generated in the oxygen minimum zone as an electron source, effectively decoupling carbon fixation from

16

oxygen generation. In consequence, fluxes of organic matter into the OMZ exceeded oxygen fluxes, although O_2 would remain the oxidant of choice in respiration. This imbalance perpetuated oxygen depletion and, in consequence, allowed bacterial sulfate reduction in the OMZ to regenerate the H_2S required to sustain anoxygenic photosynthesis.

Such biogeochemical feedbacks could, and apparently did, sustain Earth's middle age for more than a billion years. What, then, initiated Neoproterozoic transition to a more modern biosphere? In combination, erosion of the surface sulfur reservoir (Canfield 2004) and increased iron fluxes associated with supercontinental breakup apparently tipped the geochemical balance in subsurface water masses back to iron some 150 million years (or more; Johnston, Poulton et al. 2010) before latest Proterozoic oxygen increase (Canfield et al. 2007, 2008). As alternative electron donors declined, oxygenic photosynthesis came to dominate primary production throughout the photic zone, and for the first time in Earth history, organic carbon burial came to be balanced quantitatively by oxygen production. Thus, enhanced burial of organic matter associated with high latest Proterozoic rates of sediment accumulation resulted in higher PO_2.

The foregoing suggests that biological processes played a critical role in sustaining Earth's long middle age, but that the physical Earth facilitated environmental transition. Setting the fossil record within the framework of late Proterozoic environmental change, what might we conclude about the evolutionary consequences of these environmental transitions? First, sulfide is generally toxic to eukaryotes—among other effects, it interferes with cytochrome oxidase activity in mitochondria. For this reason, the widespread presence of sulfidic water masses just a few tens of meters below the sea surface should have dampened eukaryotic evolution in all but the shallowest reaches of the ocean (Martin et al. 2003). Existing paleontological and biomarker geochemical data do, in fact, suggest that Neoproterozoic collapse of OMZ euxinia coincided in time with a major increase in eukaryotic diversity in the oceans (fig. 1.4). Some 90 percent of the protistan species described from well-documented 800–713 Ma fossil assemblages have no representation in older rocks (e.g., Allison and Hilgert 1986; Butterfield et al. 1994; Butterfield 2005; Porter et al. 2003; Knoll et al. 2006).

Steranes, molecular fossils sourced mostly by eukaryotes, become abundant features of the organic geochemical record at this time, as well (Brocks 2006; Knoll et al. 2007), and stromatolites decline in abundance (Grotzinger and Knoll 1999). We further note that, besides sulfide loss, increased nitrate and Mo availability in the photic zone may have facilitated algal expansion (Anbar and Knoll 2002).

Of course, the defining biological feature of Earth's third biospheric state is complex multicellularity (figures 1.3b and 1.3c). Three-dimensional multicellularity in which only a subset of cells are in direct contact with the environment must have been limited to small sizes by PO_2 in organisms dependent on diffusion to oxygenate interior cells (Runnegar 1991). With rising oxygen levels, organismal size would increase, leading to increasingly strong oxygen, nutrient, and molecular signal gradients within tissues. In a functional sense, complex multicellularity can be understood as the circumvention of diffusion by active transport of nutrients and signaling molecules (Knoll and Hewitt 2010), and this is the breakthrough that permitted the radiations of bilaterian animals; complex red, green, and brown algae; and fungi. Consistent with this scenario, biomarkers and microfossils record nascent animals in rocks 632–650 Ma, or older (Yin et al. 2007; Love et al. 2009), but macroscopic metazoans and motile bilaterian animals with high rates of exercise metabolism enter the record only at the beginning and culmination of the Ediacaran oxygen transition, respectively. Equally, whereas simple multicellular red algae evolved 1200 Ma, three-dimensionally complex florideophyte reds first occur in 580–560 Ma rocks from China (fig. 1.3b; Xiao et al. 2004). And, while simple coenocytic and multicellular green and stramenopile algae occur in earlier deposits, macroscopic seaweeds likely to record complex multicellularity in these groups appear only near the Proterozoic-Cambrian boundary (Xiao et al. 2002).

Nothing in the foregoing downplays the importance of organism-organism interactions in large-scale evolutionary pattern. Indeed, Butterfield (2007) has argued compellingly that the introduction of animals permanently changed ecosystem structure and the nature of selective pressures in the oceans. To paraphrase Hutchinson, however, the evolutionary play takes place in an environmental as well as an ecological theater, and one with continually shifting sets and props. Especially when considering the striking temporal coincidence, it is difficult to account for the deep history of life without placing it firmly in the context of Earth's environmental evolution.

SUMMARY: EVOLUTION ON A DYNAMIC PLANET

Paleontology's chief contribution to the Neodarwinian Synthesis, George Gaylord Simpson's (1944) *Tempo and Mode in Evolution*, hardly mentioned environmental history and devoted less than a page to mass extinction; Simpson's objective was to show how evolutionary processes illuminated by population biologists could account for evolutionary pattern in the geologic record. By the 1970s, however, it had become clear that paleontological pattern is not simply the product of population genetics played out on million-year time scales, an intellectual shift most famously summarized by Stanley's (1975) epigram: "macroevolution is decoupled from microevolution." Today, this "decoupling" is discussed in terms of evolutionary hierarchies and, key in our estimation, Earth's unpredictable environmental history. Indeed, to a first approximation, macroevolutionary pattern in the fossil record may largely reflect microevolutionary processes played out on an environmentally dynamic planet.

This fundamental dynamic, in which populations continually track moving environmental targets, provides a required framework for assessing the evolutionary consequences of current global change. It helps us to understand the rich evolutionary history recorded in Phanerozoic rocks, with plankton to pachyderms responding to shifting continents, oscillating climate, large but transient environmental perturbation and, of course, changes in biological components of the environment perceived by organisms. On the time scale of Phanerozoic history, such environmental interactions are as important as population genetics to an understanding of how the biological diversity of our present moment came to be.

And, in the twenty-first century, the tools are finally in place to understand the coevolution of Earth and life across the entirety of Earth history, documenting environmental events that Hutton could scarcely have imagined, a pre-Cambrian evolutionary history that Darwin predicted but never expected to establish, and a long-term directionality to Earth's planetary history that Lowell would have appreciated. Indeed, in the new millennium, outcrop-scale investigation of environmental history has been extended to Mars, confirming that directional change is the rule there, too (e.g., Squyres and Knoll 2005). The big picture underscores the view that life has truly been a planetary phenomenon, shaping and being shaped by Earth's dynamic surface, from the moment it began.

REFERENCES

Allison, C. W., and J. W. Hilgert. 1986. Scale microfossils from the early Cambrian of northwest Canada. *J. Paleontol.* 60: 973–1015.

Anbar, A. D., and A. H. Knoll. 2002. Proterozoic ocean chemistry and evolution: A bioinorganic bridge? *Science* 297: 1137–1142.

Anbar, A. D., et al. 2007. A whiff of oxygen before the Great Oxidation Event? *Science* 317: 1903–1906.

Arnold, G. L., A. D. Anbar, J. Barling, and Lyons, T.W. 2004. Molybdenum isotope evidence for widespread anoxia in mid-Proterozoic oceans. *Science* 304: 87–90.

Brocks, J. J. 2006. Proterozoic ocean chemistry and the evolution of complex life. *Geochim. Cosmochim. Acta* 70: A68.

Brocks, J. J., G. D. Love, R. R. Summons, A. H. Knoll, G. A. Logan, and S. Bowden. 2005. Biomarker evidence for green and purple sulfur bacteria in an intensely stratified Paleoproterozoic ocean. *Nature* 437: 866–870.

Buick, R. 2008. When did oxygenic photosynthesis evolve? *Phil. Trans. Royal Soc. B* 363: 2731–2743.

Butterfield, N. J. 2000. *Bangiomorpha pubescens* n. gen., n. sp.: implications for the evolution of sex, multicellularity, and the Mesoproterozoic/Neoproterozoic radiation of eukaryotes. *Paleobiology* 26: 386–404.

———. 2005. Reconstructing a complex early Neoproterozoic eukaryote, Wynniatt Formation, arctic Canada. *Lethaia* 38: 155–169.

———. 2007. Macroevolution and macroecology through deep time. *Palaeontology* 50: 41–55.

Butterfield, N. J., A. H. Knoll, and K. Swett. 1994. Paleobiology of the Upper Proterozoic Svanbergfjellet Formation, Spitsbergen. *Fossils and Strata* 34: 1–84.

Canfield, D. E. 1998. A new model for Proterozoic ocean chemistry. *Nature* 396: 45453.

———. 2004. The evolution of the Earth surface sulfur reservoir. *Am. J. Sci.* 304: 839–861.

Canfield D. E., S. W. Poulton, A. H. Knoll, G. M. Narbonne, G. Ross, T. Goldberg, and H. Strauss. 2008. Ferruginous conditions dominated later Neoproterozoic deep-water chemistry. *Science* 321: 949–952.

Canfield, D. E., S. W. Poulton, and G. M. Narbonne. 2007. Late-Neoproterozoic deep-ocean oxygenation and the rise of animal life. *Science* 315: 92–95.

Carr, M. H. 2007. *The Surface of Mars.* Cambridge: Cambridge University Press.

Darwin, C. 1859. *On the Origin of Species by Means of Natural Selection.* London: J. Murray.

Farquhar, J., and B. A. Wing. 2003. Multiple sulfur isotopes and the evolution of the atmosphere. *Earth Planet. Sci. Lett.* 213: 1–13.

Fike, D. A., J. P. Grotzinger, L. M. Pratt, and R. E. Summon. 2006. Oxidation of the Ediacaran ocean. *Nature* 444: 744–747.

Fischer, W. W., and A. H. Knoll. 2009. An iron-shuttle for deep water silica in Late Archean and Early Paleoproterozoic iron formation. *Geol. Soc. Am. Bull.* 121: 222–235.

Grant, P. R., B. R. Grant, J. A. Markert, L. F. Keller, and K. Petren. 2004. Convergent evolution of Darwin's finches caused by introgressive hybridization and selection. *Evolution* 58: 1588–1599.

Grotzinger, J. P., and A. H. Knoll. 1999. Proterozoic stromatolites: evolutionary mileposts or environmental dipsticks? *Ann. Rev. Earth Planet. Sci.* 27: 313–358.

Hayes, J. M., and J. R. Waldbauer. 2006. The carbon cycle and associated redox processes through time. *Phil. Trans. Royal Soc. B* 361: 931–950.

Holland, H. D. 2006. The oxygenation of the atmosphere and oceans. *Phil. Trans. Royal Soc. B* 361: 903–915.

Hutton, J. 1788. Theory of the Earth; or an investigation of the laws observable in the composition, dissolution, and restoration of land upon the globe. *Trans. Royal Soc. Edinburgh* 1(II): 209–304.

Javaux, E., A. H. Knoll, and M. R. Walter. 2001. Ecological and morphological complexity in early eukaryotic ecosystems. *Nature* 412: 66–69.

———. 2004. TEM evidence for eukaryotic diversity in mid-Proterozoic oceans. *Geobiology* 2: 121–132.

Johnston, D. T., S. W. Poulton, C. Dehler, S. Porter, J. Husson, D. E. Canfield, and A. H. Knoll 2010. An emerging picture of Neoproterozoic ocean chemistry: Insights from the Chuar Group, Arizona. *Earth and Planetary Science Letters*, 290: 6473.

Johnston, D. T., F. Wolfe-Simon, A. Pearson, and A. H. Knoll. 2009. Anoxygenic photosynthesis modulated Proterozoic oxygen and sustained Earth's middle age. *Proc. Natl. Acad. Sci., USA* 106:6925–16929.

Kaufman, A. J. et al. 2007. Late Archean biospheric oxygenation and atmospheric evolution. *Science* 317: 1900–1903.

Knoll, A. H. 2003. *Life on a Young Planet: The First Three Billion Years of Evolution on Earth*. Princeton, NJ: Princeton University Press.

———. 2007. Cyanobacteria and Earth history. In A. Herrero and E. Flores, eds., *The Cyanobacteria: Molecular Biology, Genomics and* Evolution, 1–19. Heatherset, UK: Horizon Scientific Press.

Knoll, A. H., and D. Hewitt. 2010. Complex multicellularity: phylogenetic, functional and geological perspectives. In K. Sterelny and B. Calcott, eds., *The Major Transitions Revisited*. Vienna Series in Theoretical Biology. Cambridge, MA: MIT Press.

Knoll, A. H., E. J. Javaux, D. Hewitt, and P. Cohen. 2006. Eukaryotic organisms in Proterozoic oceans. *Phil. Trans. Royal Soc. B* 361: 1023–1038.

Knoll, A. H., R. E. Summons, J. Waldbauer, and J. Zumberge. 2007. The geological succession of primary producers in the oceans. In P. Falkowski and A.H. Knoll, eds., *The Evolution of Primary Producers in the Sea*, 133–163. Burlington, MA: Elsevier.

Konhauser, K. O., Hamade, T., Raiswell, R., Morris, R.C., Ferris, F.G., Southam, G., and Canfield, D.E. 2002. Could bacteria have formed the Precambrian banded iron formations? *Geology* 30: 1079–1082.

Kopp, R. E., J. L. Kirschvink, I. A. Hilburn, and C. Z. Nash. 2005. The Paleoproterozoic snowball Earth: A climate disaster triggered by the evolution of oxygenic photosynthesis. *Proc. Natl. Acad. Sci., USA* 102: 11131–11136.

Love, G. D. et al. 2009. Fossil steroids record the appearance of Demospongiae during the Cryogenian period. *Nature* 457: 718–721.

Lowell, P. 1908. *Mars as the Abode of Life*. New York: Macmillan.

Maliva, R., A. H. Knoll, and R. Siever. 1989. Secular change in chert distribution: a reflection of evolving biological participation in the silica cycle. *Palaios* 4: 519–532.

Martin, W. et al. 2003. Early cell evolution, eukaryotes, anoxia, sulfide, oxygen, fungi first (?), and a tree of genomes revisited. *IUBMB Life* 55: 193–204.

McFadden K. A., J. Huang, X. Chu, G. Jiang, A. J. Kaufman, C. Zhou, X. L. Yuan, and S. Xia. 2008. Pulsed oxidation and biological evolution in the Ediacaran Doushantuo Formation. *Proc. Natl. Acad. Sci, USA* 105: 3197–3202.

Porter, S. M., R. Meisterfeld, and A. H. Knoll. 2003. Vase-shaped microfossils from the Neoproterozoic Chuar Group, Grand Canyon: a classification guided by modern testate amoebae. *J. Paleontol.* 77: 205–225.

Poulton, S. W., P. W. Fralick, and D. E. Canfield. 2004. The transition to a sulphidic ocean similar to 1.84 billion years ago. *Nature* 431: 173–177.

Rasmussen, B., and R. Buick. 1999. Redox state of the Archean atmosphere: Evidence from detrital heavy minerals in ca. 3250–2750 Ma sandstones from the Pilbara Craton, Australia. *Geology* 27: 115–118.

Rocap, G., et al. 2003. Genome divergence in two *Prochlorococcus* ecotypes reflects oceanic niche differentiation. *Nature* 424: 1042–1047.

Runnegar, B. 1991. Precambrian oxygen levels estimated from the biochemistry and physiology of early eukaryotes. *Palaeogeogr. Palaeoclimatol. Palaeoecol.* 97: 97–111.

Rye, R., and H. D. Holland. 1998. Paleosols and the evolution of atmospheric oxygen: A critical review. *Am. J. Sci.* 98: 621–672.

Scott, C., T. W. Lyons, A. Bekker, Y. Shen, S. W. Poulton, X. Chu, and A. D. Anbar. 2008. Tracing the stepwise oxygenation of the Proterozoic ocean. *Nature* 452: 456–458.

Shen, Y. N., D. E. Canfield, and A. H. Knoll. 2002. Middle Proterozoic ocean chemistry: Evidence from the McArthur Basin, northern Australia. *Am. J. Sci.* 302: 81–109.

Shen, Y., A. H. Knoll, and M. R. Walter. 2003. Evidence for low sulphate and anoxia in a mid-Proterozoic marine basin. *Nature* 423: 632–635.

Simpson, G. G. 1944. *Tempo and Mode in Evolution*. New York: Columbia University Press.

Slack, J. F., T. Grenne, and A. Bekker. 2009. Seafloor-hydrothermal Si-Fe-Mn exhalites in the Pecos greenstone belt, New Mexico, and the redox state of ca. 1720 Ma deep seawater. *Geosphere* 5: 302–314.

Squyres, S., and A. H. Knoll. 2005. Outcrop geology at Meridiani Planum: Introduction. *Earth Planet. Sci. Lett.* 240: 1–10.

Stanley, S. M. 1975. A theory of evolution above the species level. *Proc. Natl. Acad. Sci., USA* 72: 646–650.

Tomitani, A. H. Knoll, C. M. Cavanaugh, and T. Ohno. 2006. The evolutionary diversification of cyanobacteria: molecular phylogenetic and paleontological perspectives. *Proc. Natl. Acad. Sci., USA* 103: 5442–5447.

Walter, M. R., ed. 1976. *Stromatolites*. Amsterdam, Netherlands: Elsevier.

Wille, M., J. D. Kramers, T. F. Nagler, N. J. Beukes, S. Schröder, T. Meisel, J. P. Lacassie, and A. R. Vögelin. 2007. Evidence for a gradual rise of oxygen between 2.6 and 2.5 Ga from Mo isotopes and Re-PGE signatures in shales. *Geochim. Cosmochim. Acta* 71: 2417–2435.

Xiao, S., A. H. Knoll, X. Yuan, and C. Pueschel. 2004. Phosphatized multicellular algae in the Neoproterozoic Doushantuo Formation, China, and the early evolution of florideophyte red algae. *Am. J. Bot.* 91: 214–227.

Xiao, S., X. Yuan, M. Steiner, and A. H. Knoll. 2002. Carbonaceous macrofossils in a terminal Proterozoic shale: a systematic reassessment of the Miaohe biota, South China. *J. Paleontol.* 76: 347–376.

Yin, L., M. Zhu, A. H. Knoll, X. Yuan, J. Zhang, and J. Hu. 2007. Doushantuo embryos preserved within diapause egg cysts. *Nature* 446: 661–663.

Chapter Two

Macroevolutionary Trends in Time and Space

David Jablonski

Among the most compelling large-scale evolutionary patterns are long-term trends in phenotypic traits, such as increases in suture complexity of Paleozoic ammonoids, in body size of Cenozoic horses, or in armor of post-Paleozoic molluscan shells. Such striking patterns can arise through a variety of dynamics, for example via active, protracted directional shifts or via passive diffusion away from a minimum value (e.g., Stanley 1973; McShea 1994; Gould 2002). The unfolding of large-scale trends has increasingly been appreciated as a multi-level process, with sorting simultaneously operating at the level of organisms, species, and perhaps even clades to shape the distribution of phenotypes over time (e.g., Jablonski 2008a, 2008b).

In this chapter I underscore the role that speciation and extinction of species and clades play in large-scale trends. Selection on organisms in population is important, of course, but an exclusive focus at the organismal level is clearly incomplete. I will show how a general conceptual framework that emphasizes the envelope of variation occupied by a clade over time can help both to clarify and to unify a wide variety of large-scale trending patterns, from fairly conventional phenotypic variables like body size, to environmental distributions as in the spread of clades from shallow to deep water, to latitudinal distributions as in the spread of clades from the tropics to the poles. All of these dynamics

I thank Rosemary and Peter Grant for inspiration, and the invitation to participate in this symposium; Peter R. Grant, Gene Hunt, and Susan M. Kidwell for valuable discussions and reviews; my collaborators James W. Valentine, Kaustuv Roy, and Andrew Z. Krug for many interactions, and NSF, NASA, the John Simon Guggenheim Foundation for support.

involve a strong, albeit not universal, tendency for large-scale pattern to be shaped by the diffusion of a clade along one or more axes—size, or depth or latitude, for example—and generally with only a minor contribution from sustained, directional transformation within their constituent units such as species.

MULTI-LEVEL APPROACHES TO EVOLUTIONARY TRENDS

Multi-level processes are readily detected in large-scale trends because individual species tend to be relatively static, or at least to show non-directional change, over geologic timescales. Even evolution within a strictly gradualistic system is subject to multi-level evolution, although partitioning effects among levels is operationally more difficult (Slatkin 1981; Jablonski 2008a). The fossil record shows every possible combination of phenotypic tempo and mode at the species level, but a recent analysis shows that sustained directional change is rare (~5%), so that the fossil record of morphologically defined species is heavily dominated by random walks and evolutionary stasis (Hunt 2007; see also Matilla and Bokma 2008). These results corroborate several earlier assessments (reviews in Stanley 1979; Jackson and Cheetham 1999; Gould 2002), suggesting that the dynamics of large-scale trends are not underlain purely by within-species transformation, but often include a higher-level component such as differential extinction or origination of species and clades.

When viewed at the clade level, the simplest evolutionary dynamic that could generate a trend is directional speciation (fig. 2.1a), which would produce an active trend sensu Wagner (1996): a clade shifts from its starting point at time 1 to a new phenotypic state at time 2, abandoning its earlier phenotypic state, in this simplified instance by speciating only in one direction. (McShea's [1994, 2000] "driven trend" includes directional speciation but excludes differential speciation or extinction as an active mechanism; see also Wang 2001.) An alternative is a diffusive process, often called a passive trend, wherein speciation occurs in both directions, but changes are limited by a boundary at one margin, such as a minimum viable body size for the clade (fig. 2.1b). In this instance, the maximum value still increases from time 1 to time 2, but other morphologies persist, reflecting a very different

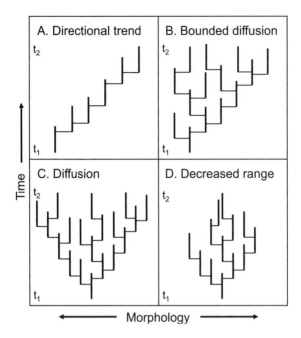

Figure 2.1. Clade-level patterns of phenotypic change from time 1 to time 2. Vertical lines are individual species. (a) Preferential speciation in the direction of the overall trend. (b) Increase over time maximum value to the right, bounded on the left side. (c) Unbounded, unbiased speciation, producing an increase in the maximum value and a decrease in the minimum. (d) Narrowing of variation by a decrease in the maximum value, and an increase in the minimum.

evolutionary dynamic from that in figure 2.1a. Unbounded diffusion (fig. 2.1c), where the clade expands freely away from the initial starting point, will still move the leading edge to greater values as the range of occupied or traversed phenotypic space expands, but as Gould (1988, 2002) emphasized, this pattern is not a trend in the classic sense of the term. Conversely, clades may decrease in range as they lose species or as speciation is concentrated near a modal value (fig. 2.1d). The kind of trend that unfolds over large timescales, then, depends in part on the starting point relative to barriers. Such barriers can be permanent constraints, as in limits to miniaturization or simplification (e.g., Stanley's [1973] "left wall"), or they can be temporary obstacles such as incumbents on an adaptive peak, as apparently occurred with mammals and their post-Cretaceous size increase (e.g., Alroy 1998). Additional

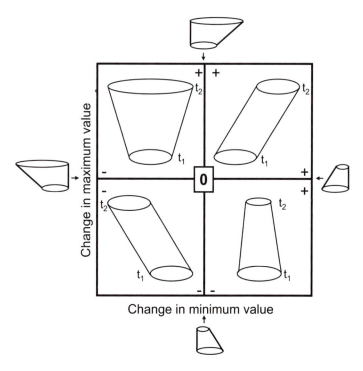

Figure 2.2. Bivariate approach to clade dynamics in a phenotypic space, with cartoons showing changes in minimum and maximum from time 1 to time 2. Vertical axis is change in the maximum value, horizontal axis is change in the minimum value, clades can be plotted as points in this space, the position of each determined by the difference between the initial and final sizes of its smallest and largest species, i.e., the leading and trailing edges of its size envelope. Cases of bounded diffusion, with static minimum or maximum, fall on the axes of the diagram as shown; origin of axes corresponds to a lack of net change in both minimum and maximum. Modified after Jablonski (1997).

complexities can be added by local attractors within the phenotype space (e.g., Brown et al. 1993; Alroy 1998, 2000), which may or may not equate to adaptive peaks: high speciation rates can increase the density of taxa within a morphological space even in the absence of a local selective optimum.

These different trajectories can be analyzed in a single space where the horizontal axis is the change in the minimum phenotypic value for a clade, and the vertical axis is the change in the maximum value for that clade (fig. 2.2). Under this scheme, each quadrant corresponds to a different macroevolutionary dynamic, with taxa plotted individually or grouped as frequencies for each quadrant (Jablonski 1996, 1997): clades falling in the upper right quadrant undergo a directional increase in the value under study, corresponding roughly to figure 2.1a; in the upper left an expansion in range as in figure 2.1c, in the lower left a directional decrease, and in the lower right a decrease in range, where the maximum value decreases and the minimum increases so that the total range constricts as in figure 2.1d. Clades where the minimum remains fixed (as in fig. 2.1b) will plot on the vertical axis, and clades with a fixed maximum will fall on the horizontal one. Although usually applied to analyses of body size (Jablonski 1996, 1997; Roy et al. 2000; Dommergues et al. 2002; Hone et al. 2005; Schmidt et al. 2006; Novack-Gottshall and Lanier 2008), this approach can be used to assess any change in a measurable value over evolutionary time. For many questions, it is more revealing than simply tracking means or medians over time (and see Solow and Wang 2008), which will often fail to capture key details of clade dynamics, such as fixed bounds or increasing ranges. Pitfalls remain because minima and maxima are not statistically well-behaved, being sensitive both to sampling and to diversity, and so I emphasize robust measures such as frequencies in quadrants, but evolutionary models for trends in clade minima and maxima as a function of time and diversity would be valuable as a vehicle for more detailed among-clade comparisons. One approach might be to analyze quartile or quintile values, which are statistically more robust than extremes (G. Hunt, personal communication 2009).

BODY-SIZE TRENDS

For the classic issue of body-size evolution, the horizontal axis in the state-space is the change in the minimum size in the clade, that is, the size of its smallest species at time 2 relative to the starting value, and the vertical axis is the change in the maximum size of the clade, the size of its largest species at time 2 relative to the starting value. Thus, Cope's Rule in the strict sense falls in the upper right quadrant: channeled, directional change of the clade, such that both the minimum and maximum sizes within the clade have increased—not only is the largest species in the clade larger in time 2 than in time 1, the smallest

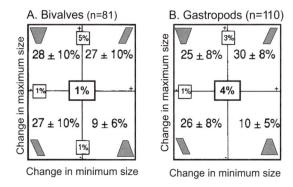

Figure 2.3. Percentages of Late Cretaceous (a) bivalves and (b) gastropods showing different patterns of changes in body size using the approach shown in figure 2.2 (with 95% binomial confidence intervals). Numbers in small boxes show percentages showing changes on one axis but stability in the other, suggesting a bounded increase in range. For raw data, with each genus plotted as a point on these axes, see Jablonski (1996, 1997). Analyses using quartiles rather than minima and maxima may be more robust, though still subject to the effects of small numbers, and give similar results, e.g., directional size increase occurs in 28% of both bivalves and gastropods, and directional size decreases are 25% and 19%, respectively.

species in the clade is also larger in time 2 than in time 1, yielding a net evolutionary trend.

Analyzed in this way, marine bivalve and gastropod genera from the Late Cretaceous can be seen to follow a variety of body-size trajectories, with Cope's Rule no more common than several other modes of change (fig. 2.3). A lack of net change (operationally defined as <5% change in size for the leading and trailing edge) and decreases in range show the lowest frequencies, perhaps unsurprising if size is free to change as clades diversify. Analyses using quartiles rather than minima and maxima, as mentioned above, give qualitatively similar results (see fig. 2.3 caption). If Cope's Rule is a pervasive tendency for directional increase in body size, then it does not hold for these mollusks. Other groups also vary significantly in their frequency of directional size increase, being relatively high in dinosaurs (Hone et al. 2005) and seen in 56 percent of species pairs for the Mammalia as a whole (Alroy 1998), but low in many other vertebrate and invertebrate clades in the fossil record (see Jablonski 1996 for review) and in the extant biota (Moen 2006 and references therein; McClain and Boyer 2009).

Many size trends appear to be more closely related to environmental context than to a pervasive selective pressure toward size increase as was originally held (and again argued by Kingsolver and Pfenning 2004). For example, size increases in fossil deep-sea ostracode crustaceans show a significant inverse correlation with the long-term temperature trend, a macroevolutionary analog of Bergmann's Rule that cooler temperatures promote larger body sizes (see also Ashton and Feldman 2003; Schmidt et al. 2006; Finkel et al. 2007; Roy 2008). Trends in body size can also be discordant across taxonomic levels, as in the increase in mean size seen in Paleozoic brachiopod orders but not within their constituent families, suggesting that differential origination and survival are important here (Novack-Gottshall and Lanier 2008, and see Gillman's [2007] nested analysis of primate clades). Such hierarchical discordances suggest that simple extrapolations from short-term observations can break down at macroevolutionary scales (Jablonski 2007, 2008a), or can be seen in terms of cross-level conflict, as when organismic selection for large size produces species that are more extinction-prone (Jablonski 1996; Van Valkenburgh et al. 2004; Clauset and Erwin 2008; Liow et al. 2008). As Roy (2008) points out, however, general models for the evolution of body size will have to account for among-clade differences in covariation among factors. For example, mollusks and mammals contrast in the relation between body size and fecundity, and apparently in the relation between body size and extinction risk (Jablonski 1996; Roy 2008). Even within mammals, however, the relation between body size and extinction risk in the geologic past may be confounded by other factors; hypercarnivores, which tend to be large, may be extinction-prone for reasons related to trophic level rather than body size, for example (Van Valkenburgh et al. 2004).

Environmental Trends

A similar clade-level dynamic can be seen in long-term environmental trends, at least in marine invertebrates where the fossil record has been examined in most detail. Many groups found today in deep water, exclusively or as part of their total bathymetric range, began in much shallower environments. Clades usually ranked as orders that originated in

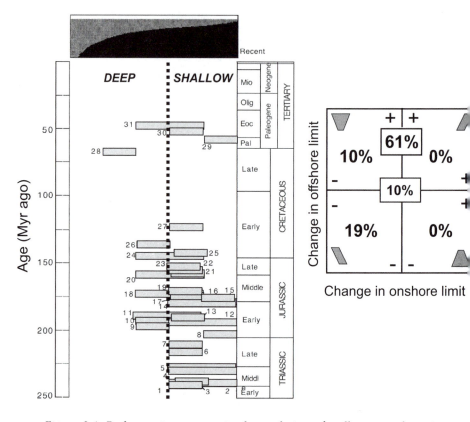

Figure 2.4. Bathymetric patterns in the evolution of well-preserved marine invertebrate orders since the start of the Mesozoic. *Left*: First occurrences of orders, after Jablonski (2005), who provides details. *Right*: Environmental dynamics of those orders. Most genera start in shallow water (and so cannot expand in the onshore direction) and expand their range into deeper water.

shallow water but now occur only at depths >100 m include the lychniscosidan sponges, the millericrinid and isocrinid crinoids, and the calycinid (salenioid) and holasteroid echinoids (Jablonski 2005). Tracking all thirty-one of the post-Paleozoic marine invertebrate orders having good preservation potential to their oldest known fossil species, we see that twenty-five started in shallow water, and then expanded across the continental shelf over millions of years (fig. 2.4). This result appears to be robust to preservation and sampling (Jablonski et al. 1997; Jablonski 2005). Further, it is not simply a function of taxonomic definitions, because the pattern was strong in analyses performed a decade

apart using different classifications, the first appearance of functionally important character states also tends to be onshore, and divergence in multivariate morphospace of the first members of major
groups tends to be greater than expected by a random draw from a frequency distribution of morphological distances or disparity among all
the other species in their ancestral clade (Jablonski and Bottjer 1990;
Jablonski et al. 1997; Eble 2000; Jablonski 2005). Something about
onshore environments preferentially elicits the origin of major novelties, or conversely, something about offshore environments tends to
damp novelty production.

The environmental trends of these orders can be analyzed using the
same approach as for body size. Here, instead of a symmetrical increase in range, 61 percent of the orders show a lopsided one: they
cannot go onto land, and so are constrained to exhibit bounded diffusion, expanding to the deeper continental shelf without losing their
hold on onshore settings. At this taxonomic level, not one clade starts
deep to become exclusively shallow water. Conversely, about 20 percent of the clades abandon their shallow-water beginnings for exclusively deep-water occurrences. Viewed in this way, they are anomalous
relative to the 61 percent that retain shallow-water occupancy, and
thus are interesting targets for comparative analysis, as are clades now
assigned lower rank, such as neogastropods and stylasterine corals,
that originated in deep water and widened their bathymetric range to
include onshore environments (Jablonski and Bottjer 1990; Lindner
et al. 2008).

These major trends are not simple extrapolations of evolutionary
events at lower levels. The marine orders start onshore, but novelties
within orders of bryozoans do not (Jablonski et al. 1997), nor do genera within echinoid and crinoid orders (Jablonski 2005). Further, no
onshore-offshore trend is seen when environmental transitions are
traced within a detailed phylogeny, as is available for the echinoid
order Calycina (=Salenioida): 32 percent of its nineteen genera originated in the environment of their putative ancestors, approximately
26 percent shifted onshore relative to the ancestor, approximately
20 percent shifted offshore, 10 percent expanded offshore, and 10 percent occupied a narrower depth range than their putative ancestral
species, at least at first. Once they originated, 52 percent of the genera

were static in their initial environments throughout their histories, approximately 16 percent moved onshore, approximately 16 percent moved offshore, and approximately 16 percent are so short-lived that their environmental history cannot be assessed. Numbers are painfully small, but lack any overall pattern to fuel the order-level dynamic (Jablonski and Smith 1990). Analyses of additional groups in such a combined environmental and phylogenetic framework are needed.

As with body size, a major component of these large-scale environmental trends is a diffusive increase in range. Moreover, the large-scale patterns are evidently shaped by differential survival and production of subtaxa, rather than as extensions of the environmental dynamics of constituent species or genera. Many hypotheses have been proposed for this overarching process, with none strongly corroborated (e.g., Jablonski 2005; Lindner et al. 2008). Definitive work will require a partnership between evolutionary paleobiology and ecological aspects of evolutionary developmental biology.

BIOGEOGRAPHIC TRENDS

A similar multi-level dynamic evidently shapes major biogeographic patterns, including the latitudinal diversity gradient. This latitudinal trend is perhaps the most prevalent biodiversity pattern, with dramatic increases in taxonomic richness from poles to equator observed in hundreds of studies on land and sea (Hillebrand 2004). Because extinction and range shifts undermine the use of present-day distributions to infer past spatial dynamics, paleontological data are especially valuable for analyzing the origin and maintenance of such large-scale biogeographic patterns, although caution is needed owing to strong sampling biases against tropical occurrences during the Cenozoic Era (Jablonski et al. 2006).

Analyses of first occurrences and subsequent spread of clades in the fossil record indicate an "out-of-the-tropics" diffusive process in marine organisms at several taxonomic levels. The well-preserved post-Paleozoic marine invertebrate orders originate in the tropics far more often than expected given the observed biases in sampling and preservation (Jablonski 1993, 2005). Most of those orders are still present in the tropics and a subset of them have spread to higher latitudes, just

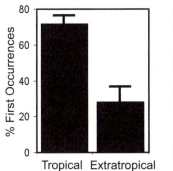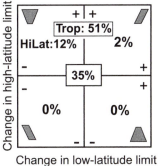

Figure 2.5. Latitudinal patterns in the evolution of marine bivalve genera. *Left*: Proportion starting in tropics since the start of the Late Miocene (11.6 Myr ago), see Jablonski et al. (2006) for details. *Right*: Latitudinal dynamics of those genera, with 51% starting in the tropics (and thus with a fixed lower bound by this protocol) and expanding to higher latitudes.

as they have spread to greater depths (Jablonski 2005; Martin et al. 2007).

We can dissect this pattern further by tracking marine bivalve genera, targeting this group because they are the macrofossils with the densest, richest, and taxonomically most standardized Cenozoic fossil record (Valentine et al. 2006; Jablonski 2008b), and because their temporal and spatial diversity patterns mirror those of the overall biota. Over the past 12 Myr, significantly more bivalve genera started in the tropics than in extratropical regions, whether analyzed as a single time bin (fig. 2.5) or partitioned into Late Miocene, Pliocene, and Pleistocene intervals (Jablonski et al. 2006). Because the fossil record is better in the temperate zones, this count of tropical starts must be an underestimate: some of the genera that seem to start in the temperate zone probably originated in the tropics but were not recorded until they reached the better sampled extratropical areas. Viewed in bivariate space, 51 percent of these bivalve genera first appeared in the tropics and spread poleward, but only 12 percent first appeared at high latitudes and spread into the tropics (and, again, sampling biases mean that this is an overestimate); 35 percent of the genera evidently stayed in their climate zone or origin, and only 2 percent actually vacated the tropics. Thus, contrary to the dichotomy emphasized for the past thirty-five years (framed by Stebbins 1974 but dating back much further),

35

the tropics are both a cradle and a museum (Jablonski et al. 2006; Krug et al. 2009). Like orders, genera preferentially originate in the tropics and spread to higher latitudes without losing their tropical members, so that the tropics are both generators and accumulators of biodiversity.

This out-of-the-tropics dynamic breaks down at the species level, however. For marine invertebrates, at least, individual species do not spread from equator to pole. Fewer than 5 percent of extant bivalve species cover such a wide range, with most apparently remaining in their climate zone of origin, or perhaps exploiting coastal conditions (such as cool upwelling water or warm protected bays) to extend a bit further. Instead, the latitudinal spread of clades (here, genera) appears to be underlain by speciation: a species in one climatic zone stays there while giving rise to another in a neighboring zone. The most species-rich genera in each climate zone also occupy the greatest latitudinal range: the more prolific speciators are the most effective range-expanders, presumably by occasionally speciating across provincial boundaries (Krug et al. 2008). This dynamic is not a simple time-dependent process—latitudinal ranges of clades are better predicted by species-richness than by clade age—but is evidently fueled by differential diversification rates among clades. More work is needed to decompose diversification rates more precisely into origination and extinction, but the basic message is clear: as with body size and onshore-offshore trends, the tropical to polar trend is shaped not by the behavior of individual species, but by speciation events.

DISCUSSION

Although populations can evolve rapidly and selection can hone remarkable adaptations from finch beaks to butterfly wings, large-scale trends appear to be molded at least in part by differential speciation and extinction; trends unfold as a multi-level process. Selection of organisms in populations may help to determine directions or patterns of speciation or extinction, but species-level properties such as geographic range and genetic population structure also influence extinction risk and speciation propensity (reviewed by Jablonski 2008a), again imposing a hierarchical structure to long-term processes. Processes at different levels may be mutually reinforcing, but cross-level conflicts must also arise, as when short-term organismic selection for large body size produces species that are more extinction-prone. And by the very nature of multi-level processes, phenotypic traits will hitch-hike on the differential speciation and extinction rates that propel many large-scale trends (see Wagner 1996; Jablonski 2008a, 2008b). Thus, Cooper et al. (2008) found that body size and fecundity hitch-hike on geographic range in present-day frogs, owing to the extinction vulnerability of species having narrow geographic ranges (and see Reynolds et al. 2005 for a similar result in fishes). Similar effects must operate in the environmental and latitudinal trends discussed here. The nodal rings of cirri that distinguish isocrinid crinoids or the enlarged anterior ventral plates in holasteroid echinoids probably did not necessitate shifts to exclusively offshore habitats, and we know from the fossil record that those features once functioned effectively in shallow water. Nevertheless, those traits now occur only in offshore settings, presumably because they were shifted there by forces operating on other aspects of their clades, which could have ranged from predation vulnerability of constituent organisms to narrow geographic ranges of constituent species.

The approach to macroevolutionary trends advocated here pinpoints some key questions concerning the following three topics that deserve greater attention.

NON-RANDOM STARTING POSITIONS

The position of the initiator of a trend is crucial to understanding all that follows (Stanley 1973; Gould 1988): if trends start at a phenotypic extreme, such as the smallest viable body size, the shallowest water, or the lowest latitudes, then simple diffusion will do the rest. Reconstruction of ancestral states for continuous characters in extant taxa is difficult, and can yield well-supported but inaccurate values, so that paleontological data are particularly valuable for defining initial points (Webster and Purvis 2002; Finarelli and Flynn 2006; Wiens et al. 2007; note that Pagel's (2002, p. 276) directional GLS method for inferring ancestral values for continuous characters such as body size requires data on character states at differing total path lengths from the root—such as data from fossils or the operational equivalent). For example, the fossil record clearly shows that mammals started at small body sizes. Whether the first Mesozoic mammals were small for intrinsic reasons related to the threshold to true endothermy, or because they were confined there by dinosaurs and other Mesozoic dominants, remains unclear, however. On the other hand, the first (Devonian) tetrapods are ~1 m long, large relative to the median or modal sizes of modern tetrapods, and *Archaeopteryx* is small for a Jurassic dinosaur but well above the median size of birds today. Similarly, the preferential initiation of major marine clades in shallow water is now seen as an intriguing intersection of ecology and developmental biology that bears further investigation. And the preferential initiation of clades in the tropics can now be seen as a problem where a simple probabilistic result of greater standing diversity needs to be tested against a higher per-taxon origination rate.

LOSS OF ANCESTRAL STATES

Directional trends, as opposed to increases in range, become especially interesting from this standpoint. Under true directionality (the upper-right quadrant in fig. 2.1) clades abandon their primitive states, for example, small-bodied species become extinct or, more rarely, transform phyletically into large species; onshore members of clades disappear; and a (minute) fraction of clades that reach the temperate zones lose their initial tropical foothold. How often is this loss deterministic, and how often purely stochastic? The small number of species within most clades, particularly relative to the number of organisms within those species, makes those clades subject to what has been termed phylogenetic drift (Stanley 1979; Gould 2002). Better analytic tools are needed to take this hierarchical effect into account.

CROSSING THRESHOLDS

Some trends involve a dramatic break with primitive character states, or a breaking of the statistical property of heritability among related organisms or species. Body size is "heritable" across species (e.g., Smith et al. 2004); several studies have shown that geographic range, a species-level trait, is heritable in the sense that closely related species have more similar range sizes, and positions of range endpoints, than expected by chance (Jablonski 2008a; Roy et al. 2009); and comparative analyses have found phylogenetic signal in a wide range of organism and species attributes (Freckleton et al. 2002) and ecological require-ments (Losos 2008). Nonetheless, large-scale analyses show that clades do occasionally break these phylogenetic correlations and enter new areas, and not simply in terms of phenotypes, such as upward and downward changes in body size (which in fact seems to be a relatively labile character, e.g., Hunt 2007). Similarly, clades leave turbulent, thermally variable, nutrient-rich water for quiet offshore zones, a sig-nificant transition physiologically, and clades leave the tropics for more strongly seasonal temperate zones, again a significant transition for an ectothermic marine invertebrate. The latter, latitudinal, dynamic seems to require that generalists evolve from specialists, in that lin-eages speciating into more seasonal settings tend to have broader tol-erances and exploit broader ranges of trophic resources than those in the tropics (e.g., Valentine et al. 2008), a perhaps counterintuitive pro-cess that needs to be tested against an exaptation hypothesis that only generalist ancestors give rise to barrier-crossing species (but see Nosil and Mooers 2005 on other transitions from generalist to specialist). Such a non-random draw might be suspected if widespread species preferentially cross barriers, but of course the widespread ones are also the most likely to encounter provincial boundaries simply by vir-tue of their broad ranges. More work is needed on when and how the phylogenetic similarities among species are disrupted to create macro-evolutionary trends that traverse new developmental and ecological terrain.

CONCLUSIONS

The most powerful approach to macroevolutionary problems is to search for generalities by comparing systems, and to integrate present-day and fossil data as done here. Placing diverse kinds of large-scale trends into a single analytical framework can reveal unrecognized commonalities, and suggest new approaches and new questions for a long-standing set of issues. When size—and other traits as different as bathymetric and latitudinal distributions—are considered in terms of envelopes of variation rather than as isolated transitions, phylogenetic increases in range are seen as an important part of the evolutionary process, and directional trends can be seen often to involve sorting among species over long periods of time (and may be a combination of different evolutionary modes, e.g., Wang 2001; Simpson 2009). A multi-level approach that treats clades as clouds of related species, where the edges of the clouds are key variables, can provide new insights and raise new questions for one of the most dramatic and least understood features of macroevolution.

References

Airoy, J. 1998. Cope's rule and the dynamics of body mass evolution in North American fossil mammals. *Science* 280: 731–734.

———. 2000. Understanding the dynamics of trends within evolving lineages. *Paleobiology* 26: 319–329.

Ashton, K. G., and C. R. Feldman. 2003. Bergmann's rule in non-avian reptiles: Turtles follow it, lizards and snakes reverse it. *Evolution* 57: 1151–1163.

Brown, J. H., P. A. Marquet, and M. L. Taper. 1993. Evolution of body size: Consequences of an energetic definition of fitness. *Am. Nat.* 142: 573–584.

Clauset, A., and D. H. Erwin. 2008. The evolution and distribution of species body size. *Science* 321: 399–401.

Cooper, N., J. Bielby, G. H. Thomas, and A. Purvis. 2008. Macroecology and extinction risk correlates of frogs. *Global Ecol. Biogeogr.* 17: 211–221.

Dommergues, J.-L., S. Montuire, and P. Neige, 2002. Size patterns through time: the case of the Early Jurassic ammonite radiation. *Paleobiology* 28: 423–434.

Eble, G. J. 2000. Contrasting evolutionary flexibility in sister groups: disparity and diversity in Mesozoic atelostomate echinoids. *Paleobiology* 26: 56–79.

Finarelli, J. A. 2007. Mechanisms behind active trends in body size evolution of the Canidae (Carnivora: Mammalia). *Am. Nat.* 170: 876–885.

Finarelli, J. A., and J. J. Flynn. 2006. Ancestral state reconstruction of body size in the Caniformia (Carnivora, Mammalia): The effects of incorporating data from the fossil record. *Syst. Biol.* 55: 301–313.

Finkel, Z. V., J. Sebbo, S. Feist-Burkhardt, A. J. Irwin, M. E. Katz, O.M.E. Schofield, J. R. Young, and P. G. Falkowski. 2007. A universal driver of macroevolutionary change in the size of marine phytoplankton over the Cenozoic. *Proc. Natl. Acad. Sci. USA* 104: 20416–20420.

Freckleton, R. P., P. H. Harvey, and M. Pagel. 2002. Phylogenetic analysis and comparative data: A test and review of evidence. *Am. Nat.* 160: 712–726.

Gillman, M. P. 2007. Evolutionary dynamics of vertebrate body mass range. *Evolution* 61: 685–693.

Gould, S. J. 1988. Trends as changes in variance: a new slant on progress and directionality in evolution. *J. Paleontol.* 62: 319–329.

———. 2002. *The Structure of Eevolutionary Theory.* Cambridge, MA: Harvard University Press.

Hillebrand, H. 2004. On the generality of the latitudinal diversity gradient. *Am. Nat.* 163: 192–211.

Hone, D.W.E., T. M. Keesey, D. Pisani, and A. Purvis. 2005. Macroevolutionary trends in the Dinosauria: Cope's rule. *J. Evol. Biol.* 18: 587–595.

Hunt, G. 2007. The relative importance of directional change, random walks, and stasis in the evolution of fossil lineages. *Proc. Natl. Acad. Sci. USA* 104: 18404–18408.

Jablonski, D. 1993. The tropics as a source of evolutionary novelty: The post-Palaeozoic fossil record of marine invertebrates. *Nature* 364: 142–144.

———. 1996. Body size and macroevolution. In D. Jablonski, D. H. Erwin, and J. H. Lipps, eds., *Evolutionary Paleobiology*, 256–289. Chicago: University of Chicago Press.

———. 1997. Body-size evolution in Cretaceous molluscs and the status of Cope's rule. *Nature* 385: 250–252.

———. 2005. Evolutionary innovations in the fossil record: The intersection of ecology, development and macroevolution. *J. Exp. Zool.* 304B: 504–519.

———. 2007. Scale and hierarchy in macroevolution. *Palaeontology* 50: 87–109.

———. 2008a. Species selection: Theory and data. *Ann. Rev. Ecol. Evol. Syst.* 39: 20–42.

———. 2008b. Extinction and the spatial dynamics of biodiversity. *Proc. Natl. Acad. Sci. USA* 105 (Suppl. 1): 11528–11535.

Jablonski, D., and D. J. Bottjer. 1990. The origin and diversification of major groups: Environmental patterns and macroevolutionary lags. In P. D. Taylor and G. P. Larwood, eds., *Major Evolutionary Radiations*, 17–57. Oxford: Clarendon Press.

Jablonski, D., S. Lidgard, and P. D. Taylor. 1997. Comparative ecology of bryozoan radiations: Origin of novelties in cyclostomes and cheilostomes. *Palaios* 12: 505–523.

Jablonski, D., K. Roy, and J.W. Valentine. 2006. Out of the Tropics: Evolutionary dynamics of the latitudinal diversity gradient. *Science* 314: 102–106.

Jablonski, D., and A. B. Smith. 1990. Ecology and phylogeny: Environmental patterns in the evolution of the echinoid order Salenioida. *Geol. Soc. Am. Abstr.* 22: A266.

Jackson, J.B.C., and A. H. Cheetham. 1999. Tempo and mode of speciation in the sea. *Trends Ecol. Evol.* 14: 72–77.

Kingsolver, J. G., and D. W. Pfennig. 2004. Individual-level selection as a cause of Cope's rule of phyletic size increase. *Evolution* 58: 1608–1623.

Krug, A. Z., D. Jablonski, and J. W. Valentine. 2008. Species-genus ratios reflect a global history of diversification and range expansion in marine bivalves. *Proc. Roy. Soc. London* B 275: 1117–1123.

Krug, A. Z., D. Jablonski, J. W. Valentine, and K. Roy. 2009. Generation of Earth's first-order biodiversity pattern. *Astrobiology* 9: 113–124.

Lindner, A., S. D. Cairns, and C. W. Cunningham. 2008. From offshore to on-shore: Multiple origins of shallow-water corals from deep-sea ancestors. *PLoS One* 3: e2429.

Liow, L. H., M. Fortelius, E. Bingham, K. Lintulaakso, H. Mannila, L. Flynn, and N. C. Stenseth. 2008. Higher origination and extinction rates in larger mammals. *Proc. Natl. Acad. Sci. USA* 105: 6097–6102.

Losos, J. B. 2008. Phylogenetic niche conservatism, phylogenetic signal and the relationship between phylogenetic relatedness and ecological similarity among species. *Ecol. Lett.* 11: 995–1007.

Martin, P. R., F. Bonier, and J. J. Tewksbury. 2007. Revisiting Jablonski (1993): Cladogenesis and range expansion explain latitudinal variation in taxonomic richness. *J. Evol. Biol.* 20: 930–936.

Matilla, T. M. and F. Bokma. 2008. Extant mammal body masses suggest punctuated equilibrium. *Proc. Roy. Soc. London* B 275: 2195–2199.

McClain, C. R., and A. G. Boyer. 2009. Biodiversity and body size are linked across metazoans. *Proc. Roy. Soc. London* B 276: 2209–2215.

McShea, D. W. 1994. Mechanisms of large-scale evolutionary trends. *Evolution* 48: 1747–1763.

———. 2000. Trends, tools, and terminology. *Paleobiology* 26: 330–333.

Moen, D. S. 2006. Cope's rule in cryptodiran turtles: Do the body sizes of extant species reflect a trend of phyletic size increase? *J. Evol. Biol.* 19: 1210–1221.

Nosil, P., and A. Ø. Mooers. 2005. Testing hypotheses about ecological specialization using phylogenetic trees. *Evolution* 59: 2256–2263.

Novack-Gottshall, P. M., and M. A. Lanier. 2008. Scale-dependence of Cope's rule in body size evolution of Paleozoic brachiopods. *Proc. Natl. Acad. Sci. USA* 105: 5430–5434.

Pagel, M. 2002. Modelling [sic] the evolution of continuously varying characters on phylogenetic trees: The case of hominid cranial capacity. In N. MacLeod and P. L. Forey, eds., *Morphology, Shape and Phylogenetics*, 269–286. London: Taylor & Francis.

Reynolds, J. D., T. J. Webb, and L. A. Hawkins. 2005. Life history and ecological correlates of extinction risk in European freshwater fishes. *Can. J. Fish. Aquatic Sci.* 62: 854–862.

Roy, K. 2008. Dynamics of body size evolution. *Science* 321: 1451–1452.

Roy, K., G. Hunt, D. Jablonski, A. Z Krug, and J. W Valentine. 2009. A macroevolutionary perspective on species range limits. *Proc. Roy. Soc. London* B 276: 1485–1493.

Roy, K., D. Jablonski, and K. K. Martien, 2000. Invariant size-frequency distributions along a latitudinal gradient in marine bivalves. *Proc. Natl. Acad. Sci. USA* 97: 13150–13155.

Schmidt, D. N., D. Lazarus, J. R. Young, and M. Kucera. 2006. Biogeography and evolution of body size in marine plankton. *Earth-Sci. Rev.* 78: 239–266.

Simpson, C. 2010. Species selection and driven mechanisms jointly generate a large-scale morphological trend in monobathrid crinoids. *Paleobiology* 36: 481–496.

Slatkin M. 1981. A diffusion model of species selection. *Paleobiology* 7: 421–425.

Smith, F. A., et al. 2004. Similarity of mammalian body size across the taxonomic hierarchy and across space and time. *Am. Nat.* 163: 672–691.

Solow, A. R., and S. C. Wang. 2008. Some problems with assessing Cope's Rule. *Evolution* 62: 2092–2096.

Stanley, S. M. 1973. An explanation for Cope's Rule. *Evolution* 27: 1–26.

———. 1979. *Macroevolution*. San Francisco: W. H. Freeman.

Stebbins, G. L. 1974. *Flowering Plants: Evolution Above the Species Level*. Cambridge, MA: Belknap Press.

Valentine, J. W., D. Jablonski, S. M. Kidwell, and K. Roy. 2006. Assessing the fidelity of the fossil record by using marine bivalves. *Proc. Natl. Acad. Sci. USA* 103: 6599–6604.

Valentine, J. W., D. Jablonski, A. Z. Krug, and K. Roy. 2008. Incumbency, diversity, and latitudinal gradients. *Paleobiology* 34: 169–178.

Van Valkenburgh, B., X. Wang, and J. Damuth. 2004. Cope's Rule, hypercarnivory, and extinction in North American canids. *Science* 306: 101–104.

Wagner, P. J. 1996. Contrasting the underlying patterns of active trends in morphologic evolution. *Evolution* 50: 990–1007.

Wang, S. C. 2001. Quantifying passive and driven large-scale evolutionary trends. *Evolution* 55: 849–858.

Webster, A. J., and A. Purvis. 2002. Testing the accuracy of methods for reconstructing ancestral states of continuous characters. *Proc. Roy. Soc. London* B 269: 143–149.

Wiens, J. J., C. A. Kuczynski, W. E. Duellman, and T. W. Reeder. 2007. Loss and re-evolution of complex life cycles in marsupial frogs: Does ancestral trait reconstruction mislead? *Evolution* 61: 1886–1899.

Chapter Three

Dinosaurs Live!

Philip J. Currie

Dinosaur Research over the Years

The scientific discovery of dinosaurs is usually pegged at 1824, when Buckland described the lower jaw of *Megalosaurus*, although the name Dinosauria was not coined for another two decades (Owen 1842). The discovery and description of dinosaurs blossomed during the latter part of the nineteenth century and during the first two decades of the twentieth century. However, following World War I and the Great Depression of the 1930s, dinosaur research was severely hampered by a lack of funding and a correspondingly low number of scientists working in the field. Although the public interest in dinosaurs remained high, many vertebrate palaeontologists considered dinosaurs as a dead-end evolutionary lineage that was too expensive to work on. Few people worked on dinosaurs anywhere in the world between the 1930s and the 1960s; consequently there were relatively few new discoveries and few research publications. The discovery and description of *Deinonychus* (Ostrom 1969) is at least a symbolic turn-around in the fortunes of dinosaur research. Research papers that followed by Bakker, Dodson, Horner, Russell and others showed that there was still much of interest to be learned about the biology of dinosaurs. The research led to increased levels of publicity about dinosaurs, which led to increased public interest, which produced more funding for collection and research. This created a feedback loop that has led to the current level of research on dinosaurs, which has never been greater (Currie 2008). More professional paleontologists and students are doing research on dinosaurs than ever before, which in turn has led to a dramatic increase in the number of scientific papers published about these animals. Even leading scientific journals like *Nature* and *Science*, which rarely published articles on dinosaurs before the 1993 release of the movie *Jurassic Park*, are now running dinosaur stories regularly. More species of dinosaurs are being described than ever before, but more importantly, the research is increasing in scope as we learn more about the biology of these animals (Currie 2008). In the past, the majority of research projects about dinosaurs were anatomical descriptions of new species and/or specimens with analyses of their relationships. The foundations of our understanding of dinosaurs continue to be built as more species are described. At last count, there are almost 700 valid generic names of dinosaurs, almost half of which were created since 1993 (Currie 2008). It is worth noting that more than 10 percent of this dinosaurian diversity comes from just two Upper Cretaceous (Campanian) sites (Dinosaur Provincial Park in Alberta and the Nemegt Basin of Mongolia). Because there would have been hundreds of different habitats around the world, and because different dinosaur species were adapted to different ecosystems, there must have been many, many more species of dinosaurs living in Campanian times that have not yet been discovered. And because different environments have different potentials for the preservation of fossils, we never will know how many dinosaur species lived in Campanian times worldwide. Furthermore, the evolution and turnover of species of hadrosaurs and ceratopsians in Dinosaur Provincial Park (Currie and Russell 2005) suggest that species of these dinosaurs lasted for less than one to two million years. Given the 150 million-year history (late Triassic to the end of the Cretaceous) of non-avian dinosaurs, there are almost certainly thousands of new dinosaur species waiting to be discovered and named.

Like all fields of vertebrate paleontology (indeed, of Science), research on dinosaurs has become progressively more multinational, multidisciplinary, invasive, and abstract, and we are learning more about the biological processes of these animals than we ever thought would have been possible ten years ago. Anatomy and phylogeny continue to be major themes of dinosaur research, but there has been a steady shift toward computer modeling (with the help of computerized tomography (CT) scanning, laser scanners, and other high-tech equipment) to analyze biomechanics and other aspects of functional morphology. Histological studies have produced surprising results in determining growth rates, physiology, and longevity (Erickson et al. 2004), and it has even become possible to determine the sex of some dinosaurs (Schweitzer et al. 2005). Although all of these things make the Dinosauria an exciting and viable focus for study, the discovery of feathered dinosaurs in northeastern China in 1996 may have been more responsible than anything else for bringing research on dinosaurs to its present levels.

ORIGIN OF BIRDS

One of the most exciting aspects of recent dinosaur studies has been their relationship with birds. *Archaeopteryx* (color plate **XX**) is generally considered to be the first (or earliest) bird (Hecht et al. 1985; Wellnhofer 2008). The first specimen was discovered in 1861 in Solnhofen, Germany, and nine additional specimens have been recovered since then (Mayr et al. 2007). Like other fossils from the lithographic limestones, the specimens are remarkably well-preserved. Most include feather impressions, which is fortunate because otherwise this animal would not have been identified as a bird when it was discovered. Although it is no bigger than a chicken, the teeth, the presence of a long, bony tail, and three free fingers with claws and many other characters make it look like a small theropod (meat-eating dinosaur). In fact, *Compsognathus* (Ostrom 1978) and *Juravenator* (Göhlich and Chiappe 2006), which are two of the smallest theropod dinosaurs known, were also recovered from the same lithographic limestones of Germany that produced the *Archaeopteryx* specimens. Ever since the first specimen was discovered, *Archaeopteryx* was recognized as the link between reptiles and birds. Thomas Huxley (1868, 1870) compared the anatomy of *Archaeopteryx* with that of other animals and came to the conclusion that birds were probably descended from dinosaurs. This was the prevailing belief until the English version of a thorough study of bird origins by a Danish bird specialist was published (Heilmann 1927). Heilmann agreed with Huxley that dinosaurs were anatomically the closest animals to *Archaeopteryx* and other early birds. However, he pointed out that dinosaurs lacked the clavicle, which is a paired bone in the shoulder girdles of most vertebrates. In *Archaeopteryx* and all birds, the clavicles have fused into a single median structure known as the furcula or wishbone. If all dinosaurs had lost their clavicles when they evolved from more primitive animals, he felt that they could not be the direct ancestors of birds that still retained clavicles. The similarities between birds and dinosaurs therefore suggested to Heilmann that they shared a common ancestor among what were known as "thecodonts" at that time. The idea that birds and dinosaurs were sister groups derived from the "Thecodontia" became the prevailing theory for the next half century. Curiously, a dinosaur known as *Oviraptor* (Osborn 1924) was found several years before the English version of Heilmann's book appeared, and it had a furcula that is almost identical to that of *Archaeopteryx*. The problem was that the bone had been identified as an interclavicle (Osborn 1924), and it was more than fifty years before it was re-identified as fused clavicles (Osmólska 1976). Once it was known that dinosaurs could have clavicles, then reexamination of specimens revealed that many theropods had furculae, including the giant tyrannosaurids (Makovicky and Currie 1998). Palaeontologists had been blinded to an extent by the "knowledge" that dinosaurs were not supposed to have clavicles, but had also missed the clavicles because they are similar in size and shape with ribs and gastralia.

A resurgence of interest in the question of bird origins developed along with the "Dinosaur Renaissance" (Bakker 1975), and two hypotheses competed with Heilmann's idea that birds evolved from "thecodonts." Walker (1972) presented evidence that birds may have been derived from early crocodylians. Although this may seem unreasonable at first glance, modern birds and crocodiles share more derived characters with each other than either shares with any other living animals. Furthermore, early crocodylians were small bipedal animals that looked somewhat similar to some "thecodonts" and early dinosaurs. With more than a dozen derived characters shared by crocodylians and birds (Whetstone and Martin 1979) and no other animals living or dead, this initially seemed like a robust theory. However, examination of well-preserved theropod fossils soon revealed that most of the "uniquely derived" characters were also present in dinosaurs (Currie 1985, 1987) and therefore had probably appeared in early archosaurs that were ultimately ancestral to crocodylians, dinosaurs, and birds.

The other hypothesis about bird origins that appeared in the 1970s was a reworking of the century-old hypothesis of Huxley (1870). In studying *Deinonychus*, Ostrom (1969) had been struck by startling similarities between this animal and *Archaeopteryx*. This led to a re-examination of the earliest bird fossils, and a new analysis of the ancestry of birds from dinosaurs (Ostrom 1973). Subsequent papers by Ostrom and others strengthened the hypothesis, but Gauthier (1986) made the strongest statement when he published a phylogenetic analysis that included more than 125 characters that were uniquely shared by theropods and birds.

WARM-BLOODED DINOSAURS

The discovery of *Deinonychus* in the 1960s also triggered another debate by inspiring images of rapidly moving, highly active dinosaurs (Ostrom 1969; Bakker 1975; Paul 1988). Warm-bloodedness in dinosaurs was not a new idea (it had been suggested independently by Frederick von Huene and Loris Russell years before), but this time it found a much wider audience. Although the debate has not been resolved, most paleontologists are willing to accept that small theropods were physiologically closer to modern birds and mammals than to extant turtles, lizards, and crocodiles. The two ideas—that theropods might be warm-blooded and that they might be the direct ancestors of birds—inevitably came together as paleontologists considered the possibility that some dinosaurs had feathers. If theropods were warm-blooded, then smaller individuals would have needed insulation to stabilize their body temperatures. And the insulation used by dinosaurs may have been some form of feather if theropods were the ancestors of birds. After all, feathers had to have developed for some reason before birds could incorporate them into their flight mechanism.

At the end of the twentieth century, the origin of birds and whether or not dinosaurs were warm-blooded were two of the biggest controversies in paleontology. Initially, more palaeontologists opposed the hypotheses of warm-blooded dinosaurs and a dinosaurian origin of birds. This has changed now, largely because of some remarkable fossils from Liaoning and adjacent provinces in northeastern China.

FEATHERED DINOSAURS

The first fossil bird was discovered in the lower Cretaceous rocks of Liaoning in 1994, but within a few short years thousands of specimens had been collected, most of which were dumped into the fossil markets of an unsuspecting world. *Confuciusornis* is undoubtedly the best-represented genus (Chiappe et al. 1999), but more than a dozen other genera have also been recovered. Most of these specimens include fossilized feathers, which only preserve under exceptional circumstances (Davis and Briggs 1995).

Given the remarkable number of well-preserved bird fossils being recovered in Liaoning, it should have been no surprise when the discovery of the first "feathered" dinosaur (*Sinosauropteryx prima*) was announced (Ji and Ji 1996). Controversy erupted immediately over whether or not this really was a "feathered" dinosaur, but the site rapidly produced additional specimens of this small, chicken-sized animal (Ji and Ji 1997b; Chen et al. 1998). All of the specimens were consistent in showing that *Sinosauropteryx* was covered by short, simple branching structures (Currie and Chen 2001) that presumably functioned to insulate the animal. Chen et al. (1998) conservatively referred to these as "integumentary structures," but everyone was aware that they were potentially protofeathers. Rather than create wider acceptance of a close relationship between birds and dinosaurs, the controversies on the ancestry of birds and warm-bloodedness in dinosaurs intensified (Brush et al. 1997; Currie 1997, 2000; Ruben et al. 1997). Such is the nature of Science—nothing should ever be accepted without question!

Sinosauropteryx is a compsognathid that is closely related to *Compsognathus* from the Solnhofen beds of Germany (Currie and Chen 2001). Numerous "feathered" dinosaurs have been subsequently discovered and described, and have been assigned to different families of theropods.

Beipiaosaurus inexpectus (Xu, Tang, and Wang 1999; Xu et al. 2009) is a larger, approximately human-sized dinosaur theropod with a relatively small head, leaf-shaped teeth, long neck, long arms, and short tail. These are characteristic of therizinosaurian theropods, although the specimen from Liaoning is the only known therizinosaurian fossil with feather-like structures similar to those that covered *Sinosauropteryx*. Long, stiff "feathers" behind its arms also distinguish it from the compsognathid.

Sinornithosaurus millenii (Xu, Wang, and Wu 1999; Xu and Wu 2001; Ji et al. 2001) is a dromaeosaurid with serrated teeth and raptorial claws similar to those of its close relative *Velociraptor*, but has feathers similar to those of *Beipiaosaurus*. *Microraptor zhaoianus* and *Microraptor gui* (Xu et al. 2000, 2003; Hwang et al. 2002) are two other, smaller dromaeosaurid species with some adaptations that suggest members of this genus may have been arboreal. A closely related form, *Hesperonykus elizabethi* (plate 1), was recently reported from Canada (Longrich and Currie 2009). Most of the "feathers" of *Microraptor* seem to have been simple branching structures like those in *Sinosauropteryx* (Currie and Chen 2001) and *Sinornithosaurus* (Xu et al. 2001), but long remiges are present behind the forearm and the tibia. This "four-winged" structure has stimulated considerable discussion about the origin of flight. *Yixianosaurus longimanus* (Xu and Wang 2003) is yet another feathered form from the region that shows characters in the hand that also indicate arboreal habits.

Protarchaeopteryx robusta (Ji and Ji 1997a) has long arms and a relatively short tail with long, quill-like feathers at the end of the tail. Each of the retrices is well enough preserved to show a central rachis, barbs and barbules (Ji et al. 1998).

At least nine specimens of *Caudipteryx zoui* (Ji et al. 1998) and *Caudipteryx dongi* (Zhou and Wang 2000) have been recovered, each with long feathers at the end of the tail, and true feathers behind the arms. A second related genus (*Similicaudipteryx*, He et al. 2008) has also been described from the same region in China. *Caudipteryx* is related to *Oviraptor* (Barsbold et al. 2000) and other oviraptorosaurs from the Late Cretaceous of China, Mongolia, and North America. Its relatively long legs suggest it was cursorial (Dyke and Norell 2005). The remiges make its arms look like rudimentary wings. However, the remiges have symmetrical vanes, and the arms are too short to have allowed it to fly. This suggests that the remiges and retrices were used for display (Currie 1998). Dinosaurs used an incredible array of ornamentation (crests, frills, horns, plates, spikes and so on) to enhance their intraspecific behavior, presumably especially during mating season. Feathers would have been relatively easy to adapt into lightweight, colorful display structures that can be shed and replaced.

Display may not have been the only function for the remiges and retrices of *Caudipteryx*. Specimens of related oviraptorosaurs have been found on nests of eggs (Dong and Currie 1996; Clark et al. 1999) in poses that suggest the remiges may have protected the eggs from the elements (Hopp and Orsen 2004). The presence of a fan of retrices at the end of the tail is presumably correlated with the reduction in the number of tail vertebrae and the development of a pygostyle in ovirap-torosaurs (Barsbold et al. 2000).

Known feathered dinosaurs are all coelurosaurian theropods, which are well represented in the Cretaceous fossil record, particularly in the Northern Hemisphere. Although feathers are rarely preserved outside of northeastern China, the majority of coelurosaurians were almost certainly covered with feathers when they were alive. Even the gigan-tic tyrannosaurids are coelurosaurians (Holtz 1994, 2000), and may have had feathers somewhere on their bodies at some stage in their lives. The idea was initially presented in a popular article in *National Geographic*, and subsequently a more primitive tyrannosauroid was discovered with feathers in Liaoning (Xu et al. 2005). *Dilong paradoxis* was a relatively small dinosaur, about the size of a German shepherd dog, and the feathers were presumably useful for insulation. However, the known Late Cretaceous tyrannosaurids are all large animals, each weighing in excess of four tonnes at maturity. Patches of skin impres-sions are known from several genera (Currie et al. 2003), and there is no indication of feathers. Such large animals would not have needed feathers as adults because their ratios of surface area to volume would have made it difficult to get rid of excess body heat if any form of insu-lation covered their bodies. Nevertheless, it is not impossible that ty-rannosaurids might have had some sort of insulating down when they hatched, or that the adults used feathers for display.

More "feathered" theropods have been found in Liaoning and are in the process of being described. In addition to feathered species from northeastern China, alvarezsaurid theropods (*Shuvuuia*) from Mongo-lia are known to have had feathers (Schweitzer et al. 1997) and several ornithomimid skeletons from Alberta also seem to preserve feather-like impressions. The described species of "feathered" dinosaurs repre-sent six different families of non-avian theropods (fig. 3.1). In addition, well-preserved footprints from the United States suggest that thero-pods may have had feather-like structures on their body as early as the Early Jurassic (Kundrat 2004). Filamentous epidermal structures in psittacosaurids (Mayr et al. 2002) and an apparent heterodontosaurid

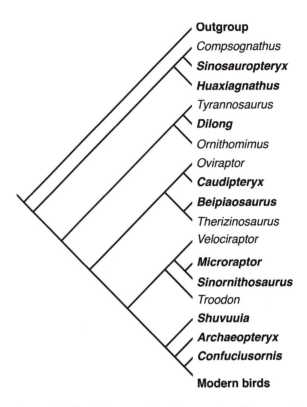

Figure 3.1. A simplified phylogeny of coelurosaurian dinosaurs showing the distribution of dinosaurs that are known to have feathers (bold italics) along with some of their better known, closest relatives.

(Zheng et al. 2009) do not seem to be particularly feather-like, but do suggest it was not just theropod dinosaurs that had a tendency to develop epidermal excrescences.

What Does It Mean to Be a Bird?

The presence of feathers does not prove that birds came from dinosaurs, but in conjunction with stronger evidence in the skeleton (Gauthier 1986; Chiappe 1995; Holtz 2000; Sumida and Brochu 2000; Chiappe and Witmer 2002), the structure of nests and eggs (Chiappe 2004), and inferred behavior (Clark et al. 1999; Chiappe 2004), it is a far more robust hypothesis than any of the alternatives for bird origins.

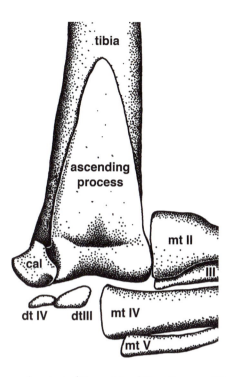

Figure 3.2. Specimen drawing of the ankle of *Caudipteryx* (National Geological Museum of China NGMC 97-9-A), showing the ascending process of the astragalus rising in front of the tibia. Note that the calcaneum is relatively small and disk-like, and does not contribute to the ascending process. Abbreviations: cal, calcaneum; dt, distal tarsal; III, metatarsal III; mt, metatarsal.

Furthermore, feathers are such complex structures that finding feathers on dinosaurs has done more to convince paleontologists and the public that birds are living representatives of the Dinosauria than all of the other lines of evidence. Nevertheless, there are a small number of ornithologists and paleontologists who strongly oppose the hypothesis (Feduccia 1996; Ruben et al. 1997; Feduccia et al. 2005; Lingham-Soliar et al. 2007). The lack of a convincing alternative for bird ancestry and the use of circular reasoning undermine their arguments, however. For example, they argue that the structure of the ankle is different in theropod dinosaurs and birds. Although they claim that *Caudipteryx* is a secondarily flightless bird (Jones et al. 2000) because it has feathers, this animal has the same ankle structure (fig. 3.2) as coelurosaurian theropods. Therefore, arguing that *Caudipteryx* is a bird because

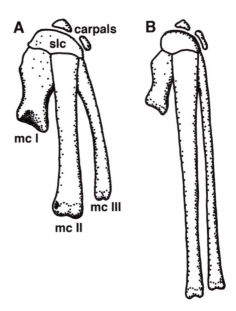

Figure 3.3. The wrists and metacarpals of dromaeosaurids (a) and early birds (b) are nearly identical in their structure with a semilunate carpal capping the three metacarpals. The most medial metacarpal (Metacarpal I according to conventional wisdom) is much shorter than the other two. However, the most lateral of the three metacarpals is much more slender than the other two. Abbreviations: mc, metacarpal; slc, semilunate carpal. (After Paul 2002.)

it has feathers eliminates the argument that birds cannot be derived from theropods because they have different ankle structure.

One of the strongest lines of argument used against the origin of birds from dinosaurs is related to apparent differences in the hand. Note that the differences being argued have nothing to do with the anatomy of the hand, because the hands of *Archaeopteryx* and other early birds are virtually identical to those of non-avian theropods (fig. 3.3). No other animals being considered as possible bird ancestors have hands that are even remotely similar to those of either group. Rather, the argument against the relationship of theropods and birds is that the fingers are developmentally different—that the three fingers of the theropod hand are I, II, and III, whereas bird fingers are II, III, and IV. Fossil evidence shows that in the evolutionary history of theropod dinosaurs, the tendency is to reduce the number of fingers from the outside of the hand. The late Triassic *Herrerasaurus* had four functional fingers, and the presence of a vestigial fifth metacarpal clearly shows that those fingers were I, II, III, and IV. Similarly, by the late Jurassic the large carnosaur *Sinraptor* had three functional fingers (I, II, III) with a vestigial fourth metacarpal. Late Cretaceous theropods like *Tyrannosaurus rex* had only two functional fingers (I, II), but still had a vestigial third metacarpal. The fingers in these theropods are identified not only by their position, but by the number of joints in each. The first finger has two phalanges, the second has three, the third has four, and the fourth has five. Embryological development of some modern birds shows that the three fingers that ossify are lateral to a cartilaginous block that might represent a vestigial finger that does not develop. This would suggest that the fingers in birds are the second, third, and fourth fingers (Larsson and Wagner 2002; Galis et al. 2003). Although this has been argued against by many (including Vargas and Fallon 2005a, b), others have suggested mechanisms whereby the apparent shift is not problematic within an evolutionary lineage (Wagner and Gauthier 1999). However, early birds like *Archaeopteryx* and *Confuciusornis* have two phalanges in the most medial finger, three in the middle finger, and four in the outside one. This is the fundamental phalangeal formula for the first three digits in the hands of not just dinosaurs, but virtually all tetrapods (including most amphibians, most primitive and modern reptiles, and primitive mammalian ancestors). And even though we have found many embryonic dinosaur fossils, the cartilages are not preserved in the wrists of any theropod dinosaurs. So how can we know that the embryological development in theropods like *Velociraptor* was different from what we see in modern birds? Maybe these animals also had a block of cartilage medial to their first finger during embryological development. We almost certainly will never know. Using embryological data to suggest that theropods cannot be ancestral to birds is therefore "negative evidence" and does not negate the vast number of derived characters shared by theropod dinosaurs and birds. We do know that the hands of most coelurosaurian dinosaurs are morphologically nearly identical to the hands of early birds, regardless of whether the three fingers represented are I, II, and III or II, III, and IV in one or both groups.

Archaeopteryx represents the dividing line between dinosaurs and birds. Related animals more derived or advanced than *Archaeopteryx* are birds, whereas more primitive species are not. At present, we know of no non-avian theropod that might have had powered flight. Birds can therefore be defined as animals that fly using wings with specialized feathers, plus their descendants that have lost the ability to fly. Although the presence of feathers separates birds from all other living animals, they do not define birds because they were also present in their ancestors that could not fly. If birds were diagnosed as all feathered animals, then it would be necessary to reclassify all feathered dinosaurs as birds, as well as all of their direct descendants. Preservation of feathers is rare, and therefore we will probably never know how widespread feathers were among theropods and other types of dinosaurs. Rather than classify dinosaurs like *Tyrannosaurus rex* as birds, it is more logical to emphasize flight, rather than feathers, in diagnosing Aves.

The fact that we are having trouble classifying *Archaeopteryx* and many theropods as either birds or non-avian theropods only emphasizes how closely related dinosaurs and birds are to each other. As we draw toward consensus on the ancestry of birds, attention is shifting to equally interesting problems—the evolution of feathers (Prum 1999), and the origin of flight (Padian and Dial 2005).

FEATHER ORIGINS

The feathers of modern birds are complex structures that serve multiple functions in living animals (Stettenheim 2000). In the absence of fossil evidence of feathers, speculation about the origin of feathers was usually creative but untestable. The apparent resolution of bird ancestry and the discovery of fossilized feathers associated with ancestral morphotypes have constrained the hypotheses of feather evolution. The most widely accepted sequence for feather evolution is that they probably appeared initially for insulation in small, endothermic dinosaurs. Some of the insulating feathers then were modified into long, stiff structures behind the arms at the end of the tail (probably for display, but perhaps also for protection of the eggs and young). The long, stiff feathers presumably gave non-avian theropods aerodynamic capabilities that could then be selected for, and eventually were modified into a flight mechanism. We can see a stepwise sequence from the simple protofeathers of *Sinosauropteryx* to the presence of remiges and retrices in *Caudipteryx* (which still apparently had more simple feathers covering its body) to the full covering of contour feathers that we see in modern birds. Clearly there were other things going on, however, as the animals "experimented" with alternative feather designs in both dinosaurs (Xu et al. 2009) and early birds.

The evolution of the complex structure of feathers has long been a source of thoughtful, often creative speculation. A recent review (Prum and Brush 2002) suggests that feathers evolved (probably from scales) through a series of adaptations in the developmental mechanisms of the follicles and feather germs in non-avian theropods. A testable model proposed by Prum (1999) envisioned that feathers originated with the first feather follicle folding the skin around the initial feather papilla into an unbranched, hollow, tubular feather. The tubular follicle collar subsequently differentiated into barb ridges to generate the barbs. At this stage, the follicle would grow a tuft of basally fused barbs. Helical growth and differentiation of the barbule plate came next, and although the order of origination of these steps is not apparent, the combination would result in a rachis with branched barbules. This would be the first bipinnate feather. After the development of the branching structure of the feather, the barbules differentiated into the locking proximal and distal segments to form the closed vane with its integrated parts. The model fits both the developmental stages of modern feathers and the phylogenetic sequence of evolutionary development of feathers in non-avian theropods.

Origin of Flight

The origin of powered flight in birds has also long been a source of fruitful investigation (Padian et al. 2001; Chatterjee and Templin 2004). Although the subject merits an entire book by itself, the question of how bird flight originated has not been, and may never be, resolved. The various hypotheses of how cursorial, non-avian theropods became airborne fall into two major categories—"ground up" or "trees down." Given that many of the non-avian, feathered theropods were long-legged, presumably ground-dwelling forms, it does make sense that these animals may have learned to fly from the "ground up" by selection working on the aerodynamic potential of the long feathers extending behind their arms (as in *Caudipteryx* and *Sinornithosaurus*). Logic has long suggested (Heilmann 1927) that active flight may have been preceded by a gliding stage, which has usually evolved from "trees down" in arboreal species (including living species of frogs, lizards, snakes, and mammals that have evolved gliding appendages). Both ideas find ample support. Dial (2003), for example, showed that galliform birds routinely flap their wings to give themselves more traction when they are running. This confirms that there is a selective advantage for a ground-dwelling animal to evolve wings, and supports the "ground up" family of hypotheses. The presence of long feathers on the hind legs of some of the smallest feathered dinosaurs, however, suggests that *Microraptor* and its closest relatives were arboreal gliders rather than cursorial (Xu et al. 2003; Chatterjee and Templin 2007). This supports the "trees down" hypothesis. The problem is that even though *Microraptor* was probably a glider, it lived in early Cretaceous times, long after the first birds were actively flying in the Late Jurassic. The presence of feathers would have allowed non-avian theropods to evolve into gliding forms any time before or after active flight had developed in one of the non-avian theropod lineages.

Indeed, it is even possible that active flight evolved in more than one lineage. If we accept that *Archaeopteryx* represents the division between non-avian and avian theropods, we may eventually find (assuming that we have not already found) non-avian theropods that were active fliers. Part of the problem, unless we can build a time-machine to transport us back in time for the purposes of observation, will always be that it is difficult to determine from even well-preserved feathered fossils exactly what the flight capabilities were in the living animals. In spite of having well-developed wings with asymmetrical feathers (Feduccia 1996), the flight capabilities of *Archaeopteryx* have been questioned.

CONCLUSIONS

The most logical ancestral forms for birds are small, meat-eating dinosaurs that lived during Jurassic times. Under a modern biological or paleontological classification (phylogenetic systematics), birds are a subdivision of the Dinosauria. As such, there are still over ten thousand species of dinosaurs alive today. That does not mean that birds are not birds, of course. It is simply that they are part of a much larger group in the same way that humans are primates, which in turn are mammals. The step-by-step transition between non-avian theropod dinosaurs and birds is well documented by a remarkable series of fossils from north-eastern China and other parts of the World. Feathers clearly appeared in non-avian theropods before birds took to the air some 150 million years ago. The exact time when the first member of this lineage became an active flier is unknown, and probably never will be known because it is impossible to observe animal behavior directly from fossils. Neither the presence of feathers, nor the ability to fly actively, can be used to diagnose birds. *Archaeopteryx* has been accepted by the vast majority of researchers as the earliest bird since its initial discovery in 1862, and is now considered by most as the dividing line between non-avian dino-saurs and birds (avian dinosaurs). Because the question of the origin of birds has now been resolved to the satisfaction of most paleontologists and ornithologists, the focus of active research has shifted more to questions of the origin of feathers and the origin of avian flight.

REFERENCES

Bakker, R. T. 1975. Dinosaur renaissance. *Scientific American* 232(4): 58–78.

Barsbold, R., H. Osmólska, M. Watabe, P. J. Currie, and K. Tsogtbaatar. 2000. A new oviraptorosaur (Dinosauria, Theropoda) from Mongolia: the first dino-saur with a pygostyle. *Acta Palaeontologica Polonica* 45: 97–106.

Brush, A., L. D. Martin, J. H. Ostrom, and P. Wellnhofer. 1997. Bird or Dino-saur?—statement of a team of specialists. *Episodes* 20: 46.

Buckland, W. 1824. Notice on *Megalosaurus* or great fossil lizard of Stones-field. *Geological Society of London, Transactions* 21: 390–397.

Chatterjee, S., and R. J. Templin. 2004. Feathered coelurosaurs from China: new light on the arboreal origin of avian flight. In P. J. Currie, E. B. Koppel-hus, M. A. Shugar and J. L. Wright, eds., *Feathered Dragons, Studies on the Transition from Dinosaurs to Birds*, 251–281. Bloomington: Indiana Univer-sity Press.

———. 2007. Biplane wing planform and flight performance of the feathered dinosaur *Microraptor gui*. *Proc. Natl. Acad. Sci. USA* 104: 1576–1580.

Chen P. J., Dong Z. M., and Zheng S. N. 1998. An exceptionally well-preserved theropod dinosaur from the Yixian Formation of China. *Nature* 391: 147–152.

Chiappe, L. M. 1995. The first 85 million years of avian evolution. *Nature* 378: 353.

———. 2004. The closest relatives of birds. *Ornitologia Neotropical* 15 (Supplement): 1–16.

Chiappe, L. M., S.-A. Ji, Q. Ji, and M. A. Norell. 1999. Anatomy and systematics of the Confuciusornithidae (Theropoda: Aves) from the Late Mesozoic of northeastern China. *American Museum of Natural History, Bulletin* 242: 1–89.

Chiappe, L. M., and L. M. Witmer. 2002. *Mesozoic Birds, Above the Heads of Dinosaurs*. Berkeley: University of California Press.

Clark, J. M., M. A. Norell, and L. M. Chiappe. 1999. An oviraptorid skeleton from the late Cretaceous of Ukhaa Tolgod, Mongolia, preserved in an avian-like brooding position over an oviraptorid nest. *American Museum Novitates* 3265: 1–36.

Currie, P. J. 1985. Cranial anatomy of *Stenonychosaurus inequalis* (Saurischia, Theropoda) and its bearing on the origin of birds. *Canad. J. Earth Sci.* 22: 1643–1658.

———. 1987. Bird-like characteristics of the jaws and teeth of troodontid theropods (Dinosauria, Saurischia). *J. Vert. Paleontol.* 7: 72–81.

———. 1997. Feathered dinosaurs. In P. J. Currie and K. Padian, eds., *The Encyclopedia of Dinosaurs*, 241. San Diego: Academic Press.

———. 1998. *Caudipteryx* Revealed. *National Geographic Magazine* 194(1): 86–89.

———. 2000. Feathered dinosaurs. In G. S. Paul, ed., *The Scientific American Book of Dinosaurs*, 183–189. New York: St. Martin's Press.

———. 2008. Fifteen years of Prehistoric Times and the cultural icon of science, the Dinosauria. *Prehistoric Times* 85: 10–11.

Currie, P. J., D. Badamgarav, and E. B. Koppelhus. 2003. The first Late Cretaceous footprints from the Nemegt locality in the Gobi of Mongolia. *Ichnos* 10: 1–13.

Currie, P. J., and P.-J. Chen. 2001. Anatomy of *Sinosauropteryx prima* from Liaoning, northeastern China. *Canad. J. Earth Sci.* 38: 1705–1727.

Currie, P. J., and D. A. Russell. 2005. The geographic and stratigraphic distribution of articulated and associated dinosaur remains. In P. J. Currie and E. B. Koppelhus, eds., *Dinosaur Provincial Park, a Spectacular Ancient Ecosystem Revealed*, 537–569. Bloomington: Indiana University Press.

Davis, P. G., and D.E.G. Briggs. 1995. Fossilization of feathers. *Geology* 23: 783–786.

Dial, K. 2003. Wing-assisted incline running and the evolution of flight. *Science* 299: 402–404.

Dong, Z. M., and P. J. Currie. 1996. On the discovery of an oviraptorid skeleton on a nest of eggs at Bayan Mandahu, Inner Mongolia, People's Republic of China. *Canad. J. Earth Sci.* 33: 631–636.

Dyke, G. J., and M. A. Norell. 2005. *Caudipteryx* as a non-avialan theropod rather than a flightless bird. *Acta Palaeontologica Polonica* 50: 101–106.

Erickson G. M., P. J. Makovicky, P. J. Currie, M. A. Norell, S. A. Yerby, and C. A. Brochu. 2004. Gigantism and comparative life-history parameters of tyrannosaurid dinosaurs. *Nature* 430: 772–775.

Feduccia, A. 1996. *The Origin and Evolution of Birds*. New Haven, CT: Yale University Press.

Feduccia, A., T. Lingham-Soliar, and J.R. Hinchliffe. 2005. Do feathered dinosaurs exist? Testing the hypothesis on neontological and paleontological evidence. *Journal of Morphology* 266: 125–166.

Galis, F., M. Kundrat, and B. Sinervo. 2003. An old controversy solved: bird embryos have five fingers. *Trends in Ecology and Evolution* 18: 7–9.

Gauthier, J. 1986. Saurischian monophyly and the origin of birds. In K. Padian, ed., *The Origin of Birds and the Evolution of Flight*, 1–55. California Academy of Sciences, San Francisco.

Göhlich, U. B., and L. M. Chiappe. 2006. A new carnivorous dinosaur from the Late Jurassic Solnhofen archipelago. *Nature* 440: 329–332.

He, T., X.-L. Wang, and Z.-H. Zhou. 2008. A new genus and species of caudipterid dinosaur from the Lower Cretaceous Jiufotang Formation of western Liaoning, China. *Vertebrata PalAsiatica* 46: 178–189.

Hecht, M. K., J. H. Ostrom, G. Viohl, and P. Wellnhofer, editors. 1985. *The Beginnings of Birds, Proceedings of the International Archaeopteryx Conference, Eichstatt, 1984*. Eichstatt, Germany: Freunde des Jura-Museums Eichstatt.

Heilmann, G. 1927. *The Origin of Birds*. London: D. Appleton and Company.

Holtz, T. R. 1994. The phylogenetic position of the Tyrannosauridae: implications for theropod systematics. *J. Paleontol.* 68: 1100–1117.

Holtz, T. R., Jr. 2000. A new phylogeny of the carnivorous dinosaurs. *Gaia* 15: 5–61.

Hopp, T. P., and M. J. Orsen. 2004. Dinosaur brooding behavior and the origin of flight feathers. In P. J. Currie, E. B. Koppelhus, M. A. Shugar, and J. L. Wright, eds., *Feathered Dragons, Studies on the Transition from Dinosaurs to Birds*, 234–250. Bloomington: Indiana University Press.

Huxley, T. H. 1868. On the animals which are most nearly intermediate between the birds and reptiles. *Annals of the Magazine of Natural History, London* 2(4): 66–75.

———. 1870. Further evidence of the affinity between the dinosaurian reptiles and birds. *Quart. J. Geol. Soc. London* 26: 12–31.

Hwang, S. H., M. A. Norell, Q. Ji, and K.-Q. Gao. 2002. New specimens of *Microraptor zhaoianus* (Theropoda: Dromaeosauridae) from northeastern China. *American Museum Novitates* 3381: 1–44.

Ji, Q., P. J. Currie, M. A. Norell, and S.-A. Ji. 1998. Two feathered dinosaurs from northeastern China. *Nature* 393: 753–761.

Ji, Q., and S.-A. Ji. 1996. On discovery of the earliest bird fossil in China and the origin of birds. *Chinese Geology* 233: 30–33 (in Chinese).

———. 1997a. Protarchaeopterygid bird (*Protarchaeopteryx* gen. nov.)— fossil remains of archaeopterygids from China. *Chinese Geology* 238: 38–41 (in Chinese).

———. 1997b. Advances in the study of the avian *Sinosauropteryx prima*. *Chinese Geology* 242: 30–32 (in Chinese).

Ji, Q., M. A. Norell, K.-Q. Gao, S.-A. Ji, and D. Ren. 2001. The distribution of integumentary structures in a feathered dinosaur. *Nature* 410: 1084–1088.

Jones, T. D., J. O. Farlow, J. A. Ruben, D. M. Henderson, and W. J. Hillenius. 2000. Cursoriality in bipedal archosaurs. *Nature* 406: 716–718.

Kundrat, M. 2004. When did theropods become feathered? Evidence for pre-*Archaeopteryx* feathery appendages. *J. Exp. Zool.* 302B: 355–364.

Larsson, H.C.E., and G. Wagner. 2002. Pentadactyl ground state of the avian wing. *J. Exp. Zool.* 294: 146–151.

Lingham-Soliar, T., A. Feduccia, and X.-L. Wang. 2007. A new Chinese specimen indicates that 'protofeathers' in the Early Cretaceous theropod dinosaur *Sinosauropteryx* are degraded collagen fibres. *Proc. R. Soc. B* 274: 1823–1829.

Longrich, N. R., and P. J. Currie. 2009. A microraptorine (Dinosauria-Dromaeosauridae) from the Late Cretaceous of North America. *Proc. Natl. Acad. Sci. USA* 106: 5002–5007.

Makovicky, P., and P. J. Currie. 1998. The presence of a furcula in tyrannosaurid theropods, and its phylogenetic and functional implications. *J. Vert. Paleontol.* 18: 143–149.

Mayr, G., D. S. Peters, G. Plodowski, and O. Vogel. 2002. Bristle-like integumentary structures at the tail of the horned dinosaur *Psittacosaurus*. *Naturwissenschaften* 89: 361–365.

Mayr, G., B. Pohl, S. Hartman, and D. S. Peters. 2007. The tenth skeletal specimen of *Archaeopteryx*. *Zool. J. Linn. Soc.* 149: 97–116.

Osborn, H. F. 1924. Three new theropoda, *Protoceratops* Zone, central Mongolia. *American Museum Novitates* 144: 1–12.

Osmólska, H. 1976. New light on skull anatomy and systematic position of *Oviraptor. Nature* 262: 683–684.

Ostrom, J. H. 1969. Osteology of *Deinonychus antirrhopus*, an unusual theropod from the Lower Cretaceous of Montana. *Peabody Museum of Natural History, Bulletin* 30, 165 pp.

———. 1973. The ancestry of birds. *Nature* 242: 136.

———. 1978. The osteology of *Compsognathus longipes* Wagner. *Zitteliana* 4: 73–118.

Owen, R. 1842. Report on British fossil reptiles. *Report of the British Association for the Advancement of Sciences, Part II*. 1841: 60–204.

Padian, K., and K. P. Dial. 2005. Could 'four-winged' dinosaurs fly? *Nature* 431: 925.

Padian, K., Q. Ji, and S.-A. Ji. 2001. Feathered dinosaurs and the origin of flight. In D. H. Tanke and K. Carpenter, eds., *Mesozoic Vertebrate Life*, 117–135. Bloomington: Indiana University Press.

Paul, G. S. 1988. *Predatory Dinosaurs of the World*. New York: Simon & Schuster.

———. 2002. *Dragons of the Air*. Baltimore: Johns Hopkins University Press.

Prum, R. O. 1999. Development and evolutionary origin of feathers. *J. Exp. Zool.* 285: 291–306.

Prum, R. O., and Brush, A. H. 2002. The evolutionary origin and diversification of feathers. *Quart. Rev. Biol.* 77: 261–295.

Ruben, J. A., T. D. Jones, and N. R. Geist. 1997. Lung structure and ventilation in theropod dinosaurs and early birds. *Science* 278: 1267–1270.

Schweitzer, M. H., J. Watt, C. Forster, M. Norell, and L. Chiappe. 1997. Keratinous structures preserved with two Late Cretaceous avian theropods from Madagascar and Mongolia. *J. Vert. Paleontol.* 17: 74A.

Schweitzer, M. H., J. L. Wittmeyer, and J. R. Horner. 2005. Gender-specific reproductive tissue in ratites and *Tyrannosaurus rex*. *Science* 308: 1456–1460.

Stettenheim, P. R. 2000. The integumentary morphology of modern birds—an overview. *American Zoologist* 40: 461– 477. 206–219.

Sumida, S. S., and C. A. Brochu. 2000. Phylogenetic context for the origin of feathers. *American Zoologist* 40: 486–503.

Vargas, A. O., and J. F. Fallon. 2005a. Birds have dinosaur wings: the molecular evidence. *J. Exp. Zool.* 304B: 86–90.

———. 2005b. The digits of the wing of birds are 1, 2 and 3: A Review. *J. Exp. Zool.* 304B: 206–219.

Wagner, G. P., and J. A. Gauthier. 1999. 1,2,3 = 2,3,4: a solution to the problem of the homology of the digits in the avian hand. *Proc. Natl. Acad. Sci. USA* 96: 5111–5116.

Walker, A. D. 1972. New light on the origin of birds and crocodiles. *Nature* 237: 257–263.

Wellnhofer, P. 2008. *Archaeopteryx*, der Urvogel von Solnhofen. Munich: Verlag Dr. Friedrich Pfeil.

Whetstone, K. N., and L. D. Martin. 1979. New look at the origin of birds and crocodiles. *Nature* 279: 234–236.

Xu, X., M. A. Norell, X.-W. Kuang, X.-L. Wang, Q. Zhao, and C.-K. Jia. 2005. Basal tyrannosauroids from China and evidence for protofeathers in tyrannosauroids. *Nature* 431: 680–684.

Xu, X., Z.-L. Tang, and X.-L. Wang. 1999. A therizinosaur dinosaur with integumentary structures from China. *Nature* 399: 350–354.

Xu, X., and X.-L. Wang. 2003. A new maniraptoran dinosaur from the early Cretaceous Yixian Formation of western Liaoning. *Vertebrata PalAsiatica* 41: 195–202.

Xu, X., X.-L. Wang, and X.-C. Wu. 1999. A dromaeosaurid dinosaur with a filamentous integument from the Yixian Formation of China. *Nature* 401: 262–266.

Xu, X., and X.-C. Wu. 2001. Cranial morphology of *Sinornithosaurus millenii* Xu et al. 1999 (Dinosauria: Theropoda: Dromaeosauridae) from the Yixian Formation of Liaoning, China. *Canad. J. Earth Sci* 38: 1739–1752.

Xu, X., Z.-H. Zhou, and R. O. Prum. 2001. Branched integumental structures in *Sinornithosaurus* and the origin of feathers. *Nature* 410: 200–204.

Xu, X., Z.-H. Zhou, and X. L. Wang. 2000. The smallest know non-avian theropod dinosaur. *Nature* 408: 705–706.

Xu, X., Zheng, X.-T., and H.-L. You. 2009. A new feather type in a nonavian theropod and the early evolution of feathers. *Proc. Natl. Acad. Sci. USA* 106: 832–834.

Xu, X., Z.-H. Zhou, X.-L. Wang, X.-W., Kuang, F.-C. Zhang, and X.-K. Du. 2003. Four-winged dinosaurs from China. *Nature* 421: 335–340.

Zheng, X.-T., H.-L. You, X. Xu, and Z.-M. Dong 2009. An Early Cretaceous heterodontosaurid dinosaur with filamentous integumentary structures. *Nature* 458: 333–336.

Zhou Z.-H., and X.-L. Wang. 2000. A new species of *Caudipteryx* from the Yixian Formation of Liaoning, northeast China. *Vertebrata PalAsiatica* 38: 111–127.

Chapter Four

Phylogeography and Phylogenetics in the Nuclear Age

Christopher N. Balakrishnan, June Y. Lee,
and Scott V. Edwards

In the 1970s and 1980s, when allozymes were firmly entrenched as the workhorse of surveys of genetic variation of natural populations, avian evolutionary genetics got off to a rocky start. The conclusions were unmistakable: typical avian species exhibited fewer genetic differences and smaller genetic distances between them than did typical species of other vertebrate groups (reviewed in Aquadro and Avise 1982; Avise and Aquadro 1982). This pattern did not imply that avian species were devoid of genetic variation within populations: measures of allozyme heterozygosity in bird populations were generally on par with those of other vertebrates. Nor did it hinder attempts to reconstruct evolutionary history: despite the smaller genetic distances between avian species, researchers were nonetheless able to make important strides in proposing and interpreting avian speciation and biogeographic patterns. The causes of these small genetic distances were difficult to discern, in part because the forces acting on allozymes were not readily apparent. On the one hand, small genetic distances could be due to reduced substitution rates in birds that caused genetic differences to accumulate slowly (Avise and Aquadro 1982; Mindell et al. 1996). Another hypothesis was that, given recent divergence of

We would like to thank past technicians, graduate students, and postdocs in the Edwards lab, particularly those working on Australian birds, as well as our many Australian friends and colleagues, for advancing our Australia work over the years. Most recently, this work has been supported by grants from the National Science Foundation and the Putnam Fund of the Museum of Comparative Zoology at Harvard University, and has been facilitated by permits from the respective wildlife departments of New South Wales, Queensland, Western Australia, and the Northern Territory.

bird species, in combination with large effective population sizes, there simply had not been sufficient time for genetic differences to accumulate via genetic drift (e.g., Barrowclough and Corbin 1978; Barrowclough et al. 1981).

In the early allozyme studies, what was widely recognized, and eventually taken as a matter of course, was that the small genetic distances of birds were driven by abundant shared polymorphisms. Such shared polymorphisms in allozymes have variously been ascribed to natural selection (e.g., balancing selection), large population sizes, hybridization, or recent divergences of avian species (Karl and Avise 1992). The remarkable evidence for fixed or nearly fixed differences at multiple allozyme loci in some groups, such as salamander populations of Northern California and the Pacific Northwest (Good and Wake 1997), was rarely observed among avian species. Shared polymorphisms were even widely observed at higher taxonomic levels in birds with many allozyme electromorphs shared between species of different genera (e.g., Barrowclough and Corbin 1978; Barrowclough et al. 1981; Avise and Aquadro 1982; Lanyon and Zink 1987). Shared polymorphisms among avian species were simply a fact of life and raised eyebrows only insofar as their quantity might differ among vertebrate groups and in particular be more common in birds than other vertebrates.

As we celebrate the retirement of Peter and Rosemary Grant, we take this opportunity to place shared genetic polymorphisms in birds in a new light based on our recent surveys of nuclear DNA sequence variation in Australian birds. This is appropriate for this festschrift; Darwin's finches, the birds that have allowed the Grants to study evolutionary processes in such detail, share abundant genetic polymorphisms among species. This conclusion is amply backed up by several studies of genetic variation among Darwin's finches by the Grants and others (e.g., Freeland and Boag 1999; Grant et al. 2005). The interpretation of these shared polymorphisms has generated alternative, if somewhat controversial, visions for the radiation of Darwin's finches (Zink 2002). Indeed, the radiation of Darwin's finches may well turn out to be exceptional among avian radiations in the degree of shared genetic polymorphisms among species and the mechanisms that have produced these shared polymorphisms. Hybridization in particular has been proffered as a potent means of generating shared polymorphisms among Darwin's finches, and this has posed challenges to identifying mechanisms that keep Darwin's finch species ultimately distinct and diverging over time (Grant and Grant 2008).

Studies of nuclear DNA sequence variation—which we have argued provide a better, clearer, and more comprehensive yardstick of avian phylogenetic history than, for example, microsatellites or mtDNA alone (Brito and Edwards 2009; Edwards and Bensch 2009; for reply, see Barrowclough and Zink 2009)—are still in their infancy in birds and our aim is not to review them in their entirety. Rather, we will focus on those patterns and conclusions that stem from our initial studies of nuclear DNA sequence variation in Australian passerines. These birds differ from Darwin's finches in numerous ways, not the least of which is the contrast between their often vast geographic ranges on a large continent and those of Darwin's finches on small oceanic islands. Nonetheless, we suspect that this work will provide a snapshot of a common pattern among many avian species, particularly continentally distributed species. We suggest, however, that the origins of shared polymorphisms in DNA sequence markers are easier to decipher than they were for allozymes. The remarkable diversity of ways in which DNA sequence data can be interpreted and analyzed—via gene trees; synonymous and nonsynonymous variation in the case of protein-coding genes; and an abundance of tests based on the so-called "site frequency spectrum" (the distribution of the frequencies of DNA sequence polymorphisms in a population)—makes DNA sequences superior to other methods of assaying genetic variation in the nuclear genome (Brito and Edwards 2009; Zink and Barrowclough 2008). This breadth of analysis provides a strong foundation for the interpretation of genetic variation in natural populations. In addition, the large number of polymorphisms that we detect help us in testing numerous hypotheses about avian phylogeography and speciation. Finally, the multiplicity of loci that we can now harness to study avian speciation provide further dimensions on which to describe and analyze avian history.

COMPARATIVE PHYLOGEOGRAPHY OF AUSTRALIAN AVIFAUNA

The fauna of Australia provides a rich system for the study of biogeography. A long history of continental isolation has generated striking patterns of endemism while more recent faunal exchange between Australia, New Guinea, and Southeast Asia has also contributed the complex biogeography of the region (reviewed in Joseph and Omland 2009). Many of the biogeographic barriers in Australia are hypothesized to have arisen during the Pleistocene, when environmental conditions in these areas are thought to have become increasingly harsh (e.g., Keast 1961; Cracraft 1986). Cladistic analyses of morphological traits indicate that a codistributed bird species were subdivided first across the Carpentarian barrier in northern Australia, and subsequently by secondary barriers to the east and west (fig. 4.1; Cracraft 1986). Pleistocene glaciation may not only have caused splitting of lineages within continental Australia, but also may have facilitated dispersal between Australia and neighboring regions by reducing the water barrier between them. Mayr (1944) described a group of over forty avian species that crossed the Timor Sea to colonize the Lesser Sunda Islands from Australia and vice versa.

The molecular tools for our recent studies have been anonymous nuclear loci. Although Karl and Avise (1993) introduced the concept of "anonymous loci" in the early 1990s, such markers have been underutilized by evolutionary geneticists until recently. Anonymous loci are non-coding DNA segments selected at random from the genome and are therefore dispersed across chromosomes. They can be easily developed for any species, and can yield a nearly unlimited number of unlinked markers that provide a better representation of genomic variation than any single locus. In addition, anonymous loci are expected to be free from possible selection pressure that other markers, such as nuclear introns, might have experienced due to their proximity to coding regions. Finally, because anonymous loci are chosen at random, they minimize the impact of ascertainment bias in marker choice (Wakeley et al. 2001). By describing the studies below, we will demonstrate that these markers are rich in information, and when combined with advances in statistical phylogeography, provide detailed insight into the history of species divergence.

Discordant Gene Trees and the Divergence of Australian Grassfinches

Grassfinches in the genus *Poephila* are distributed across northern Australia, thus spanning multiple known biogeographical barriers (fig. 4.1; Keast 1961; Cracraft 1986). The *Poephila* complex consists of three

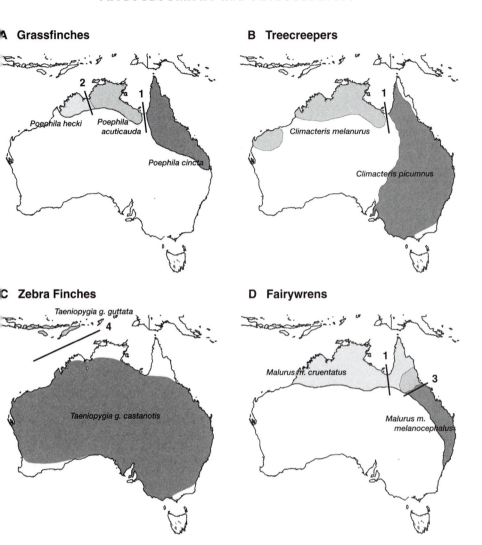

Figure 4.1. Range maps for four study groups of Australian birds. (a) *Poephila* grassfinches; (b) brown and black-tailed treecreepers (Climacteris); (c) zebra finches (Taeniopygia); and (d) red-backed fairy-wrens. Key biogeographic barriers shown are (1) the Carpentarian barrier, (2) the Kimberley plateau–Arnhem Land barrier, (3) the Cape York–Atherton Plateau barrier, and (4) the Timor Sea.

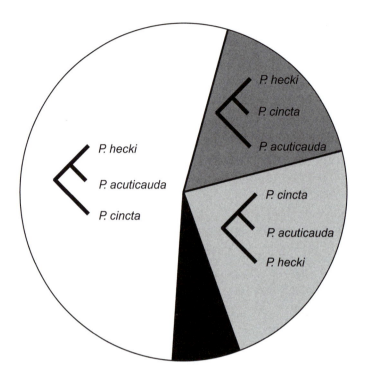

Figure 4.2. Conflicting gene tree topologies in Australian grassfinches. Shown are the frequencies of alternative tree topologies for 30 anonymous nuclear loci. The black wedge represents the frequency of unresolved topologies (2/30).

species, whose divergence has apparently been influenced by the Carpentarian barrier as well as the Kimberley Plateau -Arnhem Land barrier to the West (fig. 4.1). Cracraft (1986) hypothesized an older divergence across the Carpentarian barrier, separating the black-throated finch Poephila cincta in the east and the ancestor of two long-tailed finch species (*P. acuticauda* and *P. hecki*) in the west. To determine whether and when Pleistocene environmental change contributed to divergence among *Poephila* species, Jennings and Edwards (2005) analyzed genetic data from thirty anonymous nuclear loci. Maximum likelihood gene trees revealed the striking result that all three possible alternative tree topologies were represented in the data set (fig. 4.2) highlighting one of the consequences of ancestral polymorphism: stochasticity in the lineage sorting process will frequently cause disagreement between individual gene trees and the species tree. The favored historical scenario, in which the two long-tailed finches are sister taxa, was the most common and was recovered in sixteen of the twenty-eight (57%) resolved trees (fig. 4.2).

Topological discordance among loci, however, is not merely a nuisance. Integrating information contained in multilocus data and gene tree discordance can both yield a well-supported phylogenetic topology for the group and contribute to the estimation of historical demographic parameters (Liu and Pearl 2007). Jennings and Edwards (2005) used maximum likelihood and Bayesian approaches (Yang 2002; Rannala and Yang 2003) and estimated large ancestral populations sizes for the long-tailed finches (N_e = 384,000) and for the basal ancestor of the group (N_e = 521,000). Additionally, they estimated the divergence of long-tailed finches across the Kimberley-Arnhem Land barrier at 0.3 million years, and the divergence of black-throated and long-tailed finches across the Carpentarian barrier to roughly 0.6 million years. Confidence intervals surrounding divergence time estimates were well within the Pleistocene, supporting Pleistocene divergence among grassfinches.

Speciation in Zebra Finches Following a Founder Event

The zebra finch *Taeniopygia guttata*, well known for its role in studies of avian behavior, neurobiology, and now genomics (Zann 1996; Clayton 2004), was hypothesized by Mayr (1944) to have colonized the Lesser Sunda Islands from Australia. Two zebra finch subspecies are recognized on the basis of morphological variation: *T. guttata castanotis* on the Australian mainland, and *T. guttata guttata* on the islands (Clayton 1990). Balakrishnan and Edwards (2009) used population sampling of thirty genetic markers (including 21 anonymous loci, 4 nuclear introns, and 5 Z chromosome linked introns) to test whether the colonization of the island involved a significant founder event and to estimate divergence times among populations. Rather than the late Pleistocene as predicted by Mayr, during which sea levels were at their minimum, coalescent analyses using IM software (Hey 2005) favored a relatively deep divergence dating to the early Pleistocene with confidence intervals extending into the Pliocene (1.2–2.8 Ma). The island subspecies showed a fivefold reduction in genetic diversity relative to mainland birds, with IM analyses suggesting that only a small number

71

of individuals founded the Lesser Sundas population. By contrast, the Australian subspecies appears to have expanded dramatically over its history to a current effective population size of roughly 7 million individuals.

Despite the apparently deep historical divergence of zebra finches, the two subspecies were not reciprocally monophyletic at any of the thirty nuclear loci surveyed. IM analyses suggest that gene flow, however, is unlikely to have made an important contribution toward the genetic similarity of the two populations. Even though shared polymorphisms were the rule, the two subspecies can be readily diagnosed on the basis of genotype alone using the clustering approach employed in the software program STRUCTURE (fig. 4.3; Pritchard et al. 2000; Falush et al. 2003). Given their assortative mating in captivity, as well as behavioral (song), morphological, and genetic diagnosability (Clayton 1990; Balakrishnan and Edwards 2009), most species concepts would favor the recognition of the two zebra finch subspecies as full species. Although not as well characterized as those in the late Pleistocene, sea levels in the early Pleistocene are thought to have oscillated substantially (Lambeck and Chappell 2001; Lambeck et al. 2002). Given the broad confidence intervals around divergence time estimates (Balakrishnan and Edwards 2009), it is unclear whether the colonization of the Lesser Sundas was facilitated by reduced sea levels or was unrelated to environmental conditions. It does appear, however, that a significant founder event played a role in facilitating the divergence of zebra finches (Balakrishnan and Edwards 2009).

Conflicting Molecular and Morphological Taxonomy in the Red-backed Fairy-Wren

The red-backed fairy-wren, *Malurus melanocephalus*, is a small, sedentary Australian passerine, inhabiting northern and eastern Australia. Although the range of the red-backed fairy-wren spans the Carpentarian barrier, the ranges of two morphological subspecies, *M. m. melanocephalus* and *M. m. cruentatus*, are thought to meet along the Cape York-Atherton Tablelands barrier (fig. 4.1; Schodde and Mason 1999). In order to clarify a possible role for the Carpentarian barrier on divergence of red-backed fairy-wrens, Lee and Edwards (2008) quantified variation in twenty-nine newly developed anonymous nuclear markers,

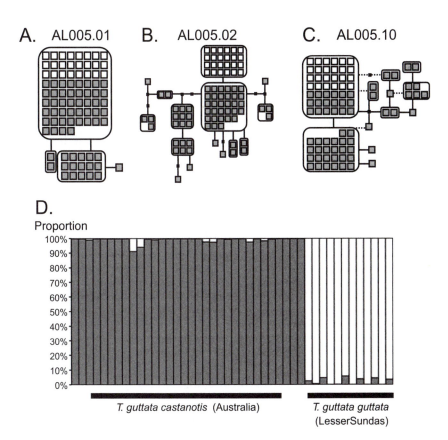

Figure 4.3. Haplotype networks for three anonymous loci (AL005.01, AL005.02, AL005.10) and two zebra finch subspecies *T. guttata guttata* (white) and *T. guttata castanotis* (gray). Squares represent individual haplotypes, and squares within a larger box share the same DNA sequence. Solid lines represent mutational steps, and small squares represent unsampled or missing haplotypes. Note that (b) represents the only case among the 30 loci where the Lesser Sundas subspecies is fixed for a unique haplotype. (d) Result of a clustering analysis in STRUCTURE. Each bar represents the proportion of each individual's ancestry attributed to one population or the other. The probabilistic assignment of individuals sampled from the different regions to distinct populations indicates strong genetic differentiation. (d) is reprinted from Balakrishnan and Edwards (2009) with permission of the Genetics Society of America.

six nuclear introns, and the mitochondrial gene NADH dehydrogenase subunit 2 (ND2). Analyses of over 15 kilobars of sequence data revealed a high degree of shared polymorphism among populations. Neither nuclear nor mitochondrial gene trees showed evidence of reciprocal monophyly among taxonomic subspecies, nor among populations on either side of the Carpentarian Barrier. In fact, shared polymorphisms were detected even between red-backed fairy-wrens and their sister species, the white-winged fairy-wren *Malurus leucopterus* (fig. 4.4), and are likely even more widespread deeper in the fairy-wren tree (Lee and Edwards unpublished). Despite traditional taxonomy considering as continuous taxa on either side of the Carpentarian barrier, Lee and Edwards (2008) found strong genetic divergence across this region (fig. 4.4), consistent with predicted area cladograms in this region (Cracraft 1986). Multilocus coalescent analysis in IM suggests that the western and eastern populations were separated approximately 0.27 million years ago with little subsequent gene flow. The traditional subspecies barrier in the east, however, is linked by gene flow (~2 migrants per generation), resulting in genetic continuity across the region.

Carpentarian Divergence of Australian Treecreepers

As a fourth test of the influence of the Pleistocene environmental change on Australia's avifauna, we have been studying divergence of black-tailed *Climacteris melanura* and brown treecreepers *Climacteris picumnus* across the Carpentarian barrier (fig. 4.1; Edwards et al. unpublished). *Climacteris* treecreepers are trunk foraging, cooperative breeders and the two species are morphologically and ecologically diagnosable. While we have yet to fully resolve the details of treecreeper population history, our preliminary results reflect some of the patterns seen in the studies described above. The two treecreeper species are not reciprocally monophyletic for any of the fifteen anonymous loci that we have sequenced (fig. 4.5). Despite this, STRUCTURE analysis recovers the two species as genetically highly distinctive groups (fig. 4.5). Lastly, coalescent estimates using the program MIGRATE (Beerli and Felsenstein 2001) of $\theta = 4N_e\mu$ (where N_e is the effective population size and μ is the mutation rate) range from 0.004–0.010, suggesting very large effective population sizes for these populations as well, in some cases an order of magnitude larger than that for the human species ($\theta \approx 0.001$).

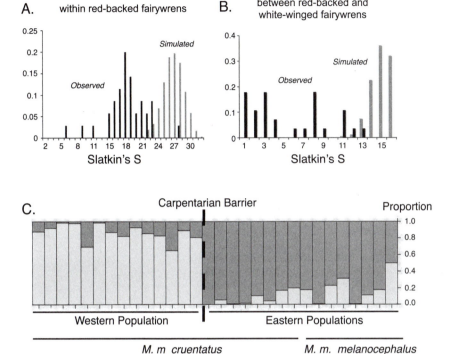

Figure 4.4. Frequency distributions of Slatkin's S in empirical and simulated datasets (a) among eastern and western populations of red-backed fairy wrens, and (b) between red-backed and white-winged fairy wrens. Slatkin's S (Slatkin and Maddison 1989) is a measure of the extent to which gene trees match geographic populations. In both (a) and (b), the distribution of empirical estimates is significantly lower than estimates based on simulated gene trees suggesting significant population structure. High estimates of Slatkin's S also emphasize extensive polymorphism sharing among populations. (c) STRUCTURE clustering diagram showing genetic differentiation across the Carpentarian barrier, but not among recognized subspecies.

Factors Influencing Genetic Diversity and the Maintenance of Ancestral Polymorphisms

The studies described above highlight the utility of multilocus genetic surveys and the pervasiveness of retained ancestral polymorphisms among Australian bird species. Even among relatively deep divergences, such as between long-tailed and black-throated finches, red-backed and white-winged fairy-wrens, and two zebra finch subspecies,

75

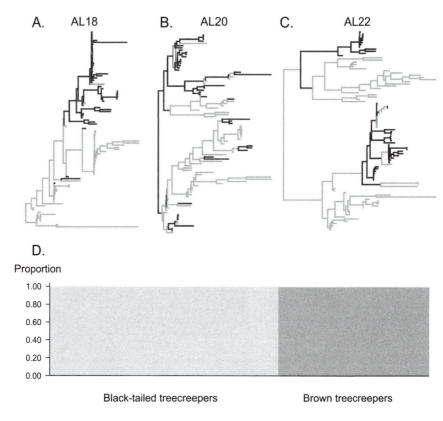

Figure 4.5. (a)–(c) Neighbor Joining gene trees from Australian treecreepers from three anonymous loci (AL18, AL20, AL22) reveal extensively shared polymorphisms among species as well as varying levels of phylogeographic structure (dark gray: brown treecreeper, *C. picumnus*; light gray: black-tailed treecreeper, *C. melanura*). (d) STRUCTURE analysis from a small number of individuals revealing substantial genetic differentiation.

ancestral polymorphisms are common, as is conflict among gene trees. The understanding of the causes of shared polymorphisms has improved dramatically in the last decade. Whereas shared polymorphisms among species were frequently attributed to gene flow, in the studies above, we have been able to apportion shared polymorphisms to either gene flow or ancestry. Rigorous field studies, exemplified by the efforts of the Grants and their colleagues in their studies of Darwin's finches, remain the best approach to quantifying patterns of hybridization and local population dynamics in ecological time scales. However, molecular approaches based on the coalescent model provide a framework within which to address the role of gene flow over evolutionary timescales. By focusing on the structure of geneologies drawn from multiple loci, the coalescent model offers the opportunity to distinguish among the signatures of recent common ancestry and gene flow that are indistinguishable using older metrics such as F_{ST}.

Ancestral population size has long been known to be a primary determinant of the timing of the transition from paraphyly to reciprocal monophyly (e.g., Hudson and Coyne 2002). Knowles and Carstens (2007) present the illustrative example that for a species with an effective population size of 100,000, and a generation time of one year, it would take over one million years of separation before fifteen sampled loci would be reciprocally monophyletic. An important trend among the populations described here is their large effective population size, reflected in the remarkable levels of genetic diversity (table 4.1; fig. 4.6). In particular, species that inhabiting the vast, arid zone of inland Australia such as zebra finches (Balakrishnan and Edwards 2009) and emus *Dromaius novaehollandiae* (Janes et al. 2009) show high genetic diversity (table 4.1) and estimates of N_e. It is therefore perhaps not surprising that among the Australian lineages described here, shared ancestral polymorphisms are rampant. As multilocus studies of avian taxa in other systems develop (such as those with smaller effective population sizes), it will be possible to determine if the extensive polymorphism sharing observed in our studies is related specifically to the biogeography of Australia.

Red-backed fairy-wren subspecies provide an example in which gene flow and ancestral polymorphism interact to promote genetic similarity. At the subspecies boundary in eastern Australia, estimates of gene flow are greater than one migrant per generation, a benchmark for the level of gene flow that would lead to genetic homogenization in the absence of divergent selection. At this interface it is of great interest to explore the selective forces influencing morphological divergence and its relationship with genetic structure. Although the Jennings and Edwards (2005) study of grassfinches did not incorporate population sampling, gene flow also is unlikely to be an important force contributing to topological discordance among gene trees in this system. The two long-tailed finches have variously been considered subspecies or species

TABLE 4.1

Descriptive statistics for four phylogeographical studies in which anonymous nuclear loci were surveyed. Presented are the number of anonymous nuclear loci amplified, their average length in base pairs, range of sample sizes per locus in alleles sequenced (n), range for the number of segregating sites (S), and the range for their nucleotide diversity (π) among loci.

	# Loci	Length (bp)	n	S	π
Red-backed Fairy-wren[1]					
M. melanocephalus	29	442	46–60	4–90	0.003–0.046
Zebra Finch[2]					
T. g. castanotis	21	228	50–64	3–42	0.002–0.025
T. g. guttata	21	228	18–24	0–14	0.000–0.010
Treecreepers					
C. melanura	15	240	42–48	4–25	0.001–0.014
C. picumnus	15	258	42–48	5–19	0.003–0.016
Emu[3]					
D. novaehollandiae	8	191	10–46	4–32	0.003–0.019

[1] Lee and Edwards (2008).
[2] Balakrishnan and Edwards (2009).
[3] Janes et al. (2009).

(Goodwin 1982) and detailed study of the contact zone between the two long-tailed finches would be an exciting extension this work.

CHALLENGES AND OPPORTUNITIES OF SHARED POLYMORPHISMS
Contrasting Patterns of Morphological and Genetic Divergence

Morphological taxonomy of red-backed fairy-wrens, based largely on plumage color, conflicts with the pattern of variation in molecular data. Divergence among the two populations in eastern Australia has been accompanied by divergence in plumage color, while splitting of eastern and western forms has not resulted in substantial morphological divergence. Genetic clustering analysis, however, groups the two morphologically divergent populations together while splitting populations

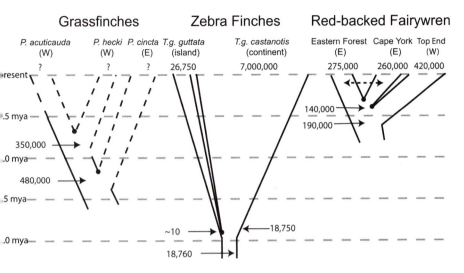

Figure 4.6. Hypotheses of historical demography for three clades of Australian songbirds. *Left*: speciation in Australian grassfinches (*Poephila*), based on Jennings and Edwards (2005). *Center*: speciation in zebra finches, based on Balakrishnan and Edwards (2009). *Right*: speciation in red-backed fairy-wrens (*Malurus melanocephalus*), based on Lee and Edwards (2008). Each scenario is placed on a common temporal scale based on mutation rates and sequence divergences discussed in detail in their respective studies. Current and ancestral effective population sizes are indicated by numbers and lineage widths. Lineages to the east (E) and west (W) of the Carpentarian barrier are indicated (see fig. 4.1). Continental and insular populations of zebra finches are indicated (see text). Filled circles at divergence events indicate speciation times. The dashed lines in the *Poephila* tree indicate that current effective population sizes are not known, since only one allele per species was sampled in that study. However, ancestral population sizes can be inferred from such data and are shown. In the case of *Malurus*, the speciation scenario is a conglomerate of results from pairs of populations, which suggested very similar divergence times for populations on the east and west of the Carpentarian barrier as well as populations only on the east side. Gene flow between eastern forest and Cape York populations is indicated by dashed arrow. These studies were all based on statistical analyses of multilocus sequence variation in 30–40 anonymous loci and introns.

that are similar in plumage color (fig. 4.4). Because of the protracted nature of the lineage sorting process (Avise et al. 1998) and the potential strength of selection, such discord between morphological and genetic divergence is commonplace. The two zebra finch subspecies, the three grassfinch species, and the treecreepers described above all display diagnostic morphological traits despite their extensive sharing of genetic polymorphisms. Indeed, the rapid divergence of morphology relative to DNA in these cases implies a role for natural and/or sexual selection in driving morphological divergence in these groups (Edwards et al. 2005).

Species Delimitation without Reciprocal Monophyly

The examples from Australian birds highlight the fact that by using nuclear gene tree monophyly as a criterion for delimiting species, we would grossly oversimplify the picture of biodiversity. Many "good" avian species, whether on islands or continents, and whether differing by plumage, song, behavior, or geography, will nonetheless routinely share nuclear polymorphisms, and in some cases will share mitochondrial polymorphisms as well. Thus, a frequent pattern observed in avian phylogeographic studies is nuclear gene tree para- or polyphyly and mitochondrial DNA reciprocal monophyly between sister species (Zink and Barrowclough 2008; Joseph and Omland 2009). Our Australian songbird studies have corroborated the pattern for nuclear genes, although only one study so far has compared these to mitochondrial DNA (Lee and Edwards 2008).

One might conclude from such patterns, as some biologists have (Zink and Barrowclough 2008), that nuclear genes are less useful for delimiting species than are mitochondrial genes. This viewpoint stems from the stringent criterion that species must be delimited based on cladistically diagnostics traits, such as the synapomorphies accumulated on gene trees. Vertebrate zoologists, and ornithologists in particular, have often favored species concepts (such as the phylogenetic species concept) that use gene tree and other types of monophyly as a means of delimiting species. This protocol has had some success, but applying it rigorously means placing undue emphasis on a single molecule (e.g., mtDNA) that may have been subject to random and deterministic forces other than those accompanying the demographic history of populations during speciation. In addition, researchers have recently pointed to powerful methods for delimiting species in the absence of gene tree monophyly. Some of these methods involve capitalizing on the coalescent patterns in gene trees (Knowles and Carstens 2007). Other methods involve standard multivariate statistical analysis of linked and unlinked SNPs, such as principal components analysis (PCA) and multidimensional scaling (MDS) (Lessa 1990; Patterson et al. 2006, Reich et al. 2008). Thus, we favor a view in which monophyly at the level of gene trees is not a criterion for species delimitation. Given the widespread occurrence of shared polymorphisms, and evolutionary biologists' growing appreciation that para- and polyphyly of gene trees are simply stages toward reciprocal monophyly after populations have become isolated (Neigel and Avise 1986), it seems reasonable to focus on the isolation of lineages and subpopulations, rather than isolation at the level of neutral genes (Edwards 2009). Once one lets go of a need to demonstrate gene tree monophyly to diagnose species, the emphasis shifts to multivariate and multilocus species delimitation as a means of delimiting species. Such approaches acknowledge that fixed differences between species are not a requirement for diagnosis and open the way forward to protocols for species delimitation that can accommodate both monophyly and shared polymorphisms in descriptions of biodiversity (Edwards et al. 2005).

The tension between cladistic interpretations of diagnosibility and statistical interpretations is reminiscent of some of the debates in the 1960s about phenetics and cladistics. Indeed, the spirit of recent applications of PCA seem very much in line with the phenetic approaches to classification and diagnosis employed by Sokal, Sneath, and others (Sokal and Sneath 1963). By contrast, other approaches to diagnosis, such as the software STRUCTURE (Pritchard et al. 2000; Falush et al. 2003), employ a modicum of a population genetic model and involve Bayesian rather than parametric statistics. However, unlike much of the literature of numerical taxonomy, some new statistical methods, such as PCA and eigenanalysis, have specific predictions as to phylogeographic process, depending on the pattern observed in the PCA plot (Patterson et al. 2006; Reich et al. 2008). The results of these techniques may not be directly translatable to rates of population genetic processes, such as rates of gene flow, as delivered by programs such as IM (Hey 2005) and MIGRATE (Beerli 2006). We note other tensions: whereas some researchers suggest that summary statistics such as F_{st} are less useful as unitary descriptors of population structure, particularly given the limits to interpretation in terms of population processes (Beerli and Felsenstein 2001), we are seeing renewed interest in F_{st} as a descriptor of population structure in other arenas (e.g., Patterson et al. 2006). Nonetheless, these are exciting developments and provide new ways to detect and quantify population structure. We cannot predict which way this debate will turn but simply note that the scale of genetic data that we can now collect to answer questions in species delimitation and phylogeography is vastly greater and more detailed than what was available when these debates were raging forty years ago.

Avian Phylogeography Looking Forward

We suggest that the size and complexity of genetic data sets, and the complexity of relationships among genes and diverging populations, particularly when hybridization is involved, will require new terms and new statistics, some of which are already emerging (Avise and Robinson 2008). Patterson et al. (2006) have developed some useful metrics for conducting power analyses aimed at determining whether a given data set can find statistical significance for a particular value of F_{ST} for two populations. Unsurprisingly, they find that with very large data sets, such as recent data sets in studies of human phylogeography in which the number of individuals times the number of loci is on the order of 1 million, F_{ST}s on the order of 0.001 can easily be deemed significant. These analyses also suggest that there is a premium on the number of loci up to a given threshold value of F_{ST}; once the number of loci is sufficient to obtain significance for this F_{ST}, then there is a premium not on loci but on numbers of individuals for detecting further structure. Of course, an equally important question for phylogeographers is whether the particular value of F_{ST} obtained is biologically significant, as well as statistically significant (Hedrick 1999). As is typical in the history of avian phylogeography, guidelines from recent trends in phylogeographic studies of humans and other model organisms are useful and point to where we will be in five or ten years.

We are also witnessing the rise of new methods in phylogenetics that address the emerging complexity of gene trees and the frequent lack of reciprocal monophyly of these same gene trees. This new class of "species tree" methods acknowledge the key distinction between gene and species trees and explicitly model the latter (Maddison and Knowles 2006; Ané et al. 2007; Liu et al. 2008; reviewed in Liu et al. 2009). They represent a promising advance over standard methods of phylogenetic analysis in the increased realism of the genealogical models they employ, as well as in their ability to handle discordance among gene trees (Edwards 2009). As we have seen in the studies reviewed here, heterogeneity in the topologies and branch lengths of gene trees is common in birds. Concatenation and supermatrix approaches can break down under such conditions, insofar as they assume a single gene tree underlying multilocus DNA sequence data sets. If such approaches do not deliver a topologically incorrect tree, then they may have difficulty assessing the confidence and probability of such trees accurately (Edwards 2009). Species trees approaches sometimes can be misled by gene flow between species or populations (Eckert and Carstens 2008), and for this reason they may be criticized as overly sensitive to such processes. But really these species tree approaches are no more sensitive to gene flow than phylogenetic approaches at other levels; for example, gene trees will also be heavily impacted by gene flow and require careful interpretation when gene flow is occurring. What does seem clear is that species tree approaches can help clarify thinking about patterns in genetic data. The fact that researchers are pointing to gene flow as a potential roadblock to such approaches is a healthy development and increases awareness of the complicating nature of this evolutionary force. As evidenced by the few applications of species tree approaches to multilocus data sets from birds and other lineages, species trees are turning out to be an excellent tool for accommodating data sets with abundant shared polymorphisms on a variety of timescales.

Conclusions

We have seen that shared polymorphisms will be common among closely related avian species and populations. These shared polymorphisms offer opportunities and challenges. Many aspects of the speciation process—the times of species divergence, the population sizes accompanying speciation, the patterns of selection and neutrality among loci—are actually facilitated when polymorphisms are shared, as opposed to when loci exhibit reciprocal monophyly. This conclusion stems from analysis of patterns of gene tree divergence in several diverging species in Australia. When reciprocal monophyly is achieved, patterns may appear clearer, but also much information on ancestral population sizes and structure is lost. Most importantly, shared polymorphisms do not mean that species cannot be easily delimited, as our analyses using STRUCTURE show.

On the other hand, shared polymorphisms present a number of challenges to the phylogeographer, although we hope that we have illustrated that many new and emerging methods can effectively deal with shared polymorphisms. Perhaps the most important dichotomy of shared polymorphisms, and the most pressing challenge, is that such polymorphisms can be shared simply to recent isolation or due to active gene flow and hybridization. Distinguishing between these two scenarios is often difficult, and our interpretations ultimately may be limited not so much by deficiencies in software or analysis tools, but by the size and complexity of our data sets. We are witnessing a gradual shift from the use of microsatellites to sequence-based markers, a shift that we think will help integrate phylogenetics and phylogeography and that will allow more nuanced interpretations of genetic data. This is because SNPs are easier to study in terms of neutrality and selection, can be compared readily between organelle and nuclear genomes, and have a less complex pattern of evolution than do microsatellites (Brito and Edwards 2009; Zink and Barrowclough 2008). Despite great advances in recent years in the size and complexity of data sets from avian and other natural populations, phylogeographic analyses still employ a relatively small number of loci, given the number of individuals and the detail with which we would like to describe phylogeographic history. On the other hand, it is likely that the number of SNPs employed in human phylogeographic studies—which now number in the hundreds of thousands—may be excessive to unravel the histories of many bird species because humans are notoriously unstructured and most avian populations will harbor more structure than will humans. Thus, we suggest that researchers push yet further to increase the number of loci employed in phylogeographic studies as much as is practical. New highly parallel sequencing techniques are emerging and are already being applied to non-model organisms, portending a new era in the collection of genomic data from natural populations (Bonneaud et al. 2008; Ellegren 2008); even further sequencing innovation is imminent (Eid et al. 2009). These techniques will enable genotyping of so many individuals that the premium will shift away from laborious lab work to the more important task of amassing suitable samples and collections from the field, and to data manipulation, computation, and statistical analyses. We predict that phylogeography of birds and other organisms will experience a continuing renaissance, one that began with the deployment of mitochondrial DNA and whose end is not yet in sight.

References

Ané, C., B. Larget, D. A. Baum, S. D. Smith, and A. Rokas. 2007. Bayesian estimation of concordance among gene trees. *Mol. Biol. Evol.* 24: 412–426.

Aquadro, C. F., and J. C. Avise. 1982. Evolutionay genetics of birds .6. a re-examination of protein divergence using varied electrophoretic conditions. *Evolution* 36: 1003–1019.

Avise, J. C., and C. F. Aquadro. 1982. A comparative summary of genetic distances in the vertebrates—patterns and correlations. *Evol. Biol.* 15: 151–185.

Avise, J. C., and T. J. Robinson. 2008. Hemiplasy: A new term in the lexicon of phylogenetics. *Syst. Biol.* 57: 503–507.

Avise, J. C., D. Walker, and G. C. Johns. 1998. Species durations and Pleistocene effects on vertebrate phylogeography. *Proc. R. Soc. Lond. B* 265: 1707–1712.

Balakrishnan, C. N., and S. V. Edwards. 2009. Nucleotide variation, linkage disequilibrium and founder-facilitated speciation in wild populations of the zebra finch (*Taeniopygia guttata*). *Genetics* 181: 1–16.

Barrowclough, G. F., and K. W. Corbin. 1978. Genetic variation and differentiation in the *Parulidae*. *Auk.* 95: 691–702.

Barrowclough, G. F., K. W. Corbin, and R. M. Zink. 1981. Genetic differentiation in the *Procellariiformes*. *Comp. Biochem. Physiol. B* 69: 629–632.

Barrowclough, G. F., and R. M. Zink. 2009. Funds enough, and time: mtDNA, nuDNA and the discovery of divergence. *Mol. Ecol.* 18: 2934–2936.

Beerli, P. 2006. Comparison of Bayesian and maximum likelihood inference of population genetic parameters. *Bioinformatics* 22: 341–345.

Beerli, P., and J. Felsenstein. 2001. Maximum likelihood estimation of a migration matrix and effective population sizes in n subpopulations by using a coalescent approach. *Proc. Nat. Acad. Sci. USA* 98: 4563–4568.

Bonneaud, C., J. Burnside, and S. V. Edwards. 2008. High-speed developments in avian genomics. *Bioscience* 58: 587–595.

Brito, P., and S. V. Edwards. 2009. Multilocus phylogeography and phylogenetics using sequence-based markers. *Genetica* 135: 439–455.

Clayton, D. F. 2004. Songbird genomics: methods, mechanisms, opportunities and pitfalls. *Ann. NY Acad. Sci.* 1016: 45–60.

Clayton, N. S. 1990. Assortative mating in zebra finch subspecies *Taeniopygia-guttata-guttata* and *T-g-castanotis*. *Phil. Trans. Roy. Soc. Lond. B.* 330: 351–370.

Cracraft, J. 1986. Origin and evolution of continental biotas—speciation and historical congruence within the Australian avifauna. *Evolution* 40: 977–996.

Eckert, A. J., and B. C. Carstens. 2008. Does gene flow destroy phylogenetic signal? The performance of three methods for estimating species phylogenies in the presence of gene flow. *Mol. Phylogenet. Evol.* 49: 832–842.

Edwards, S. V. 2009. Is a new and general theory of molecular systematics emerging? *Evolution* 63: 1–19.

Edwards, S., and S. Bensch. 2009. Looking forwards or looking backwards in avian phylogeography? A comment on Zink and Barrowclough 2008. *Mol. Ecol.* 18: 2930–2933.

Edwards, S. V., S. B. Kingan, J. D. Calkins, C. N. Balakrishnan, W. B. Jennings, W. J. Swanson, and M. D. Sorenson. 2005. Speciation in birds: Genes, geography, and sexual selection. *Proc. Nat. Acad. Sci. USA* 102: 6550–6557.

Eid, J., A. Fehr, J. Gray, K. Luong, J. Lyle, G. Otto, P. Peluso, et al. 2009. Real-Time DNA Sequencing from Single Polymerase Molecules. *Science* 323: 133–138.

Ellegren, H. 2008. Sequencing goes 454 and takes large-scale genomics into the wild. *Mol. Ecol.* 17: 1629–1631.

Falush, D., M. Stephens, and J. K. Pritchard. 2003. Inference of population structure using multilocus genotype data: Linked loci and correlated allele frequencies. *Genetics* 164: 1567–1587.

Freeland, J. R., and P. T. Boag. 1999. The mitochondrial and nuclear genetic homogeneity of the phenotypically diverse Darwin's ground finches. *Evolution* 53: 1553–1563.

Good, D. A., and D. B. Wake. 1997. Phylogenetic and taxonomic implications of protein variation in the mesoamerican salamander genus *Oedipina* (*Caudata: Plethodontidae*). *Rev. Biol. Trop.* 45: 1185–1208.

Goodwin, D. 1982. *Estrildid finches of the world*. Ithaca, NY: Cornell University Press.

Grant, P. R., and B. R. Grant. 2008. *How and Why Species Multiply: The Radiation of Darwin's Finches*. Princeton, NJ: Princeton University Press.

Grant, P. R., B. R. Grant, and K. Petren. 2005. Hybridization in the recent past. *Am. Nat.* 166: 56–67.

Hedrick, P. W. 1999. Perspective: Highly variable loci and their interpretation in evolution and conservation. *Evolution* 53: 313–318.

Hey, J. 2005. On the number of New World founders: A population genetic portrait of the peopling of the Americas. *PLoS Biol.* 3: 965–975.

Hudson, R. R., and J. A. Coyne. 2002. Mathematical consequences of the genealogical species concept. *Evolution* 56: 1557–1565.

Janes, D. E., T. Ezaz, J. A. Marshall Graves, and S. V. Edwards. 2009. Recombination and nucleotide diversity in the sex chromosomal pseudoautosomal region of the emu *Dromaius novaeholladiae. J. Hered.* 100: 125–136.

Jennings, W. B., and S. V. Edwards. 2005. Speciational history of Australian grass finches (*Poephila*) inferred from thirty gene trees. *Evolution* 59: 2033–2047.

Joseph, L., and K. E. Omland. 2009. Phylogeography: its development and impact on Australo-Papuan ornithology with special reference to paraphyly in Australian birds. *Emu* 109: 1–23.

Karl, S. A., and Avise, J. C. 1992. Balancing selection at allozyme loci in oysters: Implications from nuclear RFLPs. *Science* 256: 100–102.

———. 1993. PCR-based assays of mendelian polymorphisms from anonymous single-copy nuclear-DNA techniques and applications for population genetics. *Mol. Biol. Evol.* 10: 342–361.

Keast, A. 1961. Bird speciation on the Australian continent. *Bull. Mus. Comp. Zool.* 123: 306–495.

Knowles, L. L., and B. C. Carstens. 2007. Delimiting species without monophyletic gene trees. *Syst. Biol.* 56: 887–895.

Lambeck, K., and J. Chappell. 2001. Sea level change through the last glacial cycle. *Science* 292: 679–686.

Lambeck, K., T. M. Esat, and E. K. Potter. 2002. Links between climate and sea levels for the past three million years. *Nature* 419: 199–206.

Lanyon, S. M., and R. M. Zink. 1987. Genetic Variation in Piciform Birds Monophyly and Generic and Familial Relationships. *Auk.* 104: 724–732.

Lee, J. Y., and S. V. Edwards. 2008. Divergence across Australia's Carpentatian barrier: Statistical phylogeography of the red-backed fairywren (*Malurus melanocephalus*). *Evolution* 62: 3117–3134.

Lessa, E. P. 1990. Multidimensional Analysis of Geographic Genetic Structure. *Syst. Zool.* 39: 242–252.

Liu, L., and D. K. Pearl. 2007. Species trees from gene trees: Reconstructing Bayesian posterior distributions of a species phylogeny using estimated gene tree distributions. *Syst. Biol.* 56: 504–514.

Liu, L., D. K. Pearl, R. T. Brumfield, and S. V. Edwards. 2008. Estimating species trees using multiple-allele DNA sequence data. *Evolution* 62: 2080–2091.

Liu, L., L. Yu, L. Kubatko, D. K. Pearl, and S. V. Edwards. 2009. Coalescent methods for estimating phylogenetic trees. *Mol. Phyl. Evol.* 53: 320–328.

Maddison, W. P., and L. L. Knowles. 2006. Inferring phylogeny despite incomplete lineage sorting. *Syst. Biol.* 55: 21–30.

Mayr, E. 1944. Timor and the colonization of Australia by birds. *Emu* 44: 113–130.

Mindell, D. P., A. Knight, C. Baer, and C. J. Huddleston. 1996. Slow rates of molecular evolution in birds and the metabolic rate and body temperature hypotheses. *Mol. Biol. Evol.* 13: 422–426.

Neigel, J. E., and J. C. Avise. 1986. Phylogenetic relationships of mitochondrial DNA under various demographic models of speciation. In S. Karlin and E. Nevo, eds. *Evolutionary Processes and Theory*, 515–534. New York: Academic Press.

Patterson, N., A. L. Price, and D. Reich. 2006. Population structure and eigenanalysis. *PloS Genet.* 2: 2074–2093.

Pritchard, J. K., M. Stephens, and P. Donnelly. 2000. Inference of population structure using multilocus genotype data. *Genetics* 155: 945–959.

Rannala, B., and Z. Yang. 2003. Bayes estimation of species divergence times and ancestral population sizes using DNA sequences from multiple loci. *Genetics* 164: 1645–1656.

Reich, D., A. L. Price, and N. Patterson. 2008. Principal component analysis of genetic data. *Nat. Genet.* 40: 491–492.

Schodde, R., and I. J. Mason. 1999. *The Directory of Australian Birds: A Taxonomic and Zoogeographic Atlas of the Biodiversity of Birds in Australia and its Territories*. Canberra: CSIRO Publishing.

Slatkin, M., and W. P. Maddison. 1989. A cladistic measure of gene flow inferred from the phylogeny of alleles. *Genetics* 123: 603–613.

Sokal, R. R., and P.H.A. Sneath. 1963. *Principals of Numerical Taxonomy*. San Francisco: W. H. Freeman.

Wakeley, J., R. Nielsen, S. N. Liu-Cordero, and K. Ardlie. 2001. The discovery of single-nucleotide polymorphisms and inferences about human demographic history. *Am. J. Hum. Genet.* 69: 1332–1347.

Yang Y. 2002. Likelihood and Bayes estimation of ancestral population sizes in hominoids using data from multiple loci. *Genetics* 162: 1811–1823.

Zann, R. A. 1996. *The Zebra Finch: A Synthesis of Field and Laboratory Studies*. Oxford: Oxford University Press.

Zink, R. M. 2002. A new perspective on the evolutionary history of Darwin's finches. *Auk.* 119: 864–871.

Zink, R. M., and G. F. Barrowclough. 2008. Mitochondrial DNA under siege in avian phylogeography. *Mol. Ecol.* 17: 2107–2121.

SECTION II

MECHANISMS, MOLECULES, AND EVO-DEVO

The world is rich in biological diversity because there are so many species, and because they differ from each other in their functioning and in their appearance. We can be hypnotized by morphological diversity, to the point of believing that everything that could have been invented, evolutionarily, has been. Birds and butterflies, for example, are so varied in their colors and patterns that one wonders if any particular color combination is not found somewhere in the world. Is the supply of genetic variation essentially unlimited, resulting, given enough time, in "endless forms most beautiful," or is phenotypic variation limited by the ways in which organisms are constructed, or perhaps by strong influences from the environment? Answering such broad questions requires an understanding of how genotypes are translated into phenotypes; how genes are expressed in development, how one developmental pathway may constrain elaboration of another, and how the environment drives or denies the direction of evolution of developmental programs. In this section the authors focus on the structure of organisms and how the differences between species arise in development.

In the first chapter (chapter 5), Paul Brakefield and Mathieu Joron employ the concept of morphospace. They present the problem of explaining morphological diversity as a question of how much of the morphospace available to species within a taxon is actually occupied. The chapter is about evolvability, as determined by levels of genetic variation and covariation. Genetic covariation underlying multiple phenotypic traits may constrain the directions in which evolutionary change can occur, or can occur easily. The authors describe the results of selection experiments with an African genus of butterfly (*Bicyclus*), which show that the sizes of serially repeated eyespots on fore- and hind wings are genetically variable and relatively unconstrained by genetic correlations. Species in this genus and the related *Mycalesis* have extensively "explored" and filled available morphospace. The color of those eyespots, however, is determined by a different developmental system, under different genetic control, that is much less flexible, and variation is strongly constrained. Constraints can arise ecologically as well, and this is illustrated by *Heliconius* butterflies: a mimetic species is constrained by predation to match closely an unpalatable model with one color pattern in one area and a very different one in another. Morphospace thus has local attractors, imposed by the environment. How is this close resemblance locally, coupled with pronounced variation geographically, achieved genetically? The answer is the species deploy a small number of conserved loci that have large phenotypic effects, mutations switch the pattern abruptly from one to another, and modifier loci on other chromosomes play a secondary role, for example in effecting minor changes in the shape of a patch of color. Mimicry is something special. Perhaps switch loci with major phenotypic effects have facilitated, rather than constrained, its evolution.

The next two chapters dig deeper into genetic mechanisms. Evolutionary change takes place through mutations in the protein-coding regions of genes or in the complex regulation of those genes. As molecular geneticists grapple with the task of unraveling the mechanisms by which genes give rise to phenotypes, it is possible to take a stand on which of these sources of change is the most frequent or the most important. In chapter 6, David Kingsley reflects on how this divergence of views is a modern version of older and similarly dichotomized debates, such as the few-and-large versus many-and-small views of evolutionary change. His conclusion, as with earlier debates, is that the middle ground will eventually find greater support than the extremes. He reaches this from attempts to understand the genetic basis of change in morphology that occurred when marine sticklebacks repeatedly colonized freshwater habitats.

Most freshwater fish have lost much of the lateral plate armature that functions as a defense against vertebrate predators, are lightly armored and fast, retaining only a few plates at the anterior side of the body. Populations vary in the presence or absence of pelvic structures in different populations, in relation to the presence or absence of invertebrate predators, and they vary in pigmentation in relation to water clarity. Crossing fish from differently evolved populations and mapping the differences in measured traits in large F_2 families has revealed over a hundred different loci involved in the total phenotypic variation among fish. Importantly, they comprise a mixture of a few genes of large effect, and more numerous loci, generally unlinked, with small effects (as in chapter 5). Among those of large effect are the *Ectodysplasin* (*EDA*) locus, producing a major developmental signaling molecule that controls the formation of armor plates; *PITX1*, a major locus producing a transcription factor involved in pelvic development; and the *Kit ligand/Stem cell factor* (*KITLG*) gene, a major locus for skin color encoding a secreted signaling molecule that binds to a transmembrane receptor. From these results, two suggestions of broad relevance follow. First, evolution in natural populations occurs through fixation of hypomorphic mutations rather than loss of function mutations at each of the key developmental regulators. Second, evolution occurs through a change in where and when the gene is expressed during normal development, while at the same time normal expression at other locations is preserved. Changes in site-specific expression could come about through modification of individual enhancer sequences that control expression in some tissues but not others. We suspect that signals from neighboring tissues contribute to that control, and a series of interacting feedback loops ultimately regulate site-specific expression.

When different organisms evolve in parallel or convergently, as in mimetic butterflies (chapter 5) and sticklebacks (chapters 6 and 16), it is logical to expect a common environmental determinant (chapter 2). Traits are typically acquired or elaborated (chapter 8), such as the hairiness of leaves or spines on the stems of plants, yet sometimes they are lost; evolution goes backwards, it is regressive. As discussed in chapter 7, cavefish, as well as spiders and crustacea in similar settings, have repeatedly lost or reduced their dermal pigmentation and vision in the absence of light. Why? It is easier to understand in adaptive terms why other sensory systems (olfactory, tactile) have become elaborated in those organisms in a challenging environment than it is to understand why vision and pigmentation have been lost as opposed to being simply retained. Crossing experiments in the lab, as in chapter 6, help to pinpoint the mechanisms and sharpen the focus of this question by revealing the genetic basis of phenotypic change. They have shown, for example, that the same gene (*Oca2*) is involved in albinism in different caves, but different mutations produce the same phenotype: isolated populations have mutations in either a coding region or in a regulatory region. Joshua Gross and Cliff Tabin examine three explanations for loss: random mutations and drift, or selection through a saving of an energetic cost (direct benefit) or through a favored pleiotropic effect (indirect benefit). They find no evidence of a direct benefit. Instead, genetic analysis suggests that in every cave system eye size is subject to direct (negative) selection, whereas the number of melanophores is apparently subject to drift. It remains a challenge to determine the cause of selection against large eye size, and whether neutral mutation or selection for depigmentation is responsible for albinism.

In contrast to repeated loss of structures, ornaments that attract members of the opposite sex have repeatedly been elaborated. Extreme ornaments have arisen numerous times in the animal world, and so have weapons used by members of the same sex against each other. As Doug Emlen explains in chapter 8, the why question for both sets of traits has been answered repeatedly by applying sexual selection theory, whereas the question of how these exaggerated traits are brought about developmentally has been virtually ignored, and almost nothing is known about the genes involved. Exaggerated traits share common features. They are phenotypically plastic structures whose growth is unusually sensitive to and dependent on nutritional and social circumstances, they have unusual allometries, one trait increasing rapidly in relation to others such as body size, and they vary markedly within populations. He suggests that an explanation for all these features lies in trait-specific responses to circulating signaling molecules in the insulin pathway, and to growth factors. Change in a single trait could be produced by a mutation that affects the protein structure or function of cell surface receptors of that trait alone. Since the insulin pathway is simple, evolutionarily old and taxonomically widespread, this proposal helps us to understand why exaggeration of traits is so common. The chapter, like all the others in this section, exemplifies the field-initiated inquiry into evolutionary problems that starts with observations of conspicuous traits, probes within organisms to search for developmental mechanisms, and thereby identifies the salient genetic variation that needs to be investigated.

Chapter Five

The Flexibility of Butterfly Wing Color Patterns and Evolution in Morphospace

Paul M. Brakefield and Mathieu Joron

Biologists do not yet fully understand how the various potential processes of evolutionary change actually integrate over time to yield the patterns in which species occupy morphospace. Dramatic examples of convergent or parallel evolution appear, at least superficially, to support an all-powerful natural selection that orchestrates evolution to mold functional forms in all available ecological environments. However, natural selection can only operate in a fully unconstrained manner when there are no limitations on generating the fuel for evolution as provided by phenotypic variation in all those traits with the potential to contribute to adaptation. Perhaps, because of the properties of genetics and development, evolution tends to be biased to occur along certain trajectories rather than others (e.g., Maynard Smith et al. 1985; Schluter 1996; Brakefield 2006). History and contingency may also play a major role such that at any one time the paths traced by evolution are to some degree biased by what has gone before (Blount et al. 2008). Lynch (2007) argues persuasively, in comparing evolution in lower and higher organisms, that the more complex genomes of the latter with substantial non-coding DNA are a by-product of smaller effective population sizes with stronger stochastic influences and weaker natural selection. If future work can quantify how a radiation of related species has explored morphospace through evolutionary time (see for impressively documented examples from the fossil record, McGhee

Paul Brakefield is most grateful to Peter and Rosemary Grant for the invitation, and for their suggestion to attempt a synthesis of relevant work on the two butterfly systems. We both also thank Peter Grant for his editorial help with the manuscript.

Peter & Rosemary's Shrike-Finch

Peter & Rosemary's Chicken Finch

Figure 5.1. Two fanciful morphologies in finch morphospace. Cartoons of Henry Horn drawn at the meeting.

2006), and at the same time explain why the density of species varies through this space, we will more fully understand how evolution works. Here, we describe how work on two butterfly systems can contribute to such an approach.

Darwin's finches on the Galápagos archipelago are the paradigm for adaptive radiation. The agencies of natural selection that have led to the matching of functional morphologies in these finches to their diverse ecological environments have been revealed in an exhilarating way by Peter and Rosemary Grant (e.g., Grant and Grant 2008). Numerous distinctive morphologies of finch have evolved on the Galápagos (although not a chicken finch or a shrike finch, fig. 5.1; and cf. Grant and Grant 1989). Research on Darwin's finches has also quantified genetic variances and covariances underlying phenotypic variation in target traits for selection, and is beginning to identify developmental mechanisms that map ecologically relevant phenotypes on to genotype. However, even in this case study, the extent to which the processes are understood in sufficient detail to fully explain the pattern of occupancy in morphospace is unclear. Is this pattern primarily the result of how natural selection screens phenotypic variation, or are there important compromises established by historical contingencies and the processes that generate the phenotypic variation?

In this chapter we compare work on two systems involving butterfly wing color patterns: the marginal eyespots on the wings of *Bicyclus* butterflies and the warning colors associated with mimicry in *Heliconius* butterflies (see reviews: Beldade and Brakefield 2002; Joron et al. 2006a; Baxter, Johnston, and Jiggins 2008). Work with *B. anynana* has used a series of eyespot elements with shared development and genetics to examine the flexibility of such repeated pattern elements to follow their own individual paths in evolution and, thus, become different to each other. This issue has been explored by applying artificial selection to the eyespot pattern in different directions of morphospace together with "evo-devo" studies of morphogenesis. Knowledge of the "tool-kit" pathways of wing development in *Drosophila* flies has proved invaluable for exploring the developmental genetics of butterfly wings (see Carroll et al. 2005). The observations on the model species can now be compared with patterns of morphological diversity within radiations of related species to reveal whether the evolvability of the pattern in laboratory selection experiments may be reflected in patterns of occupancy in morphospace. We know in broad terms why eyespots matter to adult *Bicyclus* butterflies, but we have little notion about how natural selection influences the details of eyespot patterns in species in different environments. In contrast for *Heliconius*, the dynamics of mimicry in terms of both models of the evolution of the phenomenon and the action of natural selection arising through bird predation and mate choice are better understood. This broad understanding of how natural selection influences wing patterning in *Heliconius* can now be examined in the context of an ever expanding knowledge about the Mendelian loci that can switch among mimetic patterns. Thus, we can begin to develop ideas about the extent to which the genetic architecture of mimicry could bias the evolution of mimetic color patterns, even in the face of the powerful natural selection. Here, we will highlight opportunities for the future to more fully understand the balance of processes which underlie the evolution of patterns of occupancy in morphospace.

BICYCLUS EYESPOTS: EVO-DEVO AND DIVERGENCE OF SERIAL REPEATS

Research has begun to show how components of the tool-kit pathways of wing development in insects have been recruited and elaborated to yield novel structures, including the pigmented scale cells (from bristles and the *Achete-scute* pathway; Galant et al. 1998) and the eyespot pattern elements with their nested color rings (recruitment of genes such as *Distal-less, engrailed*, and *Spalt*; Carroll et al. 1994; Brunetti et al. 2001; Saenko et al. 2008). Butterfly wings are covered by sheets of scale cells that form a sort of colored mosaic like tiles on a roof. This provides the basis for the process of painting patterns on the wings in development. Eyespots and other pattern elements on butterfly wings are functional forms that contribute to variation in fitness in the wild both via interactions with predators and in mate choice (Robertson and Monteiro 2005; Stevens 2005).

DEVELOPMENT OF EYESPOTS AND ARTIFICIAL SELECTION

A reconstruction of an evolutionary "groundplan" for Nymphalid butterflies shows series or modules of different pattern elements, including bands, chevrons, and the marginal eyespots, that are arranged in anterior-posterior columns over the wings (Nijhout 1991). Both dorsal and ventral wing surfaces are subdivided by veins. In the groundplan, each section of the wing between a pair of marginal veins shows its own combination of single copies of the pattern elements, whereas in extant species some or all of these elements are often absent (e.g., the eyespots in *Heliconius*). Morphogenesis of one module, the marginal eyespots, is becoming clearer both in terms of cell-cell signaling mechanisms and candidate genetic pathways (Beldade and Brakefield 2002). Understanding pattern formation involves discovering how the epithelial cells—the scale cells to be—gain information during wing development to become fated to synthesize different color pigments shortly before adult eclosion.

The eyespots of *B. anynana* are all formed by the same developmental process (Brunetti et al. 2001; Beldade and Brakefield 2002; Reed and Serfas 2004; Saenko et al. 2008). Transplantation experiments performed in early pupae reveal that each eyespot forms around a group of organizing cells called a focus; transplanting a focus to a novel wing location yields an ectopic eyespot around the grafted tissue. Establishment of the foci occurs in late larval development, following which, in the early pupa, each focus establishes an information gradient in the surrounding epithelial cells, presumably via diffusible morphogens. These epithelial cells then respond to the gradient of the signal and, thus, become fated to subsequently synthesize a particular pigment in the late pupa. We also know from additional transplantation experiments with divergent selection lines that eyespot size depends largely on signal strength; a focus from a line with large eyespots yields larger ectopic patterns than one from a line with small eyespots irrespective of the host epithelial tissue into which it is grafted (Monteiro et al. 1994). The eyespots in *B. anynana* express the same genes in development (Brunetti et al. 2001; Reed and Serfas 2004). Also, mutant alleles typically affect all eyespots, and artificial selection on a single eyespot yields highly correlated responses for the target trait in other eyespots (Beldade et al. 2003). Shared morphogenesis led us to design artificial selection experiments in this model system to examine the potential for independent evolution of the different eyespots (fig. 5.2).

Experiments with Eyespots in Size Morphospace

The first experiments explored whether a change toward a pair of eyespots in which one became smaller and the other larger could occur as readily as one in which both eyespots were either larger or smaller. We targeted the wild type pattern for the forewing of *B. anynana* which shows a small anterior eyespot and a larger posterior one. Replicated lines were established and then selected over twenty-five generations toward each corner of the morphospace for this pattern; that is, along both the "coupled" axis for eyespot size, and an "uncoupled" axis orthogonal to a proposed genetic line of least resistance or plane of developmental drive (fig. 5.2b).

As expected, artificial selection either "up" or "down" the "coupled" axis of shared morphogenesis produced rapid responses and highly novel phenotypes, with butterflies eventually having either no eyespots

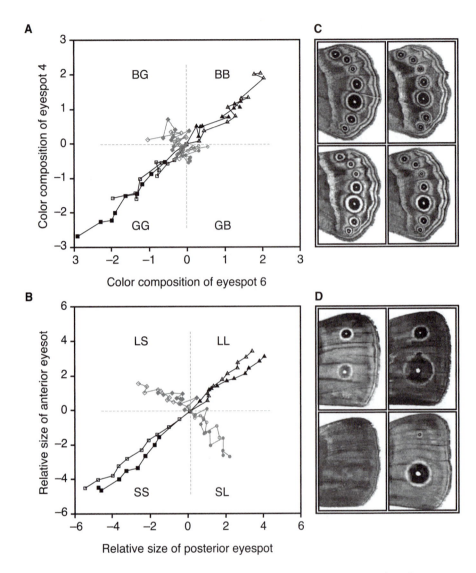

Figure 5.2. Response to artificial selection in *Bicyclus anynana* for changes through morphospace in eyespot color composition and size. (a) and (b) Responses in each generation relative to unselected control values, plotted from the starting population mean plotted at the origin. Both characters were selected in replicated populations for concerted (black points and lines) and antagonistic (gray) change in two eyespots. (a) Selection for color composition of the fourth and sixth ventral hindwing eyespots (E4 and E6): 'BB' ('Black-Black') and 'GG' ('Gold-Gold') are concerted directions; 'BG' ('Black-Gold') and 'GB' ('Gold-Black') are antagonistic directions. (b) Selection for size (relative to

at all, or two very large ones (Beldade et al. 2002). However, populations along the uncoupled axis also responded well to selection (fig. 5.2b), eventually producing highly novel phenotypes, but here with one eyespot very large, and the other absent or very small. The phenotypes resulting from this experiment were, thus, situated toward all four corners of the two-dimensional morphospace, demonstrating an overall high evolvability (Beldade et al. 2002). We suggested that this capacity for independent evolution was the product of a long legacy of natural selection and evolutionary tinkering in favor of a diversity of eyespot sizes across the wings (Beldade et al. 2003).

Experiments with Eyespots in Color Morphospace

More recent experiments on B. *anynana* have focused on making a comparison between the dynamics of the pattern of eyespot size with that of eyespot color composition (Allen et al. 2008). Like size, there is some variability in whether a particular eyespot has a comparatively narrow outer gold ring or a broader one (black or gold, respectively). Early experiments on this trait which targeted the large forewing eyespot also demonstrated a high heritability with positive genetic correlations among eyespots (Monteiro et al. 1997a). There was, however, only a very low genetic correlation between eyespot color and eyespot size indicating differing genetic architectures. Again, artificial selection along the coupled axis for a particular pair of eyespots, this time from the hindwing series, rapidly yielded novel morphologies in which both targeted eyespots have narrower or broader gold rings (as well as any flanking eyespots). However, in marked contrast to the results for eyespot size, morphological change was much more strongly limited along the uncoupling axis in which the two eyespots were selected in opposite directions in color morphospace (fig. 5.2a). This bias was emphasized by the finding that no novel "antagonistic" color phenotypes—for example,

---

Figure 5.2. (*Continued*) wing size) of the anterior and posterior eyespots on the dorsal forewing: 'LL' ('Large-Large') and 'SS' ('Small-Small') are concerted directions; 'LS' ('Large-Small') and 'SL' ('Small-Large') are antagonistic directions. (c) and (d) Representative phenotypes for each selected direction in generation 10. (c) Ventral hind wings shown for color composition lines; (d) Dorsal forewings shown for eyespot size lines with the wings arranged according to axes in (a) and (b). From Allen et al. (2008).

in gold-black or black-gold combinations—had resulted after eleven generations of selection (Allen et al. 2008).

Previous transplantation experiments among selected lines with gold or black eyespots had shown that color composition, unlike size, is determined by the threshold responses (to the signal) of the cells surrounding the focal organizer; thus, gold epithelial host tissue always yields a gold ectopic eyespot irrespective of the donor source of the signal (Monteiro et al. 1997a). This has led to a working hypothesis that whereas the pattern of relative size can readily evolve because each eyespot is determined largely by its own specific signaling organizer, the color composition over the set of eyespots on a wing is dependent on the properties of the single sheet of epithelial cells and is, therefore, much less flexible. In other words, it is possible to account for different patterns of evolvability of these eyespot traits under artificial selection in morphospace through an underlying difference in how development works. It remains unclear whether these differences are the consequence of hardwired differences in development or whether strong stabilizing selection over eyespots for color composition has resulted in little evolution of individuality among eyespots for this trait (unlike size).

Is the property of high flexibility for eyespot size reflected in the pattern of morphological disparity among extant species of *Bicyclus* and related genera? We have used museum collections to collect preliminary data on variability in the size of the two dorsal forewing eyespots among species of *Bicyclus* and *Mycalesis*, a very closely related genus in Asia within the same tribe of Mycalesina (Brakefield and Roskam 2006). When plotted together, the mean values showed a rather complete occupancy of morphospace for the pattern of relative size of the two eyespots (comparable disparity in pattern is also evident on the hind wing). In contrast, the eyespots in each of these species appear to vary little in color composition, especially when on the same wing surface; each species is characterized by eyespots of similar color, although those of different species may show comparatively narrow, intermediate, or broad gold rings (Brakefield and Roskam 2006; Allen et al. 2008). Thus, the differences in evolvability of the two eyespot traits in a single model species, as revealed in their responses to synergistic and antagonistic modes of artificial selection, appear to be reflected in patterns of disparity in morphospace that have evolved among related species. We can now move on to examining whether the generation of phenotypic variation could also influence the evolution in morphospace for *Heliconius* butterflies in which the evolution of mimicry and the ecology clearly play a dominant role.

HELICONIUS WARNING PATTERNS: COULD GENETICS OR ECOLOGY BIAS THE EVOLUTION OF MIMICRY?

Heliconius butterflies show dramatic color pattern variation among species and between geographic locations. The wing patterns of *Heliconius* are warning signals that advertise their unpalatability to predators. The variation in wing patterns is almost always linked to mimicry involving other species of *Heliconius* and/or other local butterflies and moths with which sharing the warning patterns is beneficial. The evolution of novelty, and indeed diversity, in mimetic wing patterns is somewhat unexpected since species convergence should gradually erode phenotypic diversity. Therefore, mimetic patterns are a good example of traits where both the genetics and the ecology of selection could influence the exploration of morphospace (plate 2).

Ecological Constraints on the Distribution of Mimetic Patterns in Morphospace

Natural selection on mimetic wing patterns is frequency-dependent since rare or new variants are strongly selected against because they have a low chance of being encountered by predators which have previously learned through experience to avoid them. This density and frequency-dependent mode of selection produces a rugged adaptive landscape in which abundant patterns correspond to sharp peaks of fitness, while rare and/or intermediate patterns correspond to fitness valleys. As mimicry rings often involve numerous species (Beccaloni 1997), the topography of the fitness landscape depends not only on intrinsic qualities of the warning patterns (e.g., how memorable they are) but, more importantly, on the wing patterns present locally in the species' preferred habitat (Mallet and Gilbert 1995; Joron et al. 1999; Estrada and Jiggins 2002; Joron 2005). The exploration of morphospace through this rugged fitness landscape is therefore heavily dependent on the ecological structure of the community of coexisting species.

101

Indeed, mimetic mutualistic relationships are sufficiently pervasive to directly influence niche structure in forest butterfly communities (Elias et al. 2008). Therefore, the height of fitness peaks for any specific wing morphology is strongly affected by the ecological preferences of co-mimics, and varies widely across microhabitats. Although this might appear to reflect an ecological limitation on the exploration of new wing morphologies, the distribution of diversity within mimetic butterfly clades suggests otherwise. Widespread genera, such as *Heliconius*, *Eueides* (Heliconiinae), or *Napeogenes* (Ithomiinae), have representatives in most available mimicry rings in their size class (Beccaloni 1997; Jiggins et al. 2006; Elias et al. 2009), suggesting that the ecological conditions to invade new mimicry rings are rather permissive. Rampant fluctuations of selection pressures through evolutionary time via the interplay of mimicry and habitat preferences could facilitate an efficient exploration of locally available mimetic morphologies. However, the question remains as to whether the available mimetic patterns and the areas of the morphospace actually used for mimicry are themselves in part determined by the specific details of the underlying gene-phenotype mapping, or whether these patterns can occupy any position in the morphospace and depend solely on ecological contingency of communities and predator cognition.

Qualitative Exploration of Morphospace in Mimicry

As deviant mimetic patterns are selected against (Benson 1972; Langham 2004), and gradual changes in warning signals are supported neither by theory nor by experiments (Turner 1984; Lindström et al. 1999) (although they do shift geographically—see below), the exploration of morphospace for mimicry involves shifting between distinct mimetic patterns by crossing adaptive valleys in a single step via sudden, large phenotypic changes. Major mimicry switches that provide a rough resemblance to a new model could be favored if the new mimicry ring provides improved numerical or unpalatability protection. A second phase would then involve refinement and fixation in the population by classic Fisherian selection for increased resemblance to the new mimetic pattern (Turner 1977; Baxter, Papa, et al. 2008). This form of selection is likely to sieve out specific types of genetic architectures to control wing pattern variation, a prediction which most species of

102

Heliconius studied to date seem to conform to. One of the striking features in *Heliconius* genetics is the action of Mendelian loci that switch on or off large patches of color (phase 1 loci; Brown and Benson 1974; Sheppard et al. 1985; Gilbert 2003). In most species, such as the classic examples *H. melpomene* or *H. erato* which mimic each other nearly everywhere, a small number of such large-effect loci appears to control most aspects of geographic variation in wing pattern (Mallet 1989; Jiggins and McMillan 1997).

Perhaps the most spectacular case of large-effect locus in *Heliconius* is found in the highly polymorphic *H. numata,* whose multiple "tiger-patterned" forms fly together and mimic different coexisting species of *Melinaea* Ithomiines (Brown and Benson 1974). *H. numata* polymorphism is entirely controlled by a single supergene locus (called P), with hierarchical dominance between mimetic alleles (Brown and Benson 1974; Joron, Papa, et al. 2006). This architecture enables *H. numata* simultaneously to occupy the tops of multiple, well-separated fitness peaks in morphospace, while avoiding the non-mimetic intermediate forms that might be formed by recombination between loci, or by co-expression of alleles in heterozygous genotypes. Tight linkage locks together allelic combinations and limits the occurrence of phenotypic recombinants (Joron, Papa, et al. 2006; M. Joron unpublished data), which can be considered a form of genetic bias on the phenotypic diversity available for selection in any single locality. However, color pattern recombinants are occasionally found, albeit rarely, in polymorphic populations (Brown and Benson 1974; M. Joron unpublished data), showing that other wing patterns can be formed.

Quantitative Exploration of Morphospace in Mimicry

Examination of population samples reveals substantial levels of phenotypic variation around each mimetic pattern (Brown 1976), suggesting that despite strong selection, considerable standing genetic variation is maintained that can affect wing patterns in minor ways. Genetically, quantitative variation in *Heliconius* has been classically interpreted as the result of the segregation at so-called modifier loci, which epistatically refine the resemblance provided by alleles of large-effect loci, and might have arisen through directional selection during phase 2 of mimicry evolution (Sheppard et al. 1985). In *H. melpomene*, numerous

103

races have a prominent red forewing patch, varying in size and shape. The presence of this patch is controlled by alleles of locus HmB, which does not influence its shape. Crosses between different red-forewing races of *H. melpomene* have recently shown that red patch shape is influenced by a QTL (quantitative trait locus) unlinked to HmB or to any other major-effect locus (Baxter, Papa, et al. 2008), acting as a modifier locus tuning the forewing appearance for optimal mimicry, epistatically with HmB.

Similarly, although coexisting mimetic patterns in a given locality are usually well separated from each other in the morphospace, variation can be gradual across geographic distances, as can be found in *H. numata* and *Melinaea* tiger pattern themes across the South American continent, or in the shape and size of yellow forewing patches and red hindwing rays among Amazonian races of *H. melpomene* or *H. erato* (Brown 1976, 1979). This is in fact common in many other widespread mimetic genera (e.g., *Napeogenes, Ithomia*; Jiggins et al. 2006; Elias et al. 2009). It is intriguing how local multi-species communities can gradually explore the morphospace. This may be due to some random drift in the local models, or their response to ecological contingency, in combination with an ability in the other mimetic species to quickly match those small variations (Flanagan et al. 2004). It will be of great interest to quantify the occupancy of morphospace in widespread mimetic taxa across their ranges to start describing the coverage attained via such "community drift." This however demonstrates that although mimetic selection appears to be strongly constraining in terms of how populations can shift mimetic patterns locally, their actual morphospace positions do not appear to be determined by strong constraints in pattern formation, and that there is in most cases sufficient standing genetic variation to allow for a gradual exploration of the morphospace by several species simultaneously (plate 2).

Imperfect Mimicry: Evidence for Constraints?

Exceptions to this observation of near-perfect mimicry are, however, not uncommon. For example, east-Ecuadorean co-mimics *H. erato notabilis* and *H. melpomene plesseni* show an inversed placement of pink and white areas in the proximal forewing patch (Brown 1979; Papa et al. 2008). Given the very precise mimicry of the two species nearly everywhere else, it could be considered that *H. m. plesseni* has attained a local maximum of resemblance, as if those populations were occupying a suboptimal fitness peak. This would in turn suggest that further improvement is constrained, perhaps because it requires some developmental novelty to arise or a rare favorable mutation. It is hard, however, to distinguish whether such examples of imperfect mimicry are a consequence of constraints in the machinery of wing pattern formation, or due to relaxed selection on specific pattern elements, or to population contingency and drift.

Homology of Major Loci: Evidence for Genetic Constraints?

The ecological circumstances which favor the evolution of mimicry, together with the requirement of large-effect variation to allow populations to cross adaptive valleys, might imply that the identity of the loci recruited in different species is to some degree a product of genomic contingency, that is, species-specific. Alternatively, there may be few ways of switching patterns, and genetic architectures might bias which genes must be recruited to switch to an appropriate new pattern. The rich comparative genetics data now available for *Heliconius* may help in tackling this question. The major pattern observed is the strong positional conservation of the genomic regions controlling both convergent and divergent color pattern variation in most *Heliconius* (Jiggins et al. 2005; Joron, Papa, et al. 2006; Kapan et al. 2006; Kronforst, Kapan, et al. 2006; Baxter et al. 2008b; Papa et al. 2008). For instance, the parallel geographic radiation in *H. erato* and *H. melpomene* (which are not very closely related species) involves orthologous genomic regions, on three chromosomes, in the two species. Since those species mimic each other nearly everywhere, this conserved "toolbox" of pattern genes (Gilbert 2003) might be considered evidence for biases in the patterns which can readily be evolved. However, the data suggest otherwise. Mimetic polymorphism in *H. numata*, closely related to *H. melpomene* but with very different color pattern diversity, involves a single switch locus which is a positional homologue of a cluster of loci controlling the presence of yellow patches in *H. melpomene*. The *H. numata* whole-wing variation is entirely controlled by only one of the tools in the *Heliconius* toolbox of color pattern loci (Joron, Papa, et al. 2006; Baxter, Papa, et al. 2008). The disjunct arrays of mimetic patterns in

H. melpomene and *H. numata* suggest the diversity of phenotypes controlled by this switch locus is highly flexible with few, if any, limitations. Genetic variation appears to be always available in the genome repeatedly to evolve the control of mimetic novelty by the same switch loci.

Discussion and Perspectives

Artificial selection experiments on the pattern of repeated eyespots in *B. anynana* have attempted to assay the potential role of developmental flexibility in generating phenotypic variation in the context of the evolution of occupancy of morphospace by related species. For one of the two traits examined in this way, there is evidence that it does indeed matter how development works. Thus, the mechanisms underlying variation in eyespot color composition appear to introduce a strong developmental bias that appears to be reflected in patterns of disparity among species of *Bicyclus* in Africa, and of *Mycalesis* in Asia. In contrast, eyespot size behaves in a highly flexible manner both for the response to artificial selection in the model species and for morphospace across the two lineages. Although there is a match between these experimental results and earlier interpretations of the developmental genetics of the two traits, we cannot at this stage distinguish between explanations in terms of a tight, long-term, developmental constraint, and a difference in the way in which natural selection has favored the evolution of evolvability for color and size. A combination of a wing-level developmental process and a history of strong stabilizing selection on eyespot color composition within each species in its particular environment could account for the observed absence of evolvability in this trait. This would then result in the observed lack of individuality for color composition in an eyespot module and a coupled lack of potential for independent evolution.

Another trait of interest for a related reason is eyespot shape. Here there was little evolvability for "fat" or "thin" eyespots in *B. anynana*, and what response there was to artificial selection occurred via changes in wing shape and the arrangement of scale cells across the wing (Monteiro et al. 1997b, 1997c). It is intriguing to speculate that some developmental novelty may be required to convert a circular, cone-shaped, signal to a non-circular one (although ellipsoidal eyespots occur in some butterflies, e.g., Nijhout 1991). However, using a similar reasoning to that for eyespot color, there may be valid functional explanations in terms of predation and stabilizing selection for eyespots being typically circular, and for the low evolvabilities observed for this trait.

Although these ideas are expressed here in terms of underlying developmental mechanisms and bias, they can also be set out from a genetic perspective. More data sets of this type are necessary to resolve the most appropriate terminology. Further progress will be made through the application of high throughput genetic tools, fine genetic mapping of different traits, and more detailed developmental analyses in the model species (Beldade et al. 2009), in combination with examining patterns of occupancy in morphospace in different lineages using a comparative phylogenetic framework.

The experiments and observations on eyespot patterns in the mycalesine butterflies do not directly examine evolvability in the context of the functional performance of the different forms. Although it is known that eyespot size in *Bicyclus* is influenced by natural selection and sexual selection, much more work is needed in this area. In common with many evo-devo systems, research on eyespot patterns is revealing details about the generation of phenotypic variation, but it will be much more challenging to measure fitness curves and understand exactly how natural selection works in the wild. In *B. anynana,* an attempt has been begun for patterns of allometric growth (wing:body, and forewing:hindwing size) to combine artificial selection experiments with subsequent analyses of the relative fitness of the resulting phenotypes (Frankino et al. 2005, 2007). However, *Heliconius* butterflies hold a particularly rich promise for understanding how phenotypic variation is generated in the context of a detailed picture of how natural selection influences different phenotypes in morphospace.

Unlike *Bicyclus,* knowledge about the evolvability of wing patterns in *Heliconius* is not coming from developmental genetics and artificial selection, but from genetic dissection of extant wing pattern diversity from natural populations set into the context of a rich history of theoretical and empirical studies of mimetic evolution (Turner 1977, 1984; Mallet et al. 1990). This does not directly reveal the option set of patterns that are potentially evolvable in morphospace, but rather an understanding of how the observed, somewhat discontinuous coverage of morphospace is controlled by the variation at key loci shared among all members of the clade. The overall results demonstrate an astonishing flexibility in the phenotypic action of those loci, co-opted to control a high pattern diversity as well as convergence among distantly related species (Joron, Papa, et al. 2006; Kronforst, Kapan, et al. 2006; Baxter, Papa, et al. 2008).

As was noted early on by Turner (1977), there are reasons why the ecological conditions governing the exploration of morphospace might select for specific types of genetic architectures, which in turn could limit or channel adaptation along particular directions. However, the dominant pattern is that of a lack of any obvious limitations on the evolution of new mimicry associations (although mimicry itself could be considered an inherently limited way of exploring morphospace). As the evolution of new patterns in a population must involve a significant pattern shift, it may be necessary for genetic loci of major effect to be maintained in the population as a transient polymorphism under disruptive selection (Naisbit et al. 2003). Therefore, the ecology of mimicry evolution may restrict pattern switches to be in functionally integrated regions in the genome, to be rarely disrupted by recombination, and to exhibit relatively little pleiotropy and epistasis with the rest of the genome. Switch loci can then be the prime targets of selection for large-effect variation, which could explain why the same loci are repeatedly recruited to control pattern evolution, even in locally monomorphic species. This view predicts that mimicry switch loci are eminently evolvable *units* which facilitate the exploration of morphospace. The fact that the *Heliconius* colour pattern "toolbox" is largely shared across species and composed of "plug-and-play" transferable units is further supported by the fact that foreign alleles can be introgressed into host genomes, either through natural hybridization (Kronforst, Young, et al. 2006; Mavárez et al. 2006; Dasmahapatra et al. 2007), or via artificial crosses (Gilbert 2003). They then tend to retain the same phenotypic effects as in the source species, sometimes to the point of causing hybrid speciation in certain documented cases (Mavárez et al. 2006). It appears that the switch loci recruited during early mimicry evolution could be interpreted as the evolutionary novelty allowing mimetic taxa to quickly explore and invade available mimetic niches; thus, there appears to be minimal constraint in transgressing a rugged fitness landscape. This may explain why butterfly taxa involved in color pattern mimicry (Heliconiines, Ithomiines, dismorphiine Pierids, some riodinid genera, and so on), often show spectacular adaptive radiations into numerous mimetic morphologies.

This mode of morphospace exploration does not, however, preclude quantitative variation over long distances, and one aspect that is often neglected is the richness of quantitative geographical variations in mimicry, even in species which also show highly differentiated color-pattern races. This in itself can be seen as strong support for the notion that the exact position of mimicry patterns in *Heliconius* morphospace is largely accounted for by the biology of signal-receiver interactions and ecological contingency rather than by any tight constraints introduced via developmental mechanisms. Contingency may be considered a form of constraint on morphospace exploration if the path initially taken to evolve a novel morphology diminishes the array of possible morphologies available thereafter. In this sense, given the spectacular diversity present today, the evolution of novel switch loci in the early stages of mimicry evolution could reflect the opposite of a constraint.

However, this interpretation awaits a quantitative validation through a more precise description of morphospace occupancy with the level of detail that is available for *Bicyclus*. Although *Heliconius* are not as amenable to artificial selection as *Bicyclus*, the molecular dissection of variation in their color patterns is making rapid progress toward characterizing such integrated switch loci and their regulation (Joron, Papa, et al. 2006; Baxter, Papa, et al. 2008; Papa et al. 2008). It will then become fascinating to integrate knowledge of these mechanisms of genotype-phenotype mapping with a quantitative analysis of the occupancy of morphospace for all the pattern elements modulated by such loci. Overall, the prospect for work on butterflies is to be able to examine how the developmental pathways regulating scale morphology and color patterns have been recruited in evolution to respond to the ecology of diversification in different elements of butterfly wing patterns.

REFERENCES

Allen, C., P. Beldade, B. J. Zwaan, and P. M. Brakefield. 2008. Differences in the selection response of serially repeated color pattern characters: Standing variation, development, and evolution. *BMC Evolutionary Biology* 2008, 8: 94 doi:10.1186/1471-2148-8-94.

Baxter, S.W., S. E. Johnston, and C. D. Jiggins. 2008. Butterfly speciation and the distribution of gene effect sizes during adaptation. *Heredity* 102: 57–65.

Baxter, S.W., R. Papa, N. Chamberlain, S. J. Humphray, M. Joron, R. ffrench-Constant, W. O. McMillan, and C. D. Jiggins. 2008. Parallel evolution in the genetic basis of Müllerian mimicry in *Heliconius* butterflies. *Genetics* 180: 1567–1577.

Beccaloni, G. W. 1997. Ecology, natural history and behaviour of ithomiine butterflies and their mimics in Ecuador (Lepidoptera: Nymphalidae: Ithomiinae). *Trop. Lepidoptera* 8: 103–124.

Beldade, P., and P. M. Brakefield. 2002. The genetics and evo-devo of butterfly wing patterns. *Nat. Rev. Genet.* 3: 442–452.

Beldade, P., K. Koops, and P. M. Brakefield. 2002. Developmental constraints versus flexibility in morphological evolution. *Nature* 416: 844–847.

———. 2003. Modularity, individuality, and evo-devo in butterfly wings. *Proc. Natl. Acad. Sci. USA* 99: 14262–14267.

Beldade P., S. V. Saenko, N. Pul, and A. D. Long. 2009. A gene-based linkage map for *Bicyclus anynana* butterflies allows for a comprehensive analysis of synteny with the lepidopteran reference genome. *PLoS Genetics* 5, e1000366.

Benson, W. W. 1972. Natural selection for Müllerian mimicry in *Heliconius erato* in Costa Rica. *Science* 176: 936–939.

Blount, Z. D., C. Z. Borland, and R. E. Lenski. 2008. Historical contingency and the evolution of a key innovation in an experimental population of *Escherichia coli*. *Proc. Natl. Acad. Sci. USA* 105: 7899–7906.

Brakefield, P. M. 2006. Evo-devo and constraints on selection. *Trends Ecol. Evol.* 21: 362–368.

Brakefield, P. M., and J. C. Roskam. 2006. Exploring evolutionary constraints is a task for an integrative evolutionary biology. *Am. Nat.* 168: S4–S13.

Brown, K. S. 1976. An illustrated key to the silvaniform *Heliconius* (Lepidoptera: Nymphalidae) with descriptions of new subspecies. *Trans. Amer. Entom. Soc.* 102: 373–484.

———. 1979. *Ecologia Geográfica e Evolução nas Florestas Neotropicais*. Universidade Estadual de Campinas, Campinas, Brazil.

Brown, K. S., and W. W. Benson. 1974. Adaptive polymorphism associated with multiple Müllerian mimicry in *Heliconius numata*. *Biotropica* 6: 205–228.

Brunetti, C. R., J. E. Selegue, A. Monteiro, V. French, P. M. Brakefield, and S. B. Carroll. 2001. The generation and diversification of butterfly eyespot color patterns. *Curr. Biol.* 11: 1578–1585.

Carroll, S. B., J. Gates, D. Keys, S. W. Paddock, G. F. Panganiban, J. Selegue, and J. A. Williams. 1994. Pattern formation and eyespot determination in butterfly wings. *Science* 265: 109–114.

Carroll, S. B., J. K. Grenier, and S. D. Weatherbee. 2005. *From DNA to Diversity: Molecular Genetics and the Evolution of Animal Design*. Oxford: Blackwell Scientific.

Dasmahapatra, K. K., A. Silva-Vasquez, J. W. Chung, and J. Mallet. 2007. Genetic analysis of a wild-caught hybrid between non-sister *Heliconius* butterfly species. *Biology Letters* 3: 660–663.

Elias, M., Z. Gompert, C. Jiggins, and K. Willmott. 2008. Mutualistic interactions drive ecological niche convergence in a diverse butterfly community. *PLoS Biology* 6: e300.

Elias, M., M. Joron, K. Willmott, V. Kaiser, K. Silva-Brandão, C. Arias, L. M. Gómez, et al. 2009. Out of the Andes: patterns of diversification in clearwing butterflies. *Mol. Ecol.* 18: 1716–1729.

Estrada, C., and C. D. Jiggins. 2002. Patterns of pollen feeding and habitat preference among *Heliconius* species. *Ecol. Entomol.* 27: 448–456.

Flanagan, N., A. Tobler, A. Davison, O. G. Pybus, D. D. Kapan, S. Planas, M. Linares, D. Heckel, and W. O. McMillan. 2004. The historical demography of Müllerian mimicry in the Neotropical *Heliconius* butterflies. *Proc. Natl. Acad. Sci. USA* 101: 9704–9709.

Frankino, W. A., B. J. Zwaan, D. L. Stern, and P. M. Brakefield. 2005. Natural selection and developmental constraints in the evolution of allometries. *Science* 307: 718–720.

———. 2007. Internal and external constraints in the evolution of morphological allometries in a butterfly. *Evolution* 61: 2958–2970.

Galant, R., J. B. Skeath, S. Paddock, D. L. Lewis, and S. B. Carroll. 1998. Expression of an achaete-scute homolog during butterfly scale development reveals the homology of insect scales and sensory bristles. *Curr Biol* 8: 807–813.

Gilbert, L. E. 2003. Adaptive novelty through introgression in *Heliconius* wing patterns: evidence for a shared genetic "tool box" from synthetic hybrid zones and a theory of diversification. In C. L. Boggs, W. B. Watt, and P. R. Ehrlich, ed., 14. *Ecology and Evolution Taking Flight: Butterflies as Model Systems*. Chicago: University of Chicago Press.

111

Grant, B. R., and P. R. Grant. 1989. *Evolutionary Dynamics of a Natural Population: The Large Cactus Finch of the Galápagos*. Chicago: University of Chicago Press.

———. 2008. *How and Why Species Multiply. The Radiation of Darwin's Finches*. Princeton, NJ: Princeton University Press.

Jiggins, C. D., R. Mallarino, K. R. Willmott, and E. Bermingham. 2006. The phylogenetic pattern of speciation and wing pattern change in neotropical *Ithomia* butterflies (Lepidoptera; Nymphalidae). *Evolution* 60: 1454–1466.

Jiggins, C. D., J. Mavárez, M. Beltrán, W. O. McMillan, J. S. Johnston, and E. B. Bermingham. 2005. A genetic linkage map of the mimetic butterfly, *Heliconius melpomene*. *Genetics* 171: 557–570.

Jiggins, C. D., and W. O. McMillan. 1997. The genetic basis of an adaptive radiation: warning colour in two *Heliconius* species. *Proc. R. Soc. London Ser. B* 264: 1167–1175.

Joron, M. 2005. Polymorphic mimicry, microhabitat use, and sex-specific behaviour. *J. Evol. Biol*. 18: 547–556.

Joron, M, C. D. Jiggins, A. Papanicolaou, and W. O. McMillan. 2006. *Heliconius* wing patterns: an evo-devo model for understanding phenotypic diversity. *Heredity* 97: 157–167.

Joron, M., R. Papa, M. Beltrán, N. Chamberlain, J. Mavárez, S. W. Baxter, M. Abanto, et al. 2006. A conserved supergene locus controls colour pattern diversity in *Heliconius* butterflies. *PLoS Biol*. 4: e303.

Joron, M., I. R. Wynne, G. Lamas, and J. Mallet. 1999. Variable selection and the coexistence of multiple mimetic forms of the butterfly *Heliconius numata*. *Evol. Ecol*. 13: 721–754.

Kapan, D. D., N. S. Flanagan, A. Tobler, R. Papa, R. D. Reed, J. A. Gonzalez, M. R. Restrepo, et al. 2006. Localization of Mullerian mimicry genes on a dense linkage map of *Heliconius erato*. *Genetics* 173: 735–757.

Kronforst, M. R., D. D. Kapan, and L. E. Gilbert. 2006. Parallel genetic architecture of parallel adaptive radiations in mimetic *Heliconius* butterflies. *Genetics* 174:535–539.

Kronforst, M. R., L. G. Young, L. M. Blume, and L. E. Gilbert. 2006. Multilocus analyses of admixture and introgression among hybridizing *Heliconius* butterflies. *Evolution* 60: 1254–1268.

Langham, G. M. 2004. Specialized avian predators repeatedly attack novel color morphs of *Heliconius* butterflies. *Evolution* 58: 2783–2787.

Lindström, L., R. V. Alatalo, J. Mappes, M. Riipi, and L. Vertainen. 1999. Can aposematic signals evolve by gradual change? *Nature* 397: 249–251.

Lynch, M. 2007. The frailty of adaptive hypotheses for the origins of organismal complexity. *Proc. Natl. Acad. Sci. USA* 104 (Suppl.): 8597–8604.

Mallet, J. 1989. The genetics of warning colour in Peruvian hybrid zones of *Heliconius erato* and *H. melpomene*. *Proc. R. Soc. London Ser. B* 236: 163–185.

Mallet, J., N. H. Barton, G. Lamas, J. Santisteban, C. M. Muedas, and H. Eeley. 1990. Estimates of selection and gene flow from measures of cline width and linkage disequilibrium in *Heliconius* hybrid zones. *Genetics* 124: 921–936.

Mallet, J., and L. E. Gilbert. 1995. Why are there so many mimicry rings? Correlations between habitat, behaviour and mimicry in *Heliconius* butterflies. *Biol. J. Linn. Soc.* 55: 159–180.

Mavárez, J., C. A. Salazar, E. Bermingham, C. Salcedo, C. D. Jiggins, and M. Linares. 2006. Speciation by hybridization in *Heliconius* butterflies. *Nature* 441: 868–871.

Maynard Smith, J., R. Burian, S. Kaufman, P. Alberch, J. Campbell, B. Goodwin, R. Lande, D. Raup, and L. Wolpert. 1985. Developmental constraints and evolution. *Quart. Rev. Biol.* 60: 265–287.

McGhee, G. R. 2006. *The Geometry of Evolution: Adaptive Landscapes and Theoretical Morphospaces*. Cambridge: Cambridge University Press.

Monteiro, A., P. M. Brakefield, and V. French. 1994. The evolutionary genetics and developmental basis of wing pattern variation in the butterfly *Bicyclus anynana*. *Evolution* 48: 1147–1157.

Monteiro, A., P. M. Brakefield, and V. French. 1997a. Butterfly eyespots: the genetics and development of the color rings. *Evolution* 51: 1207–1216.

———. 1997b. The genetics and development of an eyespot pattern in the butterfly *Bicyclus anynana*: Response to selection for eyespot shape. *Genetics* 146: 287–294.

———. 1997c. The relationship between eyespot shape and wing shape in the butterfly *Bicyclus anynana*: A genetic and morphometrical approach. *J. Evol. Biol.* 10: 787–802.

Naisbit, R. E., C. D. Jiggins, and J. Mallet. 2003. Mimicry: developmental genes that contribute to speciation. *Evol. Devel.* 5: 269–280.

Nijhout, H. F. 1991. *The Development and Evolution of Butterfly Wing Patterns*. Washington, DC: Smithsonian Institute Press.

Papa, R., C. Morrison, J. Walters, B. Counterman, R. Chen, G. Halder, L. Ferguson, et al. 2008. Highly conserved gene order and numerous novel repetitive elements in genomic regions linked to wing pattern variation in *Heliconius* butterflies. *BMC Genomics* 9: 345.

Reed, R. D., and M. S. Serfas. 2004. Butterfly wing pattern evolution is associated with changes in a Notch/Distal-less temporal pattern formation process. *Curr. Biol.* 14: 1159–1166.

Robertson, K. A., and A. Monteiro. 2005. Female *Bicyclus anynana* butterflies choose males on the basis of their dorsal UV-reflective eyespot pupils. *Proc. R. Soc. Lond. B* 272: 1541–1546.

Saenko S.V., V. French, P. M. Brakefield, and P. Beldade. 2008. Conserved developmental processes and the formation of evolutionary novelties: examples from butterfly wings. *Phil. Trans. Roy. Soc. B* 363: 1549–1555.

Schluter, D. 1996. Adaptive radiation along genetic lines of least resistance. *Evolution* 50: 1766–1774.

Sheppard, P. M., J.R.G. Turner, K. S. Brown, W. W. Benson, and M. C. Singer. 1985. Genetics and the evolution of Muellerian mimicry in *Heliconius* butterflies. *Phil. Trans. Roy. Soc. London Ser. B* 308: 433–610.

Stevens, M. 2005. The role of eyespots as anti-predator mechanisms, principally demonstrated in the Lepidoptera. *Biol. Rev.* 80: 573–588.

Turner, J.R.G. 1977. Butterfly mimicry: the genetical evolution of an adaptation. *Evol. Biol.* 10: 163–206.

———. 1984. Mimicry: The palatability spectrum and its consequences. In R. I. Vane-Wright and P.-R. Ackery, eds., 141–161. *The Biology of Butterflies*. London: Academic Press.

Chapter Six

Genetics, Geology, and Miracles

David Kingsley

It is a great pleasure for me to be able to participate in the Festschrift for Peter and Rosemary Grant. Ten years ago, Katie Peichel and I decided to begin a detailed genetic study of evolutionary change in naturally occurring species. That work, which focused on threespine stickleback fish, was inspired in at least three different ways by the work of Peter and Rosemary.

First, the Grants' elegant research in the Galápagos helped convince me that evolutionary change in vertebrates was actually a studiable problem. For many years, vertebrate evolution was viewed as such a slow and incremental process that there appeared to be little hope of identifying particular genes and mutations contributing to the interesting morphological adaptations seen in living or extinct species. In contrast, the Grants' work on Darwin's finches demonstrated that significant changes in functional morphology could be observed over a time course of years rather than centuries or millennia, if you were willing to roll up your sleeves and carefully measure both finches and their changing environment. Our own studies with sticklebacks began with the hope that the genetic basis of morphological change in vertebrates would also be a tractable problem, provided that we were also willing to roll up our sleeves and systematically measure the effects of different chromosomes on interesting traits in recently evolved species.

Second, before coming to my lab for postdoctoral training, Katie Peichel had been a graduate student at Princeton, and had served as teaching assistant in an evolutionary biology course taught by Rosemary. Although her own graduate thesis was in mouse molecular genetics, Katie's exposure to the Grants and their work had helped stimulate her own interest in evolution. In addition, when we decided to begin shopping for possible research organisms in summer of 1998, Katie emailed Rosemary for advice about what we were trying to do. Rosemary encouraged rather than shot down our interest in trying to tackle the genetics of vertebrate evolution, a useful psychological boost at a key early discussion stage of the project.

Finally, reading about the Grants' Galápagos work in Weiner's *Beak of the Finch* (1994) also helped us find Dolph Schluter. That book contains an interesting chapter on the work Dolph was doing in Vancouver after training with Peter and Rosemary on Darwin's finches. Dolph's subsequent studies continued the important theme that evolution was a studiable process using recently evolved species. However, after moving to Vancouver, Dolph had switched much of his research to stickleback fish, which could be experimentally crossed to generate large numbers of animals for pond ecology experiments. When Katie and I were shopping for organisms, we realized that the large clutch size of fish would also be an important advantage for tackling the actual chromosomal and molecular basis of complex traits in non-inbred organisms. As a result, we began reading about a number of different fish systems, including blind Mexican cavefish, guppies, cichlids, and sticklebacks. We were particularly impressed by the ubiquity and ease of collection of sticklebacks, their striking morphological diversity, and the depth and breadth of previous studies that had been carried out with the fish by Don McPhail, Don Hagen, Tom Reimchen, Mike Bell, Dolph Schluter, and others (Bell and Foster 1994). Katie and I scheduled a visit with Dolph in September 1998. A few weeks after that visit, Katie was already building the first comprehensive genome libraries and linkage maps for the fish, and Dolph was raising the progeny of a new intercross between divergent stickleback forms, which we would use to examine the genetics of a large number of interesting traits.

Despite the encouragement from Rosemary and Dolph, many evolutionary biologists we talked to at the beginning of the stickleback genetics project were very pessimistic about whether it would ever be possible to map and identify specific genes and mutations that contributed to the obvious morphological differences between naturally occurring species. Over and over again, we heard that most evolutionary traits were likely due to the summed effects of a huge number of genetic changes, each with infinitesimally small individual effects. The importance of small mutations for evolutionary change was actually a well-accepted tenet of the neo-Darwinian synthesis. This view can be traced all the way back to Darwin himself, who stated repeatedly in *The Origin of Species* that "Natura non facit saltum" (Nature does not make leaps), and:

> Natural selection can act only by the preservation and accumulation of infinitesimally small inherited modifications, each profitable to the preserved being. (Darwin 1859)

This view was famously questioned by "mutationists" in the early 1900s, who proposed that new traits arose all at once as single Mendelian factors that controlled important evolutionary phenotypes (De Vries 1901; Bateson 1909). In the 1940s, Goldschmidt went further and proposed that most evolution above the species level might occur by special "systemic" mutations, generating "hopeful monsters" in a single step that were then either accepted or rejected by evolution (Goldschmidt 1940). Most leaders of the neo-Darwinian synthesis countered that laboratory mutations with large phenotypic effects were invariably deleterious ("hopeless" monsters as termed by Mayr 1970). Fisher had provided an influential mathematical argument showing that mutations with the smallest phenotypic effects had the highest probability of avoiding deleterious effects during an approach to a new phenotypic optimum (Fisher 1930). Thus, according to both Darwin and the Modern Synthesis (Huxley 1942), new traits in natural populations will arise from the cumulative power of natural selection acting on very small differences, not by miraculous leaps generated all at once by the mutation process itself.

As Orr and Coyne pointed out in 1992, the long and acrimonious historical debates about the size and number of mutations underlying evolutionary change were based on surprisingly few real examples (Orr and Coyne 1992). We began the stickleback genetics project with a desire to bring new empirical data to old arguments about how organisms actually evolve under a full range of fitness constraints in natural environments. Rather than assuming that evolution was due to a few genes or to many genes, or to particular types of genes, we would simply cross fish from different evolved populations, measure traits, and map the differences in large F_2 families. This approach might reveal single

Mendelian loci that explained virtually all the variation in a given phenotype. Alternatively, genetic mapping might show that there were no genetic regions at all that had large effects, and dozens or hundreds of chromosome regions with extremely small effects. One attraction of the sticklebacks was that the same genetic mapping approach could be used to study a large number of different characters, including dramatic changes in body size, body shape, teeth, jaws, gill rakers, spines, fins, bony armor, color, behavior, physiology, and life history traits (Bell and Foster 1994). And since the same traits had evolved in many different populations, we would eventually be able to compare the genetic mechanisms that underlie many different, repeated evolutionary experiments.

We began by studying some of the most obvious differences in the physical appearance of sticklebacks (see plate 3, left). Cuvier gave different species names to marine and freshwater sticklebacks, based in part on dramatic differences in the anterior posterior patterning of bony plates along the sides of the fish (Cuvier and Valenciennes 1829). Marine fish are heavily armored and slow, with a complete row of bony plates extending from just behind the head all the way down the side of the body to the tail. In contrast, most freshwater fish are lightly armored and fast, retaining only a few plates at the anterior side of the body (Taylor and McPhail 1986; Bergstrom 2002). This was an ideal trait for genetic studies because the huge difference in number of plates provided a simple quantitative phenotype that can be easily counted and measured in crosses (Colosimo et al. 2004).

A second obvious difference was the presence or absence of pelvic structures in different populations (plate 3, middle). This trait was particularly attractive because limb structures vary extensively in vertebrates and have long been used as an important basis for phylogenetic classification (Hinchliffe and Johnson 1980). One of the most dramatic limb modifications is the complete loss of limbs that has evolved repeatedly in some species of fish, amphibians, reptiles, and mammals (Lande 1978; Bejder and Hall 2002). Marine sticklebacks resemble most fish and land animals in having two sets of paired appendages: forefins and hindfins, also known as pectoral fins and pelvic fins. The pelvic apparatus of the stickleback consists of two long, serrated bony spines that articulate in movable joints with an underlying bony pelvis.

118

The pelvis itself is a shield-shaped bone that covers much of the ventral surface. Although this elaborate pelvic structure has given sticklebacks their characteristic genus name of *Gasterosteus* (meaning bony stomach), at least two dozen populations are known around the world that have evolved partial or complete loss of pelvic structures (Bell 1974; Moodie and Reimchen 1976; Campbell and Williamson 1979; Edge and Coad 1983; Bell and Orti 1994). Crosses between marine and pelvic-reduced sticklebacks thus provide an unusual opportunity to study the genetic basis of a major change in limb patterning in natural populations.

Finally, the fish could also be used to study a number of non-skeletal traits, including obvious changes in pigmentation. Sticklebacks have colonized a large number of different lakes and streams, including freshwater locations that are either shallow or deep, and vary in water color and clarity. Several populations are now known that have substantially lighter skin and gill filaments than marine fish (plate 3, right), including the bottom-dwelling members of benthic-limnetic species pairs lakes in the Vancouver area (McPhail 1992; Miller et al. 2007). Since light skin color was present in the Paxton Lake fish used as one of the parents of the big cross set up with Dolph Schluter, it was easy to compare the genetics of skeletal and pigment changes using the very same large genetic cross that we were also using to score plates and pelvis.

Examining the genetic architecture of these traits required developing a large number of new genomic resources for threespine sticklebacks (Peichel et al. 2001; Hosemann et al. 2004; Kingsley et al. 2004; Kingsley and Peichel 2007). However, my laboratory was already well set up for high throughput sequencing, genotyping, and positional cloning studies, since we had already been using similar positional cloning methods to identify the genes responsible for classical skeletal traits in mice. Katie developed an initial first genome-wide linkage map for sticklebacks within the first eight months of our initial visit to Vancouver (Peichel et al. 2001). As fish from a "big cross" between marine and Paxton Lake sticklebacks became available, we were finally able to measure phenotypes in a large number of F_2 animals, isolate DNA from the same individuals, and type them with genome-wide linkage markers. We then systematically tested whether any chromosome regions from the original marine and freshwater grandparents were consistently present in the F_2 progeny with different plate counts, pelvic sizes, or pigmentation scores (Colosimo et al. 2004; Shapiro et al. 2004; Miller et al. 2007).

Our actual results were intermediate between the extreme positions of the previous "hopeful monsters—infinitesimal mutations" debate (see plate 3). None of the evolutionary traits we examined could be explained by a single Mendelian locus that explained all of the variation in armor plates, pelvic size, or pigmentation. Nonetheless, for each of the traits, we did detect a major locus that controlled somewhere between 40 percent and 70 percent of the variation in the F_2 cross. And for each trait, we also detected a number of unlinked loci that explained a substantially smaller proportion of the variation in phenotype (2–10%). Of course, the limited size of our cross may not have provided sufficient power to detect additional loci of even smaller effect (Beavis 1998). However, these initial genetic results suggest that evolution has occurred by a mixture of a few genes of large effect, and more numerous loci with small effects. In addition, some of the loci clearly have effect sizes that are enormous compared to classic "infinitesimal" models of evolutionary change. Similar trends are apparent in more recent analysis of numerous shape traits in sticklebacks (Albert et al. 2008); for traits analyzed in other stickleback populations (Cresko et al. 2004; Kimmel et al. 2005); and for evolutionary traits in a variety of other plant and animal systems that have now been analyzed by genome-wide linkage mapping (Orr 2001, 2005).

We have subsequently used high-resolution mapping, chromosome walking, sequencing, gene expression analysis, and transgenic studies to identify the genes of largest effect for all three of these traits. Armor plates are controlled by changes at the *Ectodysplasin* (*EDA*) locus (Colosimo et al. 2005), a major developmental signaling molecule that also controls formation of hair, teeth, and sweat glands in mice and humans, and dermal bone formation in mice, humans, and fish (Thesleff and Mikkola 2002; Harris et al. 2008)

The major locus for pelvic developmental maps to the stickleback *PITX1* gene (Shapiro et al. 2004). This homeodomain transcription factor was named for its important role in pituitary development. However, it has also been referred to as the "backfoot" gene, because it is expressed in hindlimbs but nor forelimbs of many different animals

(Shang et al. 1997). Inactivation of the gene in mice causes both pituitary abnormalities and pelvic reduction, as well as defects in jaws and craniofacial structures (Lanctot et al. 1999; Szeto et al. 1999).

Finally, the major locus for skin color maps to the stickleback *Kit ligand/Stem cell factor* (*KITLG*) gene (Miller et al. 2007). This locus encodes a secreted signaling that has been extensively studied as the product of the classic "*Steel*" coat color/sterility/anemia locus in mice. The signaling molecule binds to a transmembrane tyrosine kinase receptor that is expressed on many different stem cell populations, and is required for normal migration, proliferation, and survival of pigment, hematopoeitic, and stem cell populations during normal development (Wehrle-Haller 2003).

Although only a few evolutionary traits have now been traced to specific genes, there are interesting parallels between each of the major loci controlling large phenotypic differences in sticklebacks. Each encodes a key component of the fundamental signaling and transcription factor pathways that govern tissue formation during embryonic development. Each plays such an important role in formation of multiple different tissues that the corresponding gene is essential for normal viability, fitness, or fertility. In fact, a laboratory mouse carrying mutations in all three genes would have major hair, tooth, and gland defects, short hindlimbs, malformed jaws, defective pituitaries, absence of pigment, severe anemia, absence of mast cells, gastrointestinal motility defects, sterility, and would die at birth with cleft palate and blood cell defects. These strongly deleterious phenotypes in mice match the long-standing prediction that mutations in major developmental control genes are generally incompatible with viability, fitness, or fertility (Mayr 1970). Nonetheless, Paxton Lake sticklebacks clearly carry altered forms of all three genes, each controlling dramatic new phenotypes that have been repeatedly selected in natural freshwater environments.

How can we resolve the paradox between lethal genes in laboratory mice and new adaptive phenotypes in living sticklebacks? Our studies of the differences between marine and freshwater fish suggest that evolution in natural populations has occurred by making hypomorphic mutations rather than loss of function mutations at each of the key developmental regulators. The *EDA*, *PITX1*, and *KITLG* genes all consist of protein coding regions surrounded by large intergenic control

121

regions. The protein coding regions are either unchanged, or only slightly modified in the evolved sticklebacks, preserving many essential functions that would be lost in a null mutation at the corresponding locus. And for each of the three genes, our initial studies suggest that evolution in sticklebacks has likely occurred by changing where and when the gene is expressed at particular times and places during normal development, while preserving normal expression at other locations (Shapiro et al. 2004; Colosimo et al. 2005; Knecht et al. 2007; Miller et al. 2007).

We are currently using comparative genomics and transgenic studies to track down the actual DNA sequence changes responsible for the altered expression patterns of each gene (Chan et al. 2010). While much work remains to be done to identify the individual base pair changes responsible for morphological evolution, our ongoing studies suggest that freshwater populations have modified individual enhancer sequences that control expression in some tissues but not others. Mild protein mutations, or site-specific regulatory mutations, may make it possible to produce major morphological differences in natural populations, while preserving the other essential functions of key developmental regulators (Carroll 2000; Stern 2000; Shapiro et al. 2004).

FINDING A MIDDLE GROUND BETWEEN CATASTROPHISM AND UNIFORMITARIANISM

Geology in the nineteenth century featured strong debates between "catastrophists" arguing for periods of rapid change in Earth's history, and "uniformitarians" arguing for slow, gradual change (Whewell 1832). Charles Lyell's highly influential *Principles of Geology* in the 1830s championed the view that all major features of the geological world could be explained as the accumulated outcome of everyday processes, working slowly and steadily over time (Lyell 1830–1833). The uniformitarian view was so successful, and became so embedded in training and outlook of the field, that later proposals for rapid catastrophic events were often ridiculed as outlandish or unscientific. In the 1920s, Bretz proposed that massive sudden flooding had carved the Channeled Scablands and Columbia River basin area of Washington

State (Bretz 1923). In the 1980s, Alvarez proposed that a massive extraterrestrial impact had brought an end to the Mesozoic age (Alvarez et al. 1980). Such proposals were so far outside the accepted mainstream of uniformitarian, gradualist thinking, that it took years of additional empirical research before geologists finally began to accept both proposals.

A very similar debate has clearly occurred in the history of evolutionary biology. Darwin brought Lyell's principles of uniformitarianism to the field of biology. In Darwin's view, all the special features of organisms had arisen by the slow, gradual accumulation of small, favorable variants, rather than emerging all at once by improbable and miraculous acts of special creation, or by a mutation process that somehow caused large adaptive changes to leap out of thin air. Darwin actually drew a comparison between genetics and geology quite explicitly in *The Origin of Species*. His full sentence about "infinitesimal" modifications actually reads:

> Natural selection can act only by the preservation and accumulation of infinitesimally small inherited modifications, each profitable to the preserved being; and as modern geology has almost banished such views as the excavation of a great valley by a single diluvial wave, so will natural selection, if it be a true principle, banish belief in the continued creation of new organic beings, or of any great and sudden modification in their structure. (Darwin 1859)

Evolution by infinitesimal mutations became so well embedded in the neo-Darwinian synthesis that later proposals for new phenotypes arising by single mutations of large effect were also rejected as outlandish and unscientific. Goldschmidt's "hopeful monsters" mutations were dismissed every bit as strongly as Bretz's catastrophic flood, and Alvarez's later impact theory. In a striking echo of the earlier nineteenth-century debates about miraculous versus natural explanations in geology, Dobzhansky entitled his review of Goldschmidt's book *Catastrophism versus Evolutionism* and stated that: "in the reviewer's opinion the simplicity of Goldschmidt's theory is that of a belief in miracles" (Dobzhansky 1940).

123

Ernst Mayr concurred, stating:

The occurrence of genetic monstrosities by mutation, for instance the homeotic mutant in Drosophila, is well substantiated, but they are such evident freaks that these monsters can be designated only as "hopeless." . . . It is a general rule, of which every geneticist and breeder can give numerous examples, that the more drastically a mutation affects the phenotype, the more likely it is to reduce fitness. To believe that such a drastic mutation would produce a viable new type, capable of occupying a new adaptive zone, is equivalent to believing in miracles. (Mayr 1970)

In geology, the early acrimonious debates about sudden geological events eventually stimulated further productive research into isotope signatures, fossil transitions, and geological markers of floods and impacts. Years after Bretz's and Alvarez's original proposals, there is now general acceptance that enormous floods, eruptions, and impacts have occurred in Earth's history, and have shaped important features of the physical world. As a result, the geological history of Earth is now typically viewed as the product of *both* slow incremental processes and rare events of much larger magnitude. The alternative of insisting that all change is small and slow, or that all change is large and rapid, provides a much poorer understanding of the history of the Earth than recognition and study of both uniformitarian and catastrophic processes.

I think that many long-standing debates in evolutionary biology will also be resolved in a similar manner. Kimura and Orr have already reconsidered Fisher's theoretical arguments that the mutations favored during evolutionary adaptation will have infinitesimal effects (Kimura 1983; Orr 1998). The more recent work suggests that the distribution of effect sizes fixed during an adaptive walk will actually be exponential, with many mutations of small effect (as long proposed in the neo-Darwinian synthesis), along with some mutations with much larger effect (a possibility long dismissed as "hopeless") (Orr 2005).

Our own empirical studies of naturally evolved populations have clearly identified some genetic regions that have enormous effects on major morphological traits in sticklebacks. However, I would not want people to come away from the work thinking that stickleback genetics has confirmed an overly simplistic view of evolutionary change. We have recently found that pelvic reduction occurs by deletion of a tissue specific enhancer in the Pitx1 gene, a type of molecular lesion that could arise in a single mutational leap (Chan et al. 2010). However, we have not yet identified the actual DNA base pair lesions that have occurred in either the *EDA* or *KITLG* genes. It is thus still possible that these loci have actually accumulated a large number of individual mutations with small effects, which sum together by genetic linkage to produce a much larger phenotype. Such a phenomenon has clearly been seen in other systems (Stam and Laurie 1996; McGregor et al. 2007). It also seems likely at major genes like *EDA*, where alternative alleles for high and low plated fish turn out to be much older than the 10,000-year-old history of post-glacial lakes, and have been repeatedly fixed from ancient preexisting variants that already exist at low frequency in migratory marine ancestors (Colosimo et al. 2005; Barrett et al. 2008; see also chapter 16). In addition, our *major* genes for plates, pelvis, and pigment are clearly not the *only* genes influencing these traits. None of the morphological differences we have analyzed behave as a simple Mendelian trait, and in every case we have also found unlinked modifier loci that contribute to smaller quantitative variation in plate number, pelvic size, and pigmentation levels (plate 3). Multiple QTL have also been detected for shape traits, with many loci of small effect and a smaller number of genes with large effects (Albert et al. 2008; see also chapter 5). Finally, in overall analysis of many different skeletal and shape traits, we have already detected over a hundred different significant loci that are required to produce the total phenotypic difference seen between different fish forms (Peichel et al. 2001; Colosimo et al. 2004; Shapiro et al. 2004; Miller et al. 2007; Albert et al. 2008). Thus, evolution in sticklebacks has required a large number of different genetic changes, despite the fact that some of the major loci have effects that are much larger than expected from traditional views of infinitesimal evolutionary change.

In the last two years, a new debate has been raging about the relative importance of coding and regulatory change underlying evolutionary traits in natural populations (Carroll 2005; Hoekstra and Coyne 2007; Carroll 2008; Pennisi 2008; Stern and Orgogozo 2008). Like the monsters versus infinitesimal mutation debate, I believe that this controversy has also become over-polarized. Additional empirical studies are clearly required to determine the overall balance of coding and regulatory mutations, and I suspect that multiple types of mechanisms will likely contribute to evolutionary traits in nature.

Regardless of the overall balance of different types of genetic changes, I am encouraged that the field of evolutionary genetics has now reached the point of vigorous empirical studies in many different organisms. Importantly, no miracles are needed for future progress. Peter and Rosemary's own work shows how much can be learned from detailed studies of real species that have evolved in natural environments. We now clearly have the tools to extend such studies to the genetic level, and I look forward to the detailed answers that will come when we let fish, finches, and many other organisms reveal the molecular basis of evolutionary differences in the wild.

References

Albert, A.Y.K., S. Sawaya, T. H. Vines, A. K. Knecht, C. T. Miller, B. R. Summers, S. Balabhadra, D. M. Kingsley, and D. Schluter. 2008. The genetics of adaptive shape shift in stickleback: Pleiotropy and effect size. *Evolution* 62: 76–85.

Alvarez, L. W., W. Alvarez, F. Asaro and H. V. Michel. 1980. Extraterrestrial cause for the cretaceous-tertiary extinction. *Science* 208: 1095–1108.

Barrett, R. D., S. M. Rogers, and D. Schluter. 2008. Natural selection on a major armor gene in threespine stickleback. *Science* 322: 255–257.

Bateson, W. 1909. Heredity and variation in modern lights. In A. C. Seward, ed. *Darwin and Modern Science: Essays in Commemoration of the Centenary of the Birth of Charles Darwin and of the Fiftieth Anniversary of the Publication of "The Origin of Species."* Cambridge: Cambridge University Press.

Beavis, W. D. 1998. QTL analysis: power, precision and accuracy. In A. H. Paterson, ed., *Molecular Dissection of Complex Traits*, 145–162. Boca Raton, FL: CRC Press.

Bejder, L., and B. K. Hall. 2002. Limbs in whales and limblessness in other vertebrates: Mechanisms of evolutionary and developmental transformation and loss. *Evol. Dev.* 4: 445–458.

Bell, M. A. 1974. Reduction and loss of the pelvic girdle in gasterosteus (pisces): A case of parallel evolution. *Natural History Museum of Los Angeles County Contributions in Science* 257: 1–36.

Bell, M. A., and S. Foster. 1994. eds. *The Evolutionary Biology of the Threespine Stickleback*. Oxford: Oxford University Press.

Bell, M. A., and G. Orti. 1994. Pelvic reduction in threespine stickleback from Cook Inlet lakes: Geographical distribution and intrapopulation variation. *Copeia* 314–325.

Bergstrom, C. A. 2002. Fast-start swimming performance and reduction in lateral plate number in threespine stickleback. *Canadian Journal of Zoology-Revue Canadienne De Zoologie* 80: 207–213.

Bretz, J. H. 1923. The channeled scabland of the columbia plateau. *J. Geol.* 31: 617–649.

Campbell, R. N., and R. B. Williamson. 1979. The fishes of inland waters in the outer hebrides. *Proc Royal Soc Edinburgh* 77B: 377–393.

Carroll, S. B. 2000. Endless forms: The evolution of gene regulation and morphological diversity. *Cell* 101: 577–580.

————. 2005. Evolution at two levels: On genes and form. *PLoS Biol* 3: e245.

————. 2008. Evo-devo and an expanding evolutionary synthesis: A genetic theory of morphological evolution. *Cell* 134: 25–36.

Chan, Y. F., M. E. Marks, F. C. Jones, G. Villarreal, Jr., M. D. Shapiro, S. D. Brady, A. M. Southwick, D. M. Absher, J. Grimwood, J. Schmutz, R. M. Myers, D. Petrov, B. Jonsson, D. Schluter, M. A. Bell, and D. M. Kingsley. 2010. "Adaptive evolution of pelvic reduction in sticklebacks by recurrent deletion of a Pitx1 enhancer." *Science* 327: 302–305.

Colosimo, P. F., K. E. Hosemann, S. Balabhadra, G. Villareal, M. Dickson, J. Grimwood, J. Schmutz, R. Myers, D. Schluter, and D. M. Kingsley. 2005. Widespread parallel evolution in sticklebacks by repeated fixation of ectodysplasin alleles. *Science* 307: 1928–1933.

Colosimo, P. F., C. L. Peichel, K. S. Nereng, B. K. Blackman, M. D. Shapiro, D. Schluter, and D. M. Kingsley. 2004. The genetic architecture of parallel armor plate reduction in threespine sticklebacks. *PLoS Biology* 2: 635–641.

Cresko, W. A., A. Amores, C. Wilson, J. Murphy, M. Currey, P. Phillips, M. A. Bell, C. B. Kimmel, and J. H. Postlethwait. 2004. Parallel genetic basis for repeated evolution of armor loss in alaskan threespine stickleback populations. *Proc Natl Acad Sci U S A* 101: 6050–6055.

Cuvier, G. L., and M. A. Valenciennes. 1829. *Histoire naturelle des poissons. Tome quatrieme.* Paris: Chez F.G. Levrault.

Darwin, C. 1859. *The Origin of Species.* London: John Murray.

de Vries, H. 1901. *Die mutations theorie.* Leipzig: Veit.

Dobzhansky, T. 1940. Catastrophism versus evolutionism. *Science* 92: 356–358.

Edge, T. A., and B. W. Coad 1983. Reduction of the pelvic skeleton in the threespined stickleback gasterosteus aculeatus in 2 lakes of Quebec Canada. *Canadian Field-Naturalist* 97: 334–336.

127

Fisher, R. A. 1930. *The Genetical Theory of Natural Selection*. Oxford: Oxford University Press.

Goldschmidt, R. 1940. *The Material Basis of Evolution*. New Haven, CT: Yale University Press.

Harris, M. P., N. Rohner, H. Schwarz, S. Perathoner, P. Konstantinidis, and C. Nusslein-Volhard. 2008. Zebrafish Eda and Edar mutants reveal conserved and ancestral roles of ectodysplasin signaling in vertebrates. *PLoS Genet* 4: e1000206.

Hinchliffe, J. R., and D. R. Johnson 1980. *The Development of the Vertebrate Limb*. Oxford: Clarendon Press.

Hoekstra, H. E., and J. A. Coyne 2007. The locus of evolution: Evo devo and the genetics of adaptation. *Evolution Int J Org Evolution* 61: 995–1016.

Hosemann, K. E., P. F. Colosimo, B. R. Summers, and D. M. Kingsley 2004. A simple and efficient microinjection protocol for making transgenic stickle-backs. *Behaviour* 141: 1345–1355.

Huxley, J. S. 1942. *Evolution, The Modern Synthesis*. London: George Allen and Unwin.

Kimmel, C. B., B. Ullmann, C. Walker, C. Wilson, M. Currey, P. C. Phillips, M. A. Bell, J. H. Postlethwait, and W. A. Cresko. 2005. Evolution and development of facial bone morphology in threespine sticklebacks. *Proc Natl Acad Sci USA* 102: 5791–5796.

Kimura, M. 1983. *The Neutral Theory of Molecular Evolution*. Cambridge: Cambridge University Press.

Kingsley, D. M., B. Zhu, K. Osoegawa, P. J. de Jong, J. Schein, M. Marra, C. L. Peichel, C. Amemiya, D. Schluter, S. Balabhadra, B. Friedlander, Y. M. Cha, M. Dickson, J. Grimwood, J. Schmutz, W. S. Talbot, and R. M. Myers. 2004. New genomic tools for molecular studies of evolutionary change in stickle-backs. *Behaviour* 141: 1331–1344.

Kingsley, D. M., and C. L. Peichel 2007. The molecular genetics of evolutionary change in sticklebacks. In S. Ostlund-Nilsson, I. Mayer, and F. A. Huntingford, eds. *Biology of the Three-Spined* Stickleback, 41–81. London: CRC Press.

Knecht, A. K., K. E. Hosemann, and D. M. Kingsley 2007. Constraints on utilization of the EDA signaling pathway in threespine stickleback evolution. *Evolution & Development* 9: 141–154.

Lanctot, C., A. Moreau, M. Chamberland, M. L. Tremblay, and J. Drouin. 1999. Hindlimb patterning and mandible development require the Ptx1 gene. *Development* 126: 1805–1810.

Lande, R. 1978. Evolutionary mechanisms of limb loss in tetrapods. *Evolution* 32: 73–92.

Lyell, C. 1830–1833. *Principles of Geology*. London: John Murray.

Mayr, E. 1970. *Populations, Species and Evolution*. Cambridge, MA: Harvard University Press.

McGregor, A. P., V. Orgogozo, I. Delon, J. Zanet, D. G. Srinivasan, F. Payre, and D. L. Stern. 2007. Morphological evolution through multiple cis-regulatory mutations at a single gene. *Nature* 448: 587–590.

McPhail, J. D. 1992. Ecology and evolution of sympatric sticklebacks (gasterosteus): Evidence for a species pair in Paxton Lake, Texada Island British Columbia. *Can. J. Zool.* 70: 361–369.

Miller, C. T., S. Beleza, A. A. Pollen, D. Schluter, R. A. Kittles, M. D. Shriver, and D. M. Kingsley. 2007. Cis-regulatory changes in *Kit Ligand* expression and parallel evolution of pigmentation in sticklebacks and humans. *Cell* 131: 1179–1189.

Moodie, G.E.E., and T. Reimchen. 1976. Phenetic variation and habitat differences in gasterosteus populations of the Queen Charlotte Islands. *Systematic Zoology* 25: 49–61.

Orr, H. A. 1998. The population genetics of adaptation: The distribution of factors fixed during adaptive evolution. *Evolution* 52: 935–949.

———. 2001. The genetics of species differences. *Trends in Ecology & Evolution* 16: 343–350.

———. 2005. The genetic theory of adaptation: A brief history. *Nat Rev Genet* 6: 119–127.

Orr, H. A., and J. A. Coyne 1992. The genetics of adaptation—a reassessment. *American Naturalist* 140: 725–742.

Peichel, C. L., K. S. Nereng, K. A. Ohgi, B. L. Cole, P. F. Colosimo, C. A. Buerkle, D. Schluter, and D. M. Kingsley. 2001. The genetic architecture of divergence between threespine stickleback species. *Nature* 414: 901–905.

Pennisi, E. 2008. Evolutionary biology. Deciphering the genetics of evolution. *Science* 321: 760–763.

Shang, J., Y. Luo, and D. A. Clayton. 1997. Backfoot is a novel homeobox gene expressed in the mesenchyme of developing hind limb. *Dev. Dyn.* 209: 242–253.

Shapiro, M. D., M. E. Marks, C. L. Peichel, B. K. Blackman, K. S. Nereng, D. Schluter, B. Jonsson, and D. M. Kingsley. 2004. Genetic and developmental basis of evolutionary pelvic reduction in threespine sticklebacks. *Nature* 428: 717–723.

Stam, L. F., and C. C. Laurie. 1996. Molecular dissection of a major gene effect on a quantitative trait: The level of alcohol dehydrogenase expression in Drosophila melanogaster. *Genetics* 144: 1559–1564.

Stern, D. L. 2000. Evolutionary developmental biology and the problem of variation. *Evolution* 54: 1079–1081.

Stern, D. L., and V. Orgogozo. 2008. The loci of evolution: How predictable is genetic evolution? *Evolution* 62: 2155–2177.

Szeto, D. P., C. Rodriguez-Esteban, A. K. Ryan, S. M. O'Connell, F. Liu, C. Kioussi, A. S. Gleiberman, J. C. Izpisua-Belmonte, and M. G. Rosenfeld. 1999. Role of the bicoid-related homeodomain factor Pitx1 in specifying hindlimb morphogenesis and pituitary development. *Genes. Dev.* 13: 484–494.

Taylor, E. B., and J. D. McPhail. 1986. Prolonged and burst swimming in anadromous and fresh-water threespine stickleback, gasterosteus-aculeatus. *Canad. J. Zool.* 64: 416–420.

Thesleff, I., and M. L. Mikkola. 2002. Death receptor signaling giving life to ectodermal organs. *Sci. STKE* 131: PE22.

Wehrle-Haller, B. 2003. The role of Kit-Ligand in melanocyte development and epidermal homeostasis. *Pigment Cell Res.* 16: 287–296.

Weiner, J. 1994. *The Beak of the Finch*. New York: Knopf.

Whewell, W. 1832. Review of lyell's principles of geology. *Quarterly Review* 47: 103–132.

Chapter Seven

Evolutionary Genetics of Pigmentation Loss in the Blind Mexican Cavefish

Joshua B. Gross and Clifford J. Tabin

W hen an animal colonizes a new environment, traits that were previously important lose their utility while other traits become essential for survival and maintaining fitness. Until recently, the mechanistic and genetic basis for adaptive and neutral traits, in the context of a well-defined ecosystem, has remained elusive. The blind Mexican cavefish (*Astyanax mexicanus*) from a series of limestone subterranean caves in northeastern Mexico offers a unique opportunity to integrate a wealth of literature spanning field observations begun in the 1930s with the power of quantitative genetics. Using these contemporary tools, we can now investigate the underlying genetic basis for traits that are lost or expanded as a consequence of inhabiting the cave environment.

In general, cave animals provide a dramatic example of adaptation to a unique environment (Culver et al. 1995). Early work focusing on the phenotypes of cave animals (e.g., the "bleached" or de-pigmentation) presented a set of problems for evolutionary biologists (Eigenmann 1909). Principal among them was the question: how can the loss of tissues and organs in the cave environment be adaptive? The gradual discovery of additional cave animals ranging from isopods to salamanders

The authors wish to acknowledge Meredith Protas (University of California, Berkeley) who pioneered work on *Astyanax* in the Tabin laboratory, and our collaborator Richard Borowsky (New York University). This material is based upon work supported by the National Science Foundation under Grant No. 0821982. Any opinions, findings, and conclusions or recommendations expressed in this material are those of the authors and do not necessarily reflect the views of the National Science Foundation.

to crayfish posed an additional problem: What aspects of the cave eco-system predictably drive the loss of the same characters (namely pigmentation and the eye) in highly disparate taxa?

The question of how cave forms evolve has been limited in the past by the lack of a surrogate ancestral form to compare to the cave form. In nature, this is because oftentimes the cave-adapted (troglobitic) form and its ancestral form become geologically and genetically distinct (Renno et al. 2007), and one lineage goes extinct (Contreras-Díaz et al. 2007), precluding the possibility of hybridization and genetic mapping experiments. In the case of the blind Mexican cavefish, *Astyanax mexicanus*, both the surface (epigean) and cave (hypogean) forms thrive in the wild (Sadoğlu 1979), making them amenable to genetic investigations (plate 5; Wilkens 1988). Until very recently, contemporary techniques of linkage mapping and quantitative genetics have been limited to analyses in traditional model species. The application of these modern approaches to the biology of species that exist in the wild provides a powerful tool for clarifying the underlying genetic basis of ecologically relevant traits.

The natural habitat of the cave morphotypes of these species includes an extensive subterranean network of caves that materialized over several thousands of years from the gradual erosion of limestone bedrock in mountainous regions of northeastern Mexico (plate 6; Avise and Selander 1972). Consequently, several independent populations of fish from the genus *Astyanax* have inhabited these caves. Thus far, twenty-nine different caves have been formally described (Espinasa and Borowsky 2001), most of which were colonized toward the end of the Pleistocene era (Strecker et al. 2004). Since their discovery, these fish have been used in numerous inter-crossing experiments and phenotypic descriptions of their many cave-adapted traits (Sadoğlu and McKee 1969; Wilkens 1988; Wilkens and Strecker 2003; Protas et al. 2006; Protas et al. 2008).

In recent years, the blind Mexican cavefish has emerged as a genetic model system for the study of evolutionary and developmental biology (Jeffery 2001; Protas et al. 2006). Further, because of the widespread and independent colonization of caves by numerous conspecific cave morphs, this species is also an excellent model for the study of traits that evolve in parallel (Protas et al. 2007; see also chapters 6 and 16). Several aspects of *Astyanax* biology and ecology make it especially well suited for the study of a variety of biological processes.

Cave morphs have evolved several specialized traits as a result of perpetual life in the dark, nutrient-poor environment of the cave, including the loss of pigmentation and eyes (Dowling et al. 2002; Jeffery 2006). In order to understand how and why cave-adapted traits evolve in nature, it is essential to characterize the underlying genetic bases for these traits. Indeed, the genetic basis for these regressive traits has only recently been addressed through the use of linkage analysis in large pedigrees produced from several different caves (Protas et al. 2006; Protas et al. 2008; Gross et al. 2009). In addition to the so-called regressive traits that have evolved in this species, several progressive, or so-called constructive traits are also present as key adaptations to cave life (Hüppop 1986). Included among these is the evolution of increased numbers of taste buds and neuromasts of the lateral line system in cave morphs (reviewed in Jeffery 2001).

In this chapter, we explore the details of what is known of the genetic evolution of pigment loss in cavefish. Herein, we also delve into the question of how and why pigmentation loss occurs repeatedly in cave animals. Further, we explore whether similar genes are responsible for pigmentation alterations in other animal taxa. Finally, we examine the question of why depigmentation predictably occurs in disparate animal systems inhabiting the cave microenvironment. In this portion of the chapter, we give special attention to the two primary forces believed to play a role in regressive trait evolution, random genetic drift, and selection. Under the category of selection, we describe two potential forces of selection in the context of cave animals: direct selection and indirect selection through pleiotropy.

THE GENETIC BASIS FOR ALBINISM IN *ASTYANAX* CAVE MORPHS

The first report on the genetics of albinism in blind Mexican cavefish was published by Sadoğlu (1957), describing complete absence of melanin in ~26 percent of an F_2 pedigree between a Pachón cave x surface parental cross (plate 5). While this report focused on the presence of albinism in fish derived from the Pachón cave, albino individuals have more recently been described in additional caves including the Yerbaniz and Molino caves (Wilkens 1988; Wilkens and Strecker 2003; Wilkens 2004). It should be noted that albino individuals derived from the Japonés cave were also included in a recent genetic analysis (Protas et al. 2006); however, this cave is most likely linked to the Yerbaniz cave system.

Interestingly, in one of these caves, the Pachón cave, the presence of albinism is not fixed in every member of the population (Sadoğlu 1957; Sadoğlu and McKee 1969). While current linkage mapping projects have utilized albino individuals in various pedigree crosses, historical literature has documented the presence of both lightly pigmented ("brown") and albino individuals (plate 5). For instance, Sadoğlu and McKee (1969) reported that roughly 31.5 percent of the offspring individuals created from crosses of fish drawn directly from the wild were albino, while the remaining 68.5 percent were reported as carrying the lightly pigmented brown phenotype. Additional expeditions to the same caves of northeastern Mexico similarly reported that not every individual present in the cave populations of Pachón were albino (Kosswig 1963; Pfeiffer 1966).

The identity of the gene underlying albinism remained unknown for several years until the work of Protas et al. (2006) demonstrated that mutations at the gene *oculocutaneous albinism type II* (*Oca2*) causes albinism in the three albino populations of *Astyanax* cavefish. In two of the cave populations, Pachón and Molino, albinism is caused by coding mutations in *Oca2*. In Pachón individuals, the causative lesion in *Oca2* is a nearly complete absence of exon 24, whereas in Molino individuals *Oca2* loss of function is caused by deletion of exon 21 (Protas et al. 2006). To demonstrate that these variant alleles are responsible for the loss of *Oca2* function, the authors transfected an *Oca2*-deficient (melan-P) cell line derived from albino mice (Sviderskaya et al. 1997). The intact "wild type" allele (derived from a surface individual) was able to direct normal melanogenesis within the cell line; however, both Pachón and Molino forms of *Oca2* failed to restore pigmentation in this assay. Therefore, the different mutations present in the coding regions of the gene *Oca2* are the causative lesions mediating albinism in these two cave populations.

A third population consisting of fish derived from the Japonés cave was also included in these analyses (plate 6). These fish, along with fish from the linked Yerbaniz cave, also demonstrate albinism (Wilkens 2004). In order to determine whether the same gene is responsible for the trait in this population, the authors performed a complementation test and found that the offspring of a cross between a Pachón and Japonés individual produced only albino individuals. This cross confirmed the same gene is responsible for albinism in the Japonés cave; however, a sequence analysis of *Oca2* did not reveal presence of either the Pachón or Molino coding deletions. The Japonés allele did reveal presence of two point mutations also present in the Pachón population. Neither of the point mutations, however, mediates reduced function of the *Oca2* protein. These results implicate the same gene, *Oca2*, in the evolution of albinism in this cave population. In contrast to Pachón and Molino cave morphs, diminished *Oca2* expression is most likely caused by a *cis*-regulatory mutation (affecting transcriptional expression) rather than a coding alteration.

In sum, the simple trait of albinism affecting the ability to produce melanin has evolved in parallel in three cave populations. While the same gene mediates the same trait in all three populations, coding sequence alterations are responsible in only two of the caves. Interestingly, the causative coding mutation differs between these two caves. Further, in the third cave population, the causative mutation is not coding, but rather regulatory in nature. Therefore, *Astyanax* cavefish are susceptible to the evolution of albinism at the same locus in independent populations, albeit through distinct mutations.

THE GENETIC BASIS FOR THE BROWN MUTATION IN *ASTYANAX* CAVEFISH

A second trait affecting pigmentation was first described in the literature in 1969, in a study characterizing eye color in F_2 individuals derived from several cave populations (Sadoğlu and McKee 1969). In this study, the brown mutation was described as causing brownish eyes and a decreased number of melanophores in affected individuals. Further, when correcting for the presence of albinism, the number of affected individuals produced in an F_2 cross between Pachón cavefish and surface fish (~27%) implied the brown mutation to be a simple recessive trait.

In a series of follow-up reports, the brown mutation was more thoroughly characterized as a decrease in the amount of melanin within the melanophores of various cave populations of *Astyanax* (plate 6; Wilkens 1988; Wilkens and Strecker 2003). Wilkens and Strecker (2003) performed a series of complementation crosses between several representative cave form individuals to demonstrate that the same locus is responsible for the brown mutation in the Pachón, Yerbaniz, Piedras, and Curva caves. Therefore, the cave forms demonstrated in the literature to carry the brown mutation include the Pachón, Japonés, Yerbaniz, Sabinos, Piedras, Curva, and Chica caves (plate 6; summarized in Gross et al. 2009). Interestingly, while the brown mutation is present in the Pachón, Yerbaniz, and Japonés caves, all of which house albino individuals, it is absent from individuals drawn from the albino Molino cave (Wilkens and Strecker 2003). Thus, the brown mutation does not invariably accompany the evolution of albinism in *Astyanax* cavefish.

To determine the genetic basis for the brown phenotype, we performed a quantitative trait locus (QTL) analysis for the trait (Gross et al. 2009). We carried out a phenotypic analysis of melanophore number in the post-optic region of the head, as well as the dorsal flank, each reflecting the identical manifestation of the brown mutation. Both analyses yielded the same result, a single QTL present on linkage group P09, confirming the prediction that the brown mutation is a single gene trait. Further, concordant with the work of Wilkens and Strecker (2003), the same phenotypic analysis revealed no QTL in a backcross pedigree derived from Molino cavefish.

A genomic analysis of the sequences flanking the microsatellites housed on linkage group P09 strongly implied the affected locus resides on chromosome 18 in the zebrafish, *Danio rerio*. A search for candidate genes led directly to the gene *melanocortin receptor 1* (*Mc1r*), a gene controlling *de novo* melanin synthesis in mammalian systems (Abdel-Malek et al. 2000). When sequenced from Pachón cavefish, we identified a 2-base pair deletion in the extreme 5' end of the coding region of *Mc1r*. Further, individuals from the Yerbaniz and Japonés caves carry a point mutation (C490T) that causes an arginine-to-cysteine mutation (position 164) within the second intracellular loop of the Mc1r protein. Interestingly, the homologous arginine residue is also mutated in human individuals carrying the red hair color, pale skin (RHC) phenotype (Flanagan et al. 2000; John and Ramsay 2002; Sturm et al. 2003). Further, this arginine mutation has been shown to cause diminished Mc1r receptor function (Naysmith et al. 2004; Sánchez-Laorden et al. 2006; Beaumont et al. 2007), suggesting that both mutations lead to the phenotypic expression of the brown trait in Pachón and Yerbaniz/Japonés cave individuals.

SIMILARITIES AND DIFFERENCES IN ALBINISM AND BROWN MUTATION EVOLUTION IN *ASTYANAX* CAVEFISH

The evolution of the brown mutation is similar to the evolution of albinism in several key respects. Both mutations lead to a decrease in pigmentation and have evolved through a combination of coding and regulatory mutations. The evidence for regulatory mutations is drawn from the fact that the same locus that causes the brown mutation in Pachón, Yerbaniz, and Japonés cave individuals is responsible for the brown phenotype in an additional four caves (Chica, Curva, Piedras, and Sabinos caves). Individuals drawn from the latter four caves do not demonstrate any coding differences compared to the surface form of *Mc1r*, implying that regulatory mutations affecting the expression of *Mc1r* cause the brown phenotype in these cave populations.

Some differences do exist with respect to the expression of each trait, particularly in the geographic distribution of phenotypes. For example, albinism has been reported in only three separate cave systems: the Molino, Pachón, and linked Yerbaniz/Japonés caves. Thus, compared to albinism, the brown mutation is present within a much wider geographic distribution, being present in seven caves including: the Pachón, Yerbaniz, Japonés, Sabinos, Piedras, Curva, and Chica populations (plate 6).

Further, there are important differences in the size and structure of the genes *Mc1r* and *Oca2*. For example, *Oca2* is a very large gene, consisting of 21 exons and spanning ~119 kB in zebrafish. In contrast, *Mc1r* is a single exon gene of ~1 kB in length in most vertebrates. A higher frequency of mutations may occur in the structure of *Oca2* given its relatively large size within the genome (Protas et al. 2006). Interestingly, *Mc1r* has been invoked as an important locus for variation as a consequence of its "exclusive" role in pigmentation (Mundy 2005). Thus, coding mutations in *Mc1r* may occur without pleiotropic consequences for the organism (Carroll 2005). The gene *Oca2* similarly does not appear to have multiple (i.e., pleiotropic) functions during the development of the organism beyond its characterized role in pigmentation. Irrespective of the reason for these mutations in *Astyanax* cavefish, the same loci appear to be repeatedly involved in the evolution of de-pigmentation in distinct cave systems. This result is surprising in that it demonstrates distinct mutations arising at the same locus in independent cave forms of the same species. Thus, rather than the repeated presence of the same allele in multiple populations, in both cases the same phenotype arises via a different genotype. It is unclear why this phenomenon occurs—each cave form is subjected to identical ecological pressures (i.e., the cave environment). Therefore, if pigmentation loss is a consequence of neutral evolution, then *Mc1r* and *Oca2* may represent loci that are prone to mutate, leading to de-pigmentation in parallel cave populations of *Astyanax*. Alternatively, if these traits are being selected in the cave microenvironment, the loss-of-function mutations so far identified in both *Mc1r* and *Oca2* may rise to high frequency under strong selection among the independent populations.

EVOLUTIONARY PRESSURES LEADING TO REGRESSIVE CHANGES TO PIGMENTATION

Broadly speaking, two theories seek to explain the loss of phenotypic traits in cave animals—random genetic drift and selection (Culver 1982; Protas et al. 2007). Under neutrality, a given trait that evolved prior to colonization into the cave environment would no longer carry any benefit and, therefore, would be lost through the accumulation of loss-of-function mutations (Wilkens 1988). Alternatively, selection may drive the phenotypic evolution of certain regressive traits through either direct or indirect routes. For example, one could imagine a scenario in which the development of a trait (e.g., the presence of an eye) comes at an energetic cost to the organism, particularly those living in complete darkness (Niven 2008). This energetic cost would conceivably be directly selected against, over the course of many generations, to produce an animal without any form of an adult eye. Alternatively, "hidden" pleiotropic effects of particular genes may provide an adaptive benefit through indirect selection of a specific form of a gene (see Jeffery 2008).

Numerous features of a cave environment produce a dramatic change from the terrestrial world, including relatively constant temperatures, limited food sources, and complete absence of light (Culver 1982). In this ecosystem, many traits clearly adaptive for terrestrial organisms become useless (Plath 2004). Primary among these would be pigmentation (Weis 2002). For example, without light there is likely a relaxation of the selective pressures associated with pigmentation, including the benefits of melanin as protection from the sun's ultraviolet radiation (Schmitz et al. 1995). While pigmentation and associated ornamentation are shaped by sexual selection in many species (Badyaev et al. 2001), it carries no phenotypic benefit in the absence of light. Finally, while the value of α-melanocyte-stimulating hormone (α-MSH) signaling via the *Mc1r* receptor, and melanin-concentrating hormone (MCH) signaling, is often attributed to background adaptation and camouflage in various fish species (Logan et al. 2006), in the absence of light this value is lost.

Thus, there are numerous reasons why the value of pigmentation is diminished in the cave environment. But is pigmentation under active negative selection in the cave environment? To address this question, Protas et al. (2007) investigated the QTL polarities of two conspicuous traits in cave morphs of *Astyanax*, melanophore number and eye size. The authors found that the higher the number of cave haplotypes contributing to eye formation, the smaller the eye size. In contrast, a higher number of cave haplotypes did not always lead to lower numbers of melanophores along the body.[1] In some cases, it actually led to higher numbers of melanophores, suggesting that the melanophore number trait scored in the study is subject to drift (in either direction) in the cave environment (Protas et al. 2007).

---

[1] It should be noted that the body melanophore number trait is different from the simple, recessive "brown" mutation reported by Sadoglu and McKee (1969). In contrast to the brown phenotype, body melanophore number is a complex trait mediated by multiple loci (see Protas et al. 2007).

Alternatively, this result may indicate pleiotropic selection in one direction balancing direct selection in the other. Under this scenario, melanophore number is not directly selected; instead, the effect of a gene (or genetic network) that is positively selected in the cave environment results in pleiotropic consequences for melanophore number. Thus, the secondary effect of the selected trait in the cave environment may be to increase, or decrease, the number of melanophores. In either case, this result contrasts with the QTL polarities for eye size, which appear to be under negative selection in every cave studied.

In the specific cases of albinism and the brown mutation, it is still unclear whether these traits arose through neutral evolution or selection for de-pigmentation. A genotypic survey of two *Mc1r* variants revealed their presence exclusively in two of the cave systems (Pachón and Yerbaniz/Japonés) harboring the brown mutation (plate 6; Gross et al. 2009). Neither variant haplotype was present in any of 231 individuals sampled from surrounding surface river and stream populations. Therefore, these alleles could represent rare alleles that were strongly selected to (presumed) fixation in caves carrying the brown mutation. Alternatively, the rare *Mc1r* variants present in cave populations may also have arisen in situ and increased to high frequency through other stochastic processes, e.g., founder effects. In stickleback fish, a rare allele of *ectodysplasin* (*EDA*) present in the marine population has arisen repeatedly to high frequency in freshwater populations (chapter 6). In this case, however, repeated evolution of this trait arises from clearly identified selective forces (see Kitano et al. 2008; see also chapter 16, this volume).

A clearer understanding of this issue awaits a thorough analysis of the population genetics of the *Mc1r* variants utilizing many more individuals drawn from both cave populations and additional extant, surface populations that likely seeded the caves.

GENETIC MUTATIONS LEADING TO PIGMENTATION LOSS IN OTHER ORGANISMS

Variation at the *Mc1r* locus has been described extensively in numerous organisms including examples from natural animal populations, domesticated species, and model systems across many different taxa. In natural populations, *Mc1r* variation has been associated with plumage and coat color differences in birds and mammals. In domesticated species, *Mc1r* may have been artificially (or inadvertently) selected given specific coloration differences present in domesticated pigs, horses, and cattle. Additionally, in the wild *Mc1r* is a key locus for pigmentation evolution in a wide variety of animals including beach mice (Hoekstra et al. 2006), red-footed boobies (Baião et al. 2007), lesser snow geese (Mundy et al. 2004), and cavefish (Gross et al. 2009).

In contrast to variation at the *Mc1r* locus shared among disparate taxa, the genetic basis for albinism in other natural populations has not yet been shown to be due to common variation at the *Oca2* locus. In humans, however, ocular and cutaneous forms of albinism are most commonly caused by mutations associated with four different loci. *Oca2* is responsible for the second most common form, causing complete abrogation of melanin production in affected individuals. Interestingly, variation at the *Oca2* locus is responsible for the presence of blue eyes in humans (Duffy et al. 2007). In humans, the two key genes discovered to be important for pigmentation evolution in *Astyanax*, *Mc1r* and *Oca2*, appear to have interactive effects whereby *Mc1r* mutations modify the *Oca2* phenotype (King et al. 2003; Duffy et al. 2007). It will be interesting to investigate whether a similar modification of the *Oca2* phenotype occurs in members of the Pachón, Yerbaniz and Japonés caves—wherein albinism and brown mutations have been reported (Protas et al. 2006; Gross et al. 2009).

While albinism has arisen in other species, the affected loci are different from those shown in *Astyanax*. Most commonly, albinism in a large variety of animals, including medaka fish (Koga and Hori 1997), the domestic cat (Imes et al. 2006), ferrets (Blaszczyk 2007), and the American mink (Anistoroaei et al. 2008) is associated with different mutations in the *tyrosinase* gene. In light of the repeated evolution of albinism at the *Oca2* locus in *Astyanax*, these results are particularly interesting given that the predominant locus for mutation, in nearly every natural population of animal studied, is the *tyrosinase* gene.

CONCLUSIONS

By applying contemporary techniques of linkage mapping and quantitative genetics to the biology of cave-adapted species, we can now begin to understand the underlying genetic basis of numerous cave-specific traits. The elucidation of these genes will contribute not only to our collective understanding of how cave species evolve, but whether there are similarities or differences in the identity of the genes recruited for morphological change, irrespective of selection pressures. Furthermore, by investigating the molecular and genetic basis for morphological differences in cave and surface forms, we can shed light on the forces that lead to morphological evolution of regressive and constructive traits. In the case of eye size reduction, a QTL polarity test reveals that in every case the presence of cave alleles always leads to reduced eye size phenotype, consistent with selection for reduced eye size. In contrast, body melanophores show a variety of polarities at multiple loci—possibly indicating neutral mutation and drift as key forces for the loss of this trait.

Two of the genes so far identified as important loci in *Astyanax* are similarly important genes for variation in other species. For example, the gene *Mc1r* has been shown repeatedly to play an essential role in the evolution of plumage and coat color morphs across numerous vertebrate taxa (reviewed in Mundy 2005). In some cases, as with the rock pocket mouse (*Chaetodipus intermedius*), different coat colors confer adaptive benefits for crypsis, e.g., along lava flows in the southeastern United States (Nachman et al. 2003). Alternatively, variation at the *Mc1r* locus appears to be under sexual selection in certain bird species (Nadeau et al. 2007). In humans, *Mc1r* variation may be caused, in part, by the relaxed selection on pigmentation in humans living at higher latitudes. This is based in part on the relationship between skin color and latitude, wherein darker color is associated with proximity to equatorial geographic zones (reviewed in Barsh 2003). It should be noted, however, that different theories explain human variation in *Mc1r* as a consequence of sexual selection (Aoki 2002), as well as drift (Harding et al. 2000), without apparent consensus. Thus, despite extremely different taxa and ecological contexts, the same gene appears to be responsible for the evolution of pigmentation changes.

What have we learned thus far from genetic investigations into the evolution of cave-adapted traits? The identification of two genes (*Mc1r* and *Oca2*) underlying diminished melanin and albinism, respectively, reveals a recurring theme. In both cases, the same locus is involved repeatedly in the evolution of pigment loss. An additional pigmentation trait studied in *Astyanax*, body melanophore number, demonstrates a different scenario. Rather than cave alleles at each locus leading to a reduction of the structure (as with eye size QTL), in many instances cave alleles lead to different polarities, implying absence of selection for this trait. Amidst the background of these recent studies we can ask whether evolution of regressive traits occurs through selection or random drift. The answer depends on the trait. Eye size analyses reveal selection as a likely mechanism for eye loss in cave forms collected from the Pachón cave. De-pigmentation is a complex trait mediated by several loci. Body melanophore numbers can either increase or decrease depending on the number of cave alleles present at key loci, inconsistent with selection. Further, loss-of-function mutations in the genes *Oca2* and *Mc1r* imply that random drift may be responsible for the evolution of these simple recessive traits. At present, however, we cannot rule out the possibility of pleiotropy (i.e., indirect selection) contributing to the process of de-pigmentation. This scenario is less likely, given the reported "exclusive" roles for *Mc1r* and *Oca2* in pigmentation.

In the forthcoming years, the genetic basis for additional derived traits in the cavefish, *Astyanax*, will provide an exciting new set of data. Will we discover that the genes mutated in cavefish continue to mirror the same loci (e.g., *Oca2, Mc1r*) that cause meaningful morphological changes to other vertebrate taxa? Alternatively, will we discover the recruitment of novel or uncharacterized genes in the evolution of derived phenotypes? Continued investigations into the evolution of extreme phenotypes (eye loss) in cave-adapted species will shed light on the phenotypic evolution of taxa occupying unique environments, as well as how and which genes are important for the evolution of vertebrate-specific traits.

In summary, the evolution of structures that are lost in the cave ecosystem is a recurring theme in cave biology. Until recently, the ability to address the underlying genetic basis for these fascinating traits has been beyond technical reach. For this reason, we have adapted an emerging model system, the blind Mexican cavefish (*Astyanax mexicanus*), for genomic studies. Two principal regressive traits have been analyzed thus far, including eye loss and de-pigmentation. In each case, we observe slightly differing results. QTL polarity studies focusing on eye size measurements reveal phenotypic reduction at every cave locus, consistent with selection. In contrast, body melanophore number polarities are mixed, possibly indicating the participation of pleiotropy and/or drift in the evolution of this trait. Two simple de-pigmentation traits, brown and albinism, are due to a combination of structural and regulatory mutations in the genes *Mc1r* and *Oca2*, respectively. These results indicate that the loss of traits in cave animals occurs through a variety of mechanisms, most likely including both neutral evolution and selective forces.

References

Abdel-Malek, Z., M. C. Scott, I. Suzuki, A. Tada, S. Im, L. Lamoreux, S. Ito, G. Barsh, and V. J. Hearing. 2000. The melanocortin-1 receptor is a key regulator of human cutaneous pigmentation. *Pigment Cell Res.* 13: 156–162.

Anistoroaei, R., M. Fredholm, K. Christensen, and T. Leeb. 2008. Albinism in the American mink (*Neovison vison*) is associated with a *tyrosinase* nonsense mutation. *Anim. Genet.* 39: 645–648.

Aoki, K. 2002. Sexual selection as a cause of human skin colour variation: Darwin's hypothesis revisited. *Ann. Hum. Biol.* 29: 589–608.

Avise, J. C., and R. Selander. 1972. Evolutionary genetics of cave-dwelling fishes of the genus *Astyanax*. *Evolution* 26: 1–19.

Badyaev, A. V., G. E. Hill, P. O. Dunn, and J. C. Glen. 2001. Plumage color as a composite trait: Developmental and functional integration of sexual ornamentation. *Am. Nat.* 158: 221–235.

Baião, P. C., E. Schreiber, and P. G. Parker. 2007. The genetic basis of the plumage polymorphism in red-footed boobies (*Sula sula*): a *melanocortin-1 receptor* (*MC1R*) analysis. *J. Hered.* 98: 287–292.

Barsh, G. S. 2003. What controls variation in human skin color? *PLoS Biol.* 1: e27.

Beaumont, K. A., S. L. Shekar, R. A. Newton, M. R. James, J. L. Stow, D. L. Duffy, and R. A. Sturm. 2007. Receptor function, dominant negative activity and phenotype correlations for MC1R variant alleles. *Hum. Mol. Genet.* 16: 2249–2260.

Blaszczyk, W. M., C. Distler, G. Dekomien, L. Arning, K. P. Hoffmann, and J. T. Epplen. 2007. Identification of a *tyrosinase* (*TYR*) exon 4 deletion in albino ferrets (*Mustela putorius furo*). *Anim. Genet.* 38: 421–423.

Carroll, S. B. 2005. Evolution at two levels: On genes and form. *PLoS Biol.* 3: e245.

Contreras-Díaz, H. G., O. Moya, P. Oromí, and C. Juan. 2007. Evolution and diversification of the forest and hypogean ground-beetle genus *Trechus* in the Canary Islands. *Mol. Phylogenet. Evol.* 42: 687–699.

Culver, D. C. 1982. *Cave Life: Evolution and Ecology.* Cambridge, MA: Harvard University Press.

Culver, D. C., T. C. Kane, and D. W. Fong. 1995. *Adaptation and Natural Selection in Caves: The Evolution of* Gammarus minus*.* Cambridge, MA: Harvard University Press.

Dowling, T. E., D. P. Martasian, and W. R. Jeffery. 2002. Evidence for multiple genetic forms with similar eyeless phenotypes in the blind cavefish, *Astyanax mexicanus. Mol. Biol. Evol.* 19: 446–455.

Duffy, D. L., G. W. Montgomery, W. Chen, Z. Z. Zhao, L. Le, M. R. James, N. K. Hayward, N. G. Martin, and R. A. Sturm. 2007. A three-single-nucleotide polymorphism haplotype in intron 1 of *OCA2* explains most human eye-color variation. *Am. J. Hum. Genet.* 80: 241–252.

Eigenmann, C. H. 1909. *Cave Vertebrates of America: A Study in Degenerative Evolution.* Washington, DC: Carnegie Institution of Washington.

Espinasa, L., and R. B. Borowsky. 2001. Origins and relationship of cave populations of the blind Mexican tetra, *Astyanax fasciatus*, in the Sierra de El Abra. *Environ. Biol. Fishes* 62: 233–237.

Flanagan, N., E. Healy, A. Ray, S. Philips, C. Todd, I. J. Jackson, M. A. Birch-Machin, and J. L. Rees. 2000. Pleiotropic effects of the *melanocortin 1 receptor (MC1R)* gene on human pigmentation. *Hum. Mol. Genet.* 9: 2531–2537.

Gross, J. B., R. Borowsky, and C. J. Tabin. 2009. A novel role for *Mc1r* in the parallel evolution of depigmentation in independent populations of the cavefish, *Astyanax mexicanus. PLoS Genet.* 5: e1000326.

Harding, R. M., E. Healy, A. J. Ray, N. S. Ellis, N. Flanagan, C. Todd, C. Dixon, et al. 2000. Evidence for variable selective pressures at *MC1R. Am. J. Hum. Genet.* 66: 1351–1361.

Hoekstra, H. E., R. J. Hirschmann, R. A. Bundey, P. A. Insel, and J. P. Crossland. 2006. A single amino acid mutation contributes to adaptive beach mouse color pattern. *Science* 313: 101–104.

Hüppop, K. 1986. Oxygen consumption of *Astyanax fasciatus* (Characidae, Pisces): A comparison of epigean and hypogean populations. *Environ. Biol. Fishes* 17: 299–308.

Imes, D. L., L. A. Geary, R. A. Grahn, and L. A. Lyons. 2006. Albinism in the domestic cat (*Felis catus*) is associated with a *tyrosinase* (*TYR*) mutation. *Anim. Genet.* 37: 175–178.

145

Jeffery, W. R. 2001. Cavefish as a model system in evolutionary developmental biology. *Dev. Biol.* 231: 1–12.

———. 2006. Regressive evolution of pigmentation in the cavefish *Astyanax*. *Isr. J. Ecol. Evol.* 52: 405–422.

———. 2008. Emerging model systems in evo-devo: Cavefish and microevolution of development. *Evol. Dev.* 10: 265–272.

John, P. R., and M. Ramsay. 2002. Four novel variants in MC1R in red-haired South African individuals of European descent: S83P, Y152X, A171D, P256S. *Hum. Mutat.* 19: 461–462.

King, R. A., R. K. Willaert, R. M. Schmidt, J. Pietsch, S. Savage, M. J. Brott, J. P. Fryer, C. G. Summers, and W. S. Oetting. 2003. *MC1R* mutations modify the classic phenotype of oculocutaneous albinism type 2 (OCA2). *Am. J. Hum. Genet.* 73: 638–645.

Kitano, J., D. I. Bolnick, D. A. Beauchamp, M. M. Mazur, S. Mori, T. Nakano, and C. L. Peichel. 2008. Reverse evolution of armor plates in the threespine stickleback. *Curr. Biol.* 18: 769–774.

Koga, A., and H. Hori. 1997. Albinism due to transposable element insertion in fish. *Pigment Cell Res.* 10: 377–381.

Kosswig, C. 1963. Genetische Analyse konstruktiver und degenerativer Evolutionsprozesse. *Zeit. Zool. Syst. Evolut.* 1: 290–309.

Logan, D. W., S. F. Burn, and I. J. Jackson. 2006. Regulation of pigmentation in zebrafish melanophores. *Pigment Cell Res.* 19: 206–213.

Mundy, N. I. 2005. A window on the genetics of evolution: *MC1R* and plumage colouration in birds. *Proc. R. Soc. B* 272: 1633–1640.

Mundy, N. I., N. S. Badcock, T. Hart, K. Scribner, K. Janssen, and N. J. Nadeau. 2004. Conserved genetic basis of a quantitative plumage trait involved in mate choice. *Science* 303: 1870–1873.

Nachman, M. W., H. E. Hoekstra, and S. L. D'Agostino. 2003. The genetic basis of adaptive melanism in pocket mice. *Proc. Natl. Acad. Sci. U.S.A* 100: 5268–5273.

Nadeau, N. J., T. Burke, and N. I. Mundy. 2007. Evolution of an avian pigmentation gene correlates with a measure of sexual selection. *Proc. R. Soc. B* 274: 1807–1813.

Naysmith, L., K. Waterston, T. Ha, N. Flanagan, Y. Bisset, A. Ray, K. Wakamatsu, S. Ito, and J. L. Rees. 2004. Quantitative measures of the effect of the melanocortin 1 receptor on human pigmentary status. *J. Invest. Dermatol.* 122: 423–428.

Niven, J. E. 2008. Evolution: Convergent eye losses in fishy circumstances. *Curr. Biol.* 18: R27–R29.

Pfeiffer, W. 1966. Uber die vererbung der Schreckreaktion bei *Astyanax* (Characidae, Pisces). *Z. Vererbungsl.* 98: 97–105.

Plath, M. 2004. Cave molly females (*Poecilia mexicana*) avoid parasitised males. *Acta ethologica* 6: 47–51.

Protas, M. E., C. Hersey, D. Kochanek, Y. Zhou, H. Wilkens, W. R. Jeffery, L. I. Zon, R. Borowsky, and C. J. Tabin. 2006. Genetic analysis of cavefish reveals molecular convergence in the evolution of albinism. *Nat. Genet.* 38: 107–111.

Protas, M., M. Conrad, J. B. Gross, C. Tabin, and R. Borowsky. 2007. Regressive evolution in the Mexican cave tetra, *Astyanax mexicanus. Curr. Biol.* 17: 452–454.

Protas, M., I. Tabansky, M. Conrad, J. B. Gross, O. Vidal, C. J. Tabin, and R. Borowsky. 2008. Multi-trait evolution in a cave fish, *Astyanax mexicanus. Evol. Dev.* 10: 196–209.

Renno, J.-F., C. Gazel, G. Miranda, M. Pouilly, and P. Berrebi. 2007. Delimiting species by reproductive isolation: The genetic structure of epigean and hypogean *Trichomycterus* spp. (Teleostei, Siluriformes) in the restricted area of Torotoro (Upper Amazon, Bolivia). *Genetica* 131: 325–336.

Sadoğlu, P. 1957. A Mendelian gene for albinism in natural cave fish. *Experientia* 13: 394.

———. 1979. A breeding method for blind *Astyanax mexicanus* based on annual spawning patterns. *Copeia* 1979: 369–371.

Sadoğlu, P., and A. McKee. 1969. A second gene that affects eye and body color in Mexican blind cave fish. *J. Hered.* 60: 10–14.

Sánchez-Laorden, B. L., J. S. Sánchez-Más, E. Martínez-Alonso, A. Martínez-Menárguez, J. C. García-Borrón, and C. Jiménez-Cervantes. 2006. Dimerization of the human melanocortin 1 receptor: Functional consequences and dominant-negative effects. *J. Invest. Dermatol.* 126: 172–181.

Schmitz, S., P. D. Thomas, T. M. Allen, M. J. Poznansky, and K. Jimbow. 1995. Dual role of melanins and melanin precursors as photoprotective and phototoxic agents: Inhibition of ultraviolet radiation-induced lipid peroxidation. *Photochem. Photobiol.* 61: 650–655.

Strecker, U., V. H. Faúndez, and H. Wilkens. 2004. Phylogeography of surface and cave *Astyanax* (Teleostei) from Central and North America based on cytochrome *b* sequence data. *Mol. Phylogenet. Evol.* 33: 469–481.

Sturm, R. A., D. L. Duffy, N. F. Box, R. A. Newton, A. G. Shepherd, W. Chen, L. H. Marks, J. H. Leonard, and N. G. Martin. 2003. Genetic association and cellular function of *MC1R* variant alleles in human pigmentation. *Ann. N. Y. Acad. Sci.* 994: 348–358.

Sviderskaya, E. V., D. C. Bennett, L. Ho, T. Bailin, S.-T. Lee, and R. A. Spritz. 1997. Complementation of hypopigmentation in *p*-mutant (*pink-eyed dilu-tion*) mouse melanocytes by normal human P cDNA, and defective comple-mentation by OCA2 mutant sequences. *J. Invest. Dermatol.* 108: 30–34.

Weis, J. 2002. Blind fish may help explain human obesity. *BioScience* 52: 864.

Wilkens, H. 1988. Evolution and genetics of epigean and cave *Astyanax fascia-tus* (Characidae, Pisces): Support for the neutral mutation theory. In M. K. Hecht and B. Wallace, eds., *Evolutionary Biology,* 271–367. New York: Ple-num Publishing.

———. 2004. The *Astyanax* model (Teleostei): Neutral mutations and direc-tional selection. *Mitt. Hamb. Zool. Mus. Inst.* 101: 123–130.

Wilkens, H., and U. Strecker. 2003. Convergent evolution of the cavefish *Asty-anax* (Characidae, Teleostei): Genetic evidence from reduced eye-size and pigmentation. *Biol. J. Linn. Soc. Lond.* 80: 545–554.

Chapter Eight

A Developmental View of Exaggerated Growth and Conditional Expression in the Weapons of Sexual Selection

Douglas J. Emlen

Animals exhibit a bewildering diversity of forms, and understanding the causes of this diversity remains a fundamental objective of modern biology. Competition over access to reproduction (sexual selection, Darwin 1871) is credited with the evolution of much of this diversity, including many of nature's most extravagant structures: showy male adornments that are attractive to females (ornaments) such as the wing and tail frills of birds of paradise or the colorful dewlaps of lizards, and an arsenal of outgrowths that function in male-male combat (weapons) like the antlers of elk or the horns of beetles (Darwin 1871; Andersson 1994).

Ornaments and weapons of sexual selection can evolve incredibly rapidly. These structures typically diverge among lineages faster than other (i.e., non–sexually selected) structures, and ornaments and weapons frequently are the most diverse traits within groups of related species (Darwin 1871; Richards 1927; Björklund 1990; Andersson 1994; Prum 1997; Seehausen et al. 1999; Masta and Maddison 2002; Emlen et al. 2005; Ord and Stuart-Fox 2006; Bro Jørgensen 2007; Cardoso and Gama Mota 2008). This fantastic intrinsic potential for rapid evolution, supported also by studies indicating these traits contain unusually large quantities of genetic variation (e.g., Pomiankowski and Møller 1995; Rowe and Houle 1996; Wilkinson and Taper 1999), has fostered immense interest in their "genetic architecture," the specific genes, pathways, and developmental and physiological processes responsible for generating variation in their expression (e.g., Reinhold 1998; Lorch et al. 2003; Hunt et al. 2004; Tomkins et al. 2004; Bonduriansky and Rowe 2005; Neff and Pitcher 2005; Andersson and Simmons 2006; Bonduriansky 2007a; Radwan 2008).

This genetic architecture is complex, in part because virtually all of the most extreme ornaments and weapons are conditionally expressed: they are phenotypically plastic structures whose growth depends on larval/juvenile access to nutrition (e.g., Hill and Montgomerie 1994; Johnstone 1995; Veiga and Puerta 1996; Knell et al. 1999; Hill 2000; Kotiaho 2000; Ohlsson et al. 2002; McGraw et al. 2002; Cotton et al. 2004a; Bonduriansky and Rowe 2005; Naguib and Nemitz 2007), as well as on social and environmental factors that, like nutrition, influence the physiological condition of developing males (e.g., parasites; Hamilton and Zuk 1982; Zuk et al. 1990; Wedekind 1992; McGraw and Hill 2000; Tregenza et al. 2006). Conditional expression is especially apparent in the most elaborated structures. Indeed, many authors have suggested that increased condition-dependence and trait exaggeration should coevolve (Pomiankowski 1987; Grafen 1990; Iwasa and Pomiankowski 1994; Rowe and Houle 1996; Bonduriansky and Rowe 2005; Bonduriansky 2007a, 2007b), and numerous empirical studies have demonstrated that exaggerated ornaments and weapons are *more sensitive* to nutrition than are other structures (David et al. 1998; Knell et al. 1999; reviewed in Cotton et al. 2004b).

Consequently, the developmental mechanisms of nutrition-dependent phenotypic plasticity, and the mechanisms generating exaggerated trait growth, are likely to regulate expression of many—possibly all—of the most extreme structures of sexual selection. Genetic variation relevant to the evolution of ornaments and weapons (i.e., their genetic architecture) must therefore comprise, to a great extent, variation for the elements of these underlying developmental and physiological processes. Yet, we still know very little about how these mechanisms work, and we know almost nothing about the actual genes involved with the evolution of these magnificent structures.

Here I describe the development of an exaggerated weapon, the horns of scarab beetles, and highlight a candidate physiological mechanism for conditional expression. I focus on one specific aspect of these weapons, their extreme size, and ask the developmental question: how is disproportionate growth of a single structure achieved? Even at this early stage, it is already clear that what we know (or think we know) about these mechanisms reveals a great deal about the basic properties of sexually selected traits. Furthermore, the ubiquity of these mechanistic processes hints at why trait exaggeration may have arisen so many times in such a diversity of animal lineages and appendages.

SEXUAL SELECTION AND TRAIT EXAGGERATION: AN EVOLUTIONARY PERSPECTIVE

In the intervening years since Darwin first proposed the idea of sexual selection, a phenomenal amount has been learned about the evolution of animal ornaments and weapons (reviews: Andersson 1994; Panhuis et al. 2001; Ritchie 2007; Emlen 2008). From these studies many patterns have emerged. Here I highlight three of these, each a characteristic we now expect in the most elaborate sexually selected structures.

Extreme Relative Size

The "ornament" or "weapon" aspect of these traits is often associated with a disproportionate increase in the size of this structure relative to other body appendages (fig. 8.1). Trait exaggeration has arisen thousands of times within the animals (Darwin 1871; Richards 1927; Davitashvili 1961; Andersson 1994; Emlen 2008), and there are compelling reasons why this should be so. Competition over access to mates is often intense, generating strong directional selection for increases in ornament/weapon size (e.g., Andersson 1982; Christy and Salmon 1984; Fleming and Gross 1994; Oliveira and Custodio 1998; Hunt and Simmons 2001; Coltman et al. 2002; Preston et al. 2003; Hongo 2007). In addition, weapons can be caught up in "arms races" (Maynard Smith and Parker 1976; Parker 1979, 1983; West Eberhard 1979, 1983; Enquist and Leimar 1983; Maynard Smith and Brown 1986; Härdling 1999), and male ornaments and female preferences for those ornaments can become correlated over time, leading to escalated rates of evolution (the "Fisher" process, Fisher 1930; O'Donald 1980; Lande 1981; Kirkpatrick 1982; Kirkpatrick et al. 1990; Iwasa and Pomiankowski 1995; Kokko et al. 2002; Mead and Arnold 2004). For all of these reasons, we expect that ornaments and weapons of sexual selection will often evolve to extravagant proportions.

151

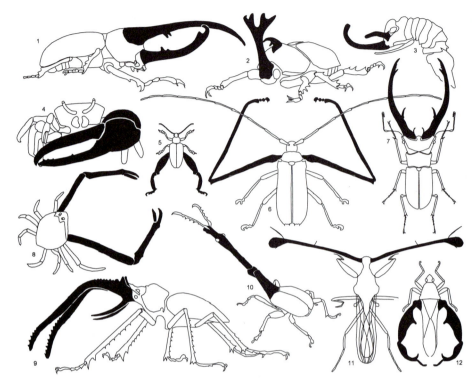

Figure 8.1. Arthropod examples of trait exaggeration. (1) Hercules beetle (*Dynastes Hercules*); (2) Japanese rhinoceros beetle (*Allomyrina dichotoma*); (3) isopod (*Dicranurus monstrosus*); (4) fiddler crab (*Uca pugnax*); (5) frog-legged leaf beetle (*Sagra bouquetti*); (6) harlequin beetle (*Acrocinus longimanus*); (7) stag beetle (*Cyclommatus elaphus*); (8) crab (*Myra fugax*); (9) stag beetle (*Chiasognathus grantii*); (10) giraffe weevil (*Lasiorhynchus barbicornis*); (11) stalk-eyed fly (*Cyrtodiopsis dalmanni*); (12) leaf-footed bug (*Esparzaniella reclusa*).

Steep Allometry Slopes

The scaling relationships between exaggerated male traits and body size tend to be steeper than those for other traits, or for the corresponding trait in females (e.g., fig. 8.2; Zeh et al. 1992; Emlen and Nijhout 2000; Palestrini et al. 2000; Baker and Wilkinson 2001; Kelly 2004; Swallow et al. 2005). Implicit within this is the fact that not all individuals express these traits to the same degree, and the range of trait sizes present among individuals within a population can be vast.

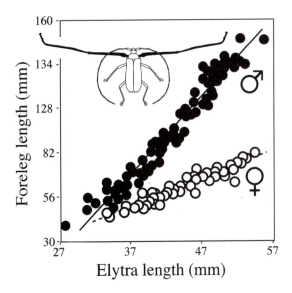

Figure 8.2. Exaggerated structures often have steep allometry slopes. Scaling relationship between foreleg length and body size of male (closed circles) and female (open circles) harlequin beetles, *Acrocinus longimanus*. Males have disproportionately longer forelegs than females, and a steeper foreleg allometry slope. Data from Zeh et al. (1992).

Evolutionary models for this phenomenon build on the idea that bigger traits are better, for either fighting or mating, and on the observation that the costs of bearing these traits are not equal for everybody. Specifically, small, poor-condition males pay a higher price for producing an ornament or weapon than do males in the best condition (Nur and Hasson 1984; Kodric-Brown and Brown 1984; Zeh and Zeh 1988; Cotton et al 2004a; Bonduriansky and Rowe 2005). When these conditions are met, directional selection for costly structures is predicted to lead to the evolution of steeper allometry slopes (Green 1992; Petrie 1992; Bonduriansky and Day 2003; Kodric-Brown et al. 2006; Bonduriansky 2007a), and a number of comparative studies have now supported this idea (e.g., Alatalo et al. 1988; Simmons and Tomkins 1996; Emlen and Nijhout 2000).

"Heightened" Conditional Expression

The growth of exaggerated traits generally is especially sensitive to the nutritional history and physiological condition of the individual males that produce them (Rowe and Houle 1996; Cotton et al. 2004a; Tomkins et al. 2004). As mentioned above, the exaggerated ornaments and weapons are often *more sensitive* to things like larval nutrition than are other traits.

Thus, ornaments and weapons of sexual selection very often reach exaggerated proportions, they are unusually variable in their expression among individuals resulting from steep allometry slopes, and their growth is especially sensitive to larval/juvenile nutrition. Despite the truly incredible breadth of shapes, styles, and types of structures that have arisen through countless independent histories of sexual selection in animals, these same three conditions almost always apply. Arguably, this trio of co-expressed characteristics is absolutely central to modern ideas of sexual selection, because together they mean that even very small differences in male body size or condition will be *amplified* into more dramatic—more visible—differences in the size of the ornament or weapon (Wallace 1987; Hasson 1989, 1991; Andersson 1994; Bondurianksy and Day 2003). For this reason, these traits are thought to be more revealing (or more honest) indicators of male quality than are other traits (Pomiankowski and Iwasa 1993; Schluter and Price 1993; Kirkpatrick 1996; Iwasa et al. 1991; Houle and Kondrashov 2001; Mead and Arnold 2004). It is not an accident, then, that exaggerated traits form the basis for male assessment of rival males, or for female choice of mates.

Although these ideas are not new, they are almost never considered from the perspective of developmental mechanisms. In the following section I revisit these three characteristics from the perspective of animal development, and I use this information to predict how exaggerated trait sizes might evolve. The traits I use to illustrate these ideas are the horns of beetles—rigid cuticular outgrowths that function as weapons in male battles over access to females.

DEVELOPMENT OF BEETLE HORNS

Beetle horns can attain extreme proportions, and as with other sexually selected structures, among-individual variation in horn expression can be profound: some males produce enormous weapons comprising 30 percent of their body weight, while other males produce only vestigial horns, or no horns at all (plate 7a). This phenotypic variation in beetle horns results primarily from heterogeneity in larval nutritional environments (Emlen 1994; Hunt and Simmons 1997, 2000; Iguchi 1998; Moczek and Emlen 1999; Karino et al. 2004). What this means, developmentally, is that the amount of growth of these structures is adjusted somehow in response to the nutritional conditions animals encounter as they develop.

Both horn size *and body size* are sensitive to variation in nutrition, with the consequence that the amount of horn growth is coupled with overall body size: larvae with access to large food amounts develop into adults with large body sizes and long horns, while larvae with small food amounts develop into adults with small body sizes and short horns. Iterated across a number of different individuals developing under a range of nutritive environments, the result of this phenotypic plasticity is a steep scaling relationship (allometry) between male horn length and body size (plate 7b, top). Other traits, such as wings, also are sensitive to nutrition, and their final dimensions also scale positively with among-individual variation in body size (plate 7b, middle). Genitalia, in contrast, are not sensitive to nutrition, and genitalia (e.g., aedeagus) size does not scale with body size (this trait has a flat allometry slope; plate 7b, bottom). This figure highlights several crucial points: (1) nutrition-dependent phenotypic plasticity and allometry are related in insects (both result from mechanisms that couple growth with nutrition; Stern and Emlen 1999; Emlen and Allen 2003; Shingleton et al. 2007, 2008; Frankino et al. 2008); (2) *traits differ* in the extent to which their growth is sensitive to the nutrition environment; and (3) *exaggerated traits* (in this case, weapons) *are the most sensitive* to the nutritional environment (e.g., horns are ten times longer in the best-fed males than they are in poorly fed males, whereas genitalia are only 1.2 times larger in these same males).

How Is Trait Growth Coupled with Nutrition in Insects?

In metamorphic insects, the adult structures (e.g., wings, horns, genitalia) develop from isolated clusters of epidermal cells called imaginal discs that are dispersed like islands within the larval body (e.g., plate 8). These discs are remarkably autonomous entities (e.g., they can be transplanted to other parts of the animal; Kopec 1922; Pohley 1965; Hadorn 1966) and each produces a specific structure, such as a left foreleg or a right hindwing (Kojima 2004; Weihe et al. 2005). Importantly, these imaginal discs do not grow at the same time as the rest of the animal (Truman and Riddiford 2002). The onset of their proliferation is delayed until late in the larval period, and much or all of their growth occurs after animals have already attained their full body size.

Yet, the amount of proliferation each disc undergoes tracks closely the nutritional condition and body size of that individual. This means that the mechanism(s) of nutrition-dependent phenotypic plasticity (and allometry) must incorporate whole-animal circulating physiological signals whose levels are sensitive to larval nutrition, and which modulate the amount of growth of the different traits—the imaginal discs—in accordance with the actual nutritional environment and overall body growth experienced by a larva. Several physiological pathways meet these criteria (e.g., the ecdysteroid pathway; Colombani et al. 2005; Mirth et al. 2005; Orme and Leevers 2005; Telang et al. 2007; or the JH pathway; Emlen and Allen 2003; Léopold and Layalle 2006; Truman et al. 2006), but the most promising of these is the insulin receptor (InR) pathway.

Most people associate the insulin receptor pathway with the regulation of ageing/lifespan (e.g., Blüher et al. 2003; Tatar et al. 2003), or with the control of body size (e.g., Edgar 1999, 2006; Nijhout 2003; Stern 2003), but increasing evidence suggests that this pathway also coordinates and regulates growth of the various body parts. In insects, insulin-like peptides secreted primarily by the brain, and probably in cooperation with growth factors secreted by the fat bodies, act as whole-animal circulating signals; when they reach the imaginal discs, they bind to the insulin receptor and activate a signal-transduction cascade that controls the rate of cell proliferation within each disc (Edgar 1999, 2006; Weinkove and Leevers 2000; Brogiolo et al. 2001; Bryant 2001; Johnston and Gallant 2002; Claeys et al. 2002).

Both insulin and growth factor levels are sensitive to nutrition (Kawamura et al. 1999; Day and Lawrence 2000; Masumura et al. 2000; Britton et al. 2002; Ikeya et al. 2002; Nijhout and Grunert 2002), and this means that because of the insulin receptor pathway, cell proliferation should occur at a *faster rate* in the imaginal discs of large, well-fed individuals than it does in the corresponding discs of smaller, poorly fed individuals. Importantly, the InR pathway is activated independently within each of the target tissues—within each imaginal disc—and this highlights one way that this pathway could contribute to the evolution of trait differences in plasticity and growth (e.g., why genitalia differ from wings or horns).

If the insulin receptor pathway works to regulate trait growth the way that we suspect it does, then the sensitivity of each imaginal disc to these insulin signals will determine, to a large extent, how that particular structure will grow: its growth rate, and perhaps more importantly, the sensitivity of its cells to the nutrition environment (how phenotypically plastic growth of that structure is). Specifically, we predict that traits sensitive to insulin signals should have pronounced nutrition-dependent phenotypic plasticity (conditional-expression), and these structures should have positive allometry slopes when scaled with body size in natural populations (as occurs in wings and horns). We predict that traits *in*sensitive to circulating insulin signals will not be phenotypically plastic, and that these structures should have flat allometry slopes (e.g., genitalia).

Beautiful tests of these ideas in *Drosophila* were provided recently by Alex Shingleton, David Stern, and colleagues (Shingleton et al. 2005). Perturbations to the insulin receptor confirmed, first, that this pathway does affect allometry, and second, that traits differ predictably in how they respond to insulin signals. Traits like wings were very sensitive both to insulin and to genetic perturbations to the insulin receptor (wing allometry was positive in control animals, but dropped flat in perturbed individuals), whereas genitalia were not sensitive to insulin, and their growth was largely unaffected by perturbations to the insulin receptor (their allometry already was flat; Shingleton et al. 2005). Laura Corley Lavine, Ian Dworkin, and I are conducting similar tests in horned beetles, and our results to date confirm that the InR pathway is active in developing beetle horn discs. Measures of relative transcription of the InR gene in these beetles suggest that horn discs in large males have greater signaling through this pathway than same-stage

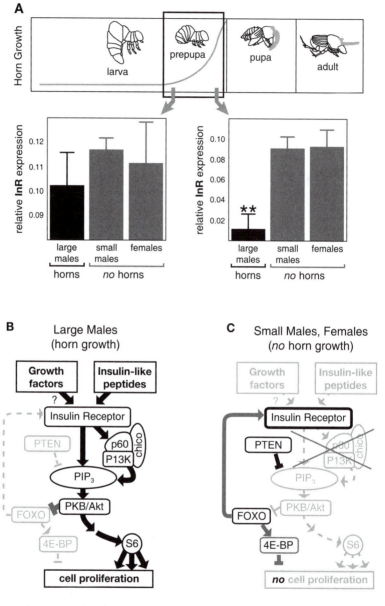

Figure 8.3. Example of insulin receptor (InR) pathway activity associated with horn growth. (a) Quantitative real-time PCR was performed using a short fragment of the *Onthophagus nigriventris* insulin receptor from a cDNA pool, using 28S as a control for overall levels of RNA. At the beginning of horn growth (left panels) there is no evidence for differential expression of the InR gene between

horn discs from small males (in fact, they suggest that activity of this pathway is truncated in small males and females, both of which do not grow horns; Emlen et al. 2006; fig. 8.3).

Similarly, measures of transcription of FOXO (another element in the InR pathway) by Emilie Snell-Rood and Armin Moczek also suggest greater pathway activity in the horn discs of large males than in small, hornless males or females (personal communication). Although these studies are preliminary, and they will need to be complemented with measures of gene expression in other traits (e.g., wings, genitalia) and with tests of pathway function, at this point in time it appears that insulin receptor pathway activity is correlated with the amount of weapon growth in beetles. Consequently, in *Drosophila* almost certainly, and possibly also in beetles, trait differences in nutrition-dependent phenotypic plasticity and allometry result at least in part from disc-specific differences in their responsiveness to circulating insulin signals, and several authors have recently proposed that the InR pathway may play an important and widespread role as a general mechanism for plasticity and allometry in the appendage growth of insects (Emlen and Allen 2003; Shingleton et al. 2007; Frankino et al. 2008). Although this hypothesis awaits further testing, we can use what we already know about the behavior of the insulin receptor pathway to revisit our original question.

Figure 8.3. (*Continued*) large males which grow horns, and small males and females who lack horns. However, during the period of maximal horn growth (right panels), there is a highly significant ($p < 0.0001$) difference in expression of the InR gene that is associated with horn growth: horn cells from small males and females had significantly higher levels of InR transcript than similar cells from large males. (b) The insulin signaling pathway, illustrating one explanation for this result: as signaling through the insulin pathway is increased overall in large males, the expression of InR decreases due to kinase-dependent inactivation of its transcriptional activator FOXO by PKB/Akt (gray bar; Kramer et al. 2003; Puig and Tijan 2005). (c) In small males and females, horn growth is blocked; we now suspect that repression of horn growth results from the truncation of pathway activity at some point downstream from the insulin receptor. Reduced signaling through this pathway would keep FOXO in an activated state, causing an up-regulation of InR transcription (Puig and Tijan 2005) and a shutting off of cell proliferation through the transcriptional inhibitor 4E-BP (gray arrows; Jünger et al. 2003). Reprinted with permission from Emlen et al. (2006).

HOW IS EXAGGERATED GROWTH OF A SINGLE
STRUCTURE ACHIEVED?

I suggest the answer could be a surprisingly simple one: *increased* sensitivity of those cells to insulin. The crucial point is that the insulin receptor pathway is activated independently within each of the different tissues. Because each trait responds independently to these circulating signals, genetic changes to the insulin receptor pathway within any specific trait would affect the growth and final size of that trait, but only that trait; the size of that structure would be changed relative to the rest of the body.

Figure 8.4 illustrates a hypothetical example. Suppose in an ancestral population of an insect (like the ancestor of modern-day harlequin beetles; figures 8.1, 8.2) a mutation arose that caused one trait, the forelegs, to become three times as sensitive to insulin as the other surrounding traits (e.g., the midlegs or hindlegs). Then, in individuals inheriting this genetic change in underlying mechanism, the forelegs would now grow three times faster than these other traits, resulting in adults with disproportionately large foreleg sizes. Note also that from this same single change in mechanism, we would get (1) well-fed individuals with enormous foreleg sizes (trait exaggeration), (2) populations with steeper allometry slopes, and (3) traits that now exhibit "heightened" conditional expression—*all three* of the defining characteristics we typically associate with nature's most extravagant structures.

Countless empirical and theoretical studies have suggested *why* sexual selection should lead to the evolution of exaggerated structures with unusual condition-sensitivity to their expression. This may be the first attempt to explain *how* these structures attain their extreme sizes and conditional expression. By examining specific physiological processes likely to be involved with the regulation of expression of beetle horns, we begin to reveal the mechanistic interconnectedness of things like trait proportion, allometry, and nutrition-dependent phenotypic plasticity. I suggest that each of these features of ornaments and weapons of sexual selection could be accounted for as a direct outcome of a change in responsiveness of their cells to a physiological signal like insulin. In fact, from a developmental perspective, all of these features

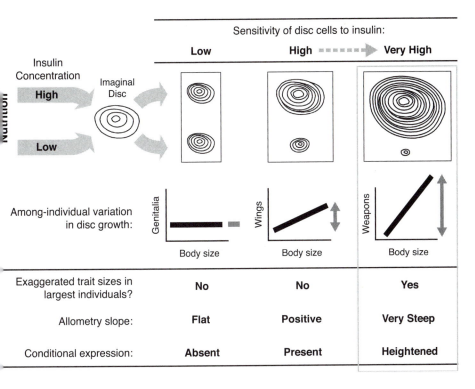

Figure 8.4. Model for the evolution of trait exaggeration. Larval nutritional state is reflected in circulating levels of insulins (and growth factors, not shown), which modulate the rate of growth of each of the trait imaginal discs. Traits whose disc cells are sensitive to these signals (e.g., wings) exhibit greater nutrition-dependent phenotypic plasticity and steeper allometry slopes than other traits whose disc cells are less sensitive to these signals (e.g., genitalia). An evolutionary increase in the sensitivity of cells within a particular trait (e.g., forelegs, see text; indicated by dashed gray arrow) would lead to disproportionately rapid growth of that trait, but only that trait (exaggerated trait sizes). This same change in mechanism would also yield steeper allometry slopes, and increased nutrition-dependent phenotypic plasticity (heightened conditional expression).

may really be the same thing: pleiotropic consequences of a single, relatively simple shift in an underlying developmental mechanism.

Why consider developmental mechanisms? Examining developmental mechanisms can reveal a great deal about the genetic architecture of complex structures: which specific genes and pathways are likely to underlie variation in their expression, and why it is that particular aspects of expression co-occur. It can also suggest how widespread, or generalizable, a biological phenomenon is likely to be, and how easy or difficult it should be to evolve. In this case, the underlying developmental mechanism in question is not unique to beetles. The insulin receptor pathway is an ancient physiological process that probably couples trait growth with nutrition in most of the metazoa (e.g., insulin-like growth factors influence antler growth; Suttie et al. 1989; Elliott et al. 1992; Price and Allen 1994; Webster et al. 1996), raising the possibility that this pathway has contributed to the evolution of many different types of ornaments and weapons. Furthermore, given what we presently understand of the properties of this pathway, gaining trait exaggeration does not appear to be difficult. Indeed, subtle genetic changes in the levels of expression of any of the genes in this insulin pathway may well be sufficient to generate extreme trait sizes, steep allometry slopes, and heightened conditional expression, all in a single step. If true, then this insight from development may help explain why trait exaggeration has arisen independently in so many different animal lineages, and in such an astonishing diversity of morphological structures.

REFERENCES

Alatalo, R. V., J. Höglund, and A. Lundberg. 1988. Patterns of variation in tail ornament size in birds. *Biol. J. Linn. Soc.* 34: 363–374.

Andersson, M. 1982. Sexual selection, natural selection and quality advertisment. *Biol. J. Linn. Soc.* 17: 375–393.

———. 1994. *Sexual Selection.* Princeton, NJ: Princeton University Press.

Andersson, M., and L. W. Simmons. 2006. Sexual selection and mate choice. *TREE* 21: 296–302.

Baker, R. H., and G. S. Wilkinson. 2001. Phylogenetic analysis of sexual dimorphism and eye-span allometry in stalk-eyed flies (Diopsidae). *Evolution* 55: 1373–1385.

Björklund, M. 1990. A phylogenetic interpretation of sexual dimorphism in body size and ornament in relation to mating system in birds. *J. Evol. Biol.* 3: 171–183.

Blüher, M., B. B. Kahn, and C. R. Kahn. 2003. Extended longevity in mice lacking the insulin receptor in adipose tissue. *Science* 299: 572–574.

Bonduriansky, R. 2007a. The evolution of condition-dependent sexual dimorphism. *Am. Nat.* 169: 9–19.

————. 2007b. The genetic architecture of sexual dimorphism: the potential roles of genomic imprinting and condition-dependence. In D. J. Fairbairn, W. U. Blanckenhorn, and T. Szekely, eds., *Sex, Size and Gender Roles: Evolutionary Studies of Sexual Size Dimorphism*, 176–184. Oxford: Oxford University Press.

Bonduriansky, R., and T. Day. 2003. The evolution of static allometry in sexually selected traits. *Evolution* 57: 2450–2458.

Bonduriansky, R., and L. Rowe. 2005. Sexual selection, genetic architecture, and the condition dependence of body shape in the sexually dimorphic fly *Prochyliza xanthostoma* (Piophilidae). *Evolution* 59: 138–151.

Britton, J. S., W. K. Lockwood, L. Li, S. M. Cohen, and B. A. Edgar. 2002. Drosophila's insulin/PI3-kinase pathway coordinates cellular metabolism with nutritional conditions. *Dev. Cell.* 2: 239–249.

Brogiolo, W., H. Stocker, T. Ikeya, F. Rintelen, R. Fernandez, and E. Hafen. 2001. An evolutionarily conserved function of the Drosophila insulin receptor and insulin-like peptides in growth control. *Curr. Biol.* 11: 213–221.

Bro-Jørgensen, J. 2007. The intensity of sexual selection predicts weapon size in male bovids. *Evolution* 61: 1316–1326.

Bryant, P. J. 2001. Growth factors controlling imaginal disc growth in Drosophila. *Novartis Found. Symp.* 237: 182–194; discussion 194–202.

Cardoso, G. C., and P. G. Mota. 2008. Speciational evolution of coloration in the genus *Carduelis*. *Evolution* 62: 753–762.

Christy, J. H., and M. Salmon. 1984. Ecology and evolution of mating systems of fiddler crabs (Genus Uca). *Biological Reviews* 59: 483–509.

Claeys, I., G. Simonet, J. Poels, T. Van Loy, L. Vercammen, A. De Loof, and J. Vanden Broeck. 2002. Insulin-related peptides and their conserved signal transduction pathway. *Peptides* 23: 807–816.

Colombani, J., L. Bianchini, S. Layalle, C. Antoniewski, C. Carre, S. Noselli, and P. Leopold. 2005. Antagonistic actions of ecdysone and insulins determine final size in Drosophila. *Science* 310: 667–670.

Coltman, D. W., M. Festa-Bianchet, J. T. Jorgenson, and C. Strobeck. 2002. Age-dependent sexual selection in bighorn rams. *Proc. Royal Soc. B* 269: 165–172.

Cotton, S., K. Fowler, and A. Pomiankowski. 2004a. Condition dependence of sexual ornament size and variation in the stalk-eyed fly *Cyrtodiopsis dalmanni* (Diptera: Diopsidae). *Evolution* 58: 1038–1046.

————. 2004b. Do sexual ornaments demonstrate heightened condition-dependent expression as predicted by the handicap hypothesis? *Proc. Royal Soc. B* 271: 771–783.

Darwin, C. 1871. *The Descent of Man and Selection in Relation to Sex.* New York: Random House, Modern Library.

David, P., A. Hingle, D. Greig, A. Rutherford, A. Pomiankowski, and K. Fowler. 1998. Male sexual ornament size but not asymmetry reflects condition in stalk-eyed flies. *Proc. Royal Soc. B* 265: 2211–2216.

Davitashvili, L. 1961. *The Theory of Sexual Selection* (In Russian). Izd. AN SSR, Moskva: Academy of Sciences Press.

Day, S. J., and P. A. Lawrence. 2000. Measuring dimensions: The regulation of size and shape. *Development* 127: 2977–2987.

Edgar, B. A. 1999. From small flies come big discoveries about size control. *Nature Cell Biol.* 1: E191–193.

———. 2006. How flies get their size: Genetics meets physiology. *Nature Rev. Gen.* 7: 907–916.

Elliott, J. L., J. M. Oldham, G. R. Ambler, J. J. Bass, G. S. Spencer, S. C. Hodgkinson, B. H. Breier, P. D. Gluckman, and J. M. Suttie. 1992. Presence of insulin-like growth factor-I receptors and absence of growth hormone receptors in the antler tip. *Endocrinology* 130: 2513–2520.

Emlen, D. J. 1994. Environmental control of horn length dimorphism in the beetle *Onthophagus acuminatus* (Coleoptera: Scarabaeidae). *Proc. Roy. Soc. B* 256: 131–136.

———. 2008. The evolution of animal weapons. *Ann. Rev. Ecol., Evol., and Syst.* 39: 387–413.

Emlen, D. J., and C. E. Allen. 2003. Genotype to phenotype: physiological control of trait size and scaling in insects. *Integr. Comp. Biol.* 43: 617–634.

Emlen, D. J., L. Corley-Lavine, and B. Ewen-Campen, B. 2007. On the origin and evolutionary diversification of beetle horns. *Proceedings of the National Academy of Sciences* 104 supplement1: 8661–8668.

Emlen, D. J., J. Marangelo, B. Ball, and C. W. Cunningham. 2005. Diversity in the weapons of sexual selection: Horn evolution in the beetle genus *Onthophagus* (Coleoptera: Scarabaeidae). *Evolution* 59: 1060–1084.

Emlen, D. J., and H. F. Nijhout. 2000. The development and evolution of exaggerated morphologies in insects. *Annu Rev. Entom.* 45: 661–708.

Emlen, D. J., Q. Szafran, L. S. Corley, and I. Dworkin. 2006. Insulin signaling and limb-patterning: candidate pathways for the origin and evolutionary diversification of beetle horns. *Heredity* 97: 179–191.

Enquist, M., and O. Leimar. 1983. Evolution of fighting behavior decision rules and assessment of relative strangth. *J. Theor. Biol.* 102: 387–410.

Fisher, R. A. 1930. *The Genetic Theory of Natural Selection.* Oxford: Oxford University Press.

Fleming, I. A., and M. R. Gross. 1994. Breeding competition in a Pacific salmon (Coho: *Oncorhynchus kisutch*): Measures of natural and sexual selection. *Evolution* 48: 637–657.

Frankino, W. A., D. J. Emlen, and A. W. Shingleton. 2008. Experimental approaches to studying the evolution of animal form: The shape of things to come. In: T. Garland and M. Rose, eds., *Experimental Evolution: Concepts, Methods, and Applications*, 419–478. Berkeley: University of California Press.

Grafen, A. 1990. Biological signals as handicaps. *J. Theor. Biol.* 144: 517–546.

Green, A. J. 1992. Positive allometry is likely with mate choice, competitive display and other functions. *Anim. Behav.* 43: 170–172.

Hadorn, E. 1966. Konstanz, Wechsel und typus der determination und differenzierung in zellen aus mannlichen genitalanlagen von *Drosophila melanogaster* nach dauerkultur *in vivo*. *Dev. Biol.* 13: 424–509.

Hamilton, W. D., and M. Zuk. 1982. Heritable true fitness and bright birds: a role for parasites? *Science* 218: 384–387.

Härdling, R. 1999. Arms races, conflict costs and evolutionary dynamics. *J. Theor. Biol.* 196: 163–167.

Hasson, O. 1989. Amplifiers and the handicap principle in sexual selection: a different emphasis. *Proc. Royal Soc. B* 235: 383–406.

———. 1991. Sexual displays as amplifiers: practical examples with an emphasis on feather decorations. *Behav. Ecol.* 2: 189–197.

Hill, G. E. 2000. Energetic constraints on expression of carotenoid-based plumage coloration in male house finches. *J. Avian Biol.* 31: 559–566.

Hill, G. E., and R. Montgomerie. 1994. Plumage color signals nutritional condition in the House Finch. *Proc. Royal Soc. B* 258: 47–52.

Hongo, Y. 2007. Evolution of male dimorphic allometry in a population of the Japanese horned beetle *Trypoxylus dichotomus septentrionalis*. *Behav. Ecol. Sociobiol.* 62: 245–253.

Houle, D., and A. S. Kondrashov. 2002. Coevolution of costly mate choice and condition-dependent display of good genes. *Proc. Royal Soc. B* 269: 97–104.

Hunt, J., and L. W. Simmons. 1997. Patterns of fluctuating asymmetry in beetle horns: an experimental examination of the honest signalling hypothesis. *Behav. Ecol. Sociobiol.* 41: 109–114.

Hunt, J., L. F. Bussière, M. Jennions, and R. Brooks. 2004. What is genetic quality? *TREE* 19: 329–333.

Hunt, J., and L. Simmons. 2001. Status-dependent selection in the dimorphic beetle *Onthophagus taurus*. *Proc. Royal Soc. B* 268: 2409–2414.

Hunt, J., and L. W. Simmons. 2000. Maternal and paternal effects on offsrping phenotype in the dung beetle *Onthophagus taurus*. *Evolution* 54: 936–941.

Iguchi, Y. 1998. Horn dimorphism in *Allomyrina dichotoma septentrionalis* (Coleoptera: Scarabaeidae) affected by larval nutrition. *Annals Entomol. Soc. America* 91: 845–847.

Ikeya, T., M. Galic, P. Belawat, K. Nairz, and E. Hafen. 2002. Nutrient-dependent expression of insulin-like peptides from neuroendocrine cells in the CNS contributes to growth regulation in Drosophila. *Curr. Biol.* 12: 1293–1300.

Iwasa, Y., and A. Pomiankowski. 1994. The evolution of mate preferences for multiple sexual ornaments. *Evolution* 48: 853–867.

———. 1995. Continual change in mate preferences. Nature 377: 420–422.

Iwasa, Y., A. Pomiankowski, and S. Nee. 1991. The evolution of costly mate preferences II. The 'handicap' principle. *Evolution* 45: 1431–1442.

Johnston, L. A., and P. Gallant. 2002. Control of growth and organ size in Drosophila. *Bioessays* 24: 54–64.

Johnstone, R. A. 1995. Sexual selection, honest advertisement and the handicap principle: Reviewing the evidence. *Biol. Rev. Cambridge Phil. Soc.* 70: 1–65.

Karino, K., N. Seki, and M. Chiba. 2004. Larval nutritional environment determines adult size in Japanese horned beetles *Allomyrina dichotoma*. *Ecol. Res.* 19: 663–668.

Kawamura, K., T. Shibata, O. Saget, D. Peel, and P. J. Bryant. 1999. A new family of growth factors produced by the fat body and active on Drosophila imaginal disc cells. *Development* 126: 211–219.

Kelly, C. D. 2004. Allometry and sexual selection of male weaponry in Wellington tree weta, *Hemideina crassidens*. *Behav. Ecol.* 16: 145–152.

Kirkpatrick, M. 1982. Sexual selection and the evolution of female choice. *Evolution* 36: 1–12.

———. 1996. Good genes and direct selection in evolution of mating preferences. *Evolution* 50: 2125–2140.

Kirkpatrick, M., T. Price, and S. J. Arnold. 1990. The Darwin-Fisher theory of sexual selection in monogamous birds. *Evolution* 44: 180–193.

Knell, R. J., N. Fruhauf, and K. A. Norris. 1999. Conditional expression of a sexually selected trait in the stalk-eyed fly *Diasemopsis aethiopica*. *Ecol. Entomol.* 24: 323–328.

Kodric-Brown, A., and J. H. Brown. 1984. Truth in advertising: The kinds of traits favored by sexual selection. *Am. Nat.* 124: 309–323.

Kodric-Brown, A., R. M. Sibly, and J. H. Brown. 2006. The allometry of ornaments and weapons. *Proc. Natl. Acad. Sci. U. S. A.* 103: 8733–8738.

Kojima, T. 2004. The mechanism of Drosophila leg development along the proximodistal axis. *Development, Growth and Differentiation* 46: 115–129.

Kokko, H., R. Brooks, J. M. McNamara, and A. Houston. 2002. The sexual selection continuum. *Proc. Royal Soc. B* 269: 1331–1340.

Kopec, S. 1922. Studies on the necessity of the brain for the inception of insect metamorphosis. *Biol. Bull.* 42: 323–341.

Kotiaho, J. S. 2000. Testing the assumptions of conditional handicap theory: costs and condition dependence of a sexually selected trait. *Behav. Ecol. Sociobiol.* 48: 188–194.

Lande, R. 1981. Models of speciation by sexual selection on polygenic traits. *Proc. Natl. Acad. Sci. U. S. A.* 78: 3721–3725.

Léopold, P., and S. Layalle. 2006. Linking nutrition and tissue growth. *Science* 312: 1317–1318.

Lorch, P. D., S. Proulx, L. Rowe, and T. Day. 2003. Condition-dependent sexual selection can accelerate adaptation. *Evol. Ecol. Research* 5: 867–881.

Masta, S. E., and W. P. Maddison. 2002. Sexual selection driving diversification in jumping spiders. *Proc. Natl. Acad. Sci. U.S.A.* 99: 4442–4447.

Masumura, M., S. I. Satake, H. Saegusa, and A. Mizoguchi. 2000. Glucose stimulates the release of bombyxin, an insulin-related peptide of the silkworm *Bombyx mori*. *General & Comparative Endocrinology* 118: 393–399.

Maynard-Smith, J., and R L.W. Brown. 1986. Competition and body size. *Theor. Pop. Biol.* 30: 166–179.

Maynard-Smith, J., and G. A. Parker. 1976. The logic of asymmetric contests. *Anim. Behav.* 24: 159–175.

McGraw, K. J., and G. E. Hill. 2000. Carotenoid-based ornamentation and status signaling in the house finch. *Behav. Ecol.* 11: 520–527.

McGraw, K. J., E. A. Mackillop, J. Dale, and M. Hauber. 2002. Different colors reveal different information: how nutritional stress affects the expression of melanin- and structurally based ornamental plumage. *J. Exp. Biol.* 205: 3747–3755.

Mead, L. S., and S. J. Arnold. 2004. Quantitative genetic models of sexual selection. *TREE* 19: 264–271.

Mirth, C., J. W. Truman, and L. M. Riddiford. 2005. The role of the prothoracic gland in determining critical weight for metamorphosis in *Drosophila melanogaster*. *Curr. Biol.* 15: 1796–1807.

Moczek, A. P., and D. J. Emlen. 1999. Proximate determination of male horn dimorphism in the beetle *Onthophagus taurus* (Coleoptera: Scarabaeidae). *J. Evol. Biol.* 12: 27–37.

Naguib, M., and A. Nemitz. 2007. Living with the past: Nutritional stress in juvenile males has immediate effects on their plumage ornaments and on adult attractiveness in Zebra finches. *PLoS One* 9: 1–5.

Neff, B. D., and T. E. Pitcher. 2005. Genetic quality and sexual selection: an integrated framework for good genes and compatible genes. *Molecular Ecology* 14: 19–38.

Nijhout, H. F. 2003. The control of body size in insects. *Dev. Biol.* 261: 1–9.

Nijhout, H. F., and L. W. Grunert. 2002. Bombyxin is a growth factor for wing imaginal disks in Lepidoptera. *Proc. Natl. Acad. Sci. U.S.A.* 99: 15446–15450.

Nur, N., and O. Hasson. 1984. Phenotypic plasticity and the handicap principle. *J. Theor. Biol.* 110: 275–298.

O'Donald, P. 1980. *Genetic Models of Sexual Selection.* Cambridge: Cambridge University Press.

Ohlsson, T., H. G. Smith, L. Råberg, and D. Hasselquist. 2002. Pheasant sexual ornaments reflect nutritional conditions during early growth. *Proc. Royal Soc. B* 269: 21–27.

Oliveira, R. F., and M. R. Custodio. 1998. Claw size, waving display and female choice in the European fiddler crab, *Uca tangeri. Ethol. Ecol. Evol.* 10: 241–251.

Ord, T. J., and D. Stuart-Fox. 2006. Ornament evolution in dragon lizards: multiple gains and widespread losses reveal a complex history of evolutionary change. *J. Evol. Biol.* 19: 797–808.

Orme, M. H., and S. J. Leevers. 2005. Flies on steroids: The interplay between ecdysone and insulin signaling. *Cell Metabolism* 2: 277–278.

Palestrini, C., A. Rolando, and P. Laiolo. 2000. Allometric relationships and character evolution in *Onthophagus taurus* (Coleoptera: Scarabaeidae). *Can. J. Zool.* 78: 1199–1206.

Panhuis, T. M., R. Butlin, M. Zuk, and T. Tregenza. 2001. Sexual selection and speciation. *TREE* 16: 364–371.

Parker, G. A. 1979. Sexual selection and sexual conflict. In M. S. Blum and N. A. Blum, eds., *Sexual Selection and Reproductive Competition in Insects*, 123–166. New York: Academic Press.

———. 1983. Arms races in evolution: An evolutionary stable strategy to the opponent-independent costs game. *Journal of Theoretical Biology* 101: 619–648.

Petrie, M. 1992. Are all secondary sexual display structures positively allometric and, if so, why? *Anim. Behav.* 43: 173–175.

Pohley, H.-J. 1965. Regeneration and the moulting cycle in *Ephestia kühniella.* In V. Kiortsis, and H.A.L. Trampusch, eds., *Regeneration in Animals and Related Problems*, 324–330. Amsterdam: North-Holland Publishing.

Pomiankowski, A. 1987. Sexual selection and the handicap principle does work sometimes. *Proc. Royal Soc. B* 231: 123–146.

Pomiankowski, A., and Y. Iwasa. 1993. Evolution of multiple sexual preferences by Fisher's runaway process of sexual selection. *Proc. Royal Soc. B* 253: 173–181.

Pomiankowski, A., and A. P. Møller. 1995. A resolution to the lek paradox. *Proc. Royal Soc. B* 260: 21–29.

Preston, B. T., I. R. Stevenson, J. M. Pemberton, D. W. Coltman, and K. Wilson. 2003. Overt and covert competition in a promiscuous mammal: the importance of weaponry and testes size to male reproductive success. *Proc. Royal Soc. B* 270: 633–640.

Price, J., and S. Allen. 2004. Exploring the mechanisms regulating regeneration of deer antlers. *Phil. Trans. R. Soc. London* 359: 809–822.

Prum, R. O. 1997. Phylogenetic tests of alternative intersexual selection mechanisms: trait macroevolution in a polygynous clade (Aves: Pipridae). *Am. Nat.* 149: 668–692.

Radwan, J. 2008. Maintenance of genetic variation in sexual ornaments: A review of the mechanisms. *Genetica* 134: 113–127.

Reinhold, K. 1998. Sex linkage among genes controlling sexually selected traits. *Behav. Ecol. Sociobiol.* 44: 1–7.

Richards, O. W. 1927. Sexual selection and allied problems in the insects. *Biological Reviews* 2: 298–364.

Ritchie, M. G. 2007. Sexual selection and speciation. *Ann. Rev. Ecol., Evol., Syst.* 38: 79–102.

Rowe, L., and D. Houle. 1996. The lek paradox and the capture of genetic variance by condition-dependent traits. *Proc. Royal Soc. B* 263: 1415–1421.

Schluter, D., and T. Price. 1993. Honesty, perception and population divergence in sexually selected traits. *Proc. Royal Soc. B* 253: 117–122.

Seehausen, O., P. J. Mayhew, and J. M. Van Alphen. 1999. Evolution of colour patterns in East African cichlid fish. *J. Evol. Biology* 12: 514–534.

Shingleton, A., A. Frankino, T. Flatt, H. F. Nijhout, and D. J. Emlen. 2007. Size and shape: the developmental regulation of static allometry in insects. *Bioessays* 29: 536–548.

Shingleton, A., C. Mirth, and P. W. Bates. 2008. Developmental model of static allometry in holometabolous insects. *Proc. R. Soc. B* 275: 1875–1885.

Shingleton, A. W., J. Das, L. Vinicius, and D. L. Stern. 2005. The temporal requirements for insulin signaling during development in Drosophila. *PLoS Biology* 3: 1607–1617.

Simmons, L. W., and J. L. Tomkins. 1996. Sexual selection and the allometry of earwig forceps. *Evol. Ecol.* 10: 97–104.

Stern, D. 2001. Body-size evolution: how to evolve a mammoth moth. *Curr. Biol.* 11: R917–919.

———. 2001. Body-size vontrol: How an insect knows it has grown enough. *Curr. Biol.* 13: R267–R269.

Stern, D. L., and D. J. Emlen. 1999. The developmental basis for allometry in insects. *Development* 126: 1091–1101.

Suttie, J. M., P. F. Fennessy, I. D. Corson, F. J. Laas, S. F. Crosbie, J. H. Butler, and P. D. Gluckman. 1989. Pulsatile growth hormone, insulin-like growth factors and antler development in red deer (*Cervus elaphus scoticus*) stags. *J. Endocrinology* 121: 351–360.

Swallow, J. G., L. E. Wallace, S. J. Christianson, P. M. Johns, and G. S. Wilkinson. 2005. Genetic divergence does not predict change in ornament expression among populations of stalk-eyed flies. *Mol. Ecol.* 14: 3787–3800.

Tatar, M., A. Bartke, and A. Antebi. 2003. The endocrine regulation of aging by insulin-like signals. *Science* 299: 1346–1351.

Telang, A., L. Frame, and M. R. Brown. 2007. Larval feeding duration affects ecdysteroid levels and nutritional reserves regulating pupal commitment in the yellow fever mosquito *Aedes aegypti* (Diptera: Culicidae). *J. Exp. Biol.* 210: 854–864.

Tomkins, J. L., J. Radwan, J. S. Kotiaho, and T. Tregenza. 2004. Genic capture and resolving the lek paradox. *TREE* 19: 323–328.

Tregenza, T., L. W. Simmons, N. Wedell, and M. Zuk. 2006. Female preference for male courtship song and its role as a signal of immune function and condition. *Anim. Behav.* 72: 809–818.

Truman, J. W., K. Hiruma, P. J. Allee, S. G. B. MacWhinnie, D. T. Champlin, and L. M. Riddiford. 2006. Juvenile hormone is required to couple imaginal disc formation with nutrition in insects. *Science* 312: 1385–1388.

Truman, J. W., and L. M. Riddiford. 2002. Endocrine insights into the evolution of metamorphosis in insects. *Ann. Rev. Entomol.* 47: 467–500.

Veiga, J., and M. Puerta. 1996. Nutritional constraints determine the expression of a sexual trait in the house sparrow, *Passer domesticus. Proc. Royal Soc. B* 263: 229–234.

Wallace, B. 1987. Ritualistic combat and allometry. *Am. Nat.* 129: 775–776.

Webster, J. R., I. D. Corson, R. P. Littlejohn, S. K. Stuart, and J. M. Suttie. 1996. Effects of season and nutrition on growth hormone and insulin-like growth factor-I in male red deer. *Endocrinology* 137: 698–704.

Wedekind, C. 1992. Detailed information about parasites revealed by sexual ornamentation. *Proc. Royal Soc. B* 247: 169–174.

Weihe, U., M. Milán, and S. M. Cohen. 2005. Drosophila limb development. *Comprehensive Molecular Insect Science* 1: 305–347.

Weinkove, D., and S. J. Leevers. 2000. The genetic control of organ growth: Insights from Drosophila. *Current Opinion in Genetics and Development* 10: 75–80.

West-Eberhard, M.-J. 1983. Sexual selection, social competition and speciation. *Quart. Rev. Biol.* 58: 155–183.

———. 1979. Sexual selection, social competition, and evolution. *Proc. Amer. Phil. Soc.* 123:222–234.

Wilkinson, G., and M. Taper. 1999. Evolution of genetic variation for condition-dependent traits in stalk-eyed flies. *Proc. Royal Soc. B* 266.

Zeh, D. W., and J. A. Zeh. 1988. Condition-dependent sex ornaments and field tests of sexual-selection theory. *Am. Nat.* 132: 454–459.

Zeh, D. W., J. A. Zeh, and G. Tavakilian. 1992. Sexual selection and sexual dimorphism in the harlequin beetle *Acrocinus longimanus*. *Biotropica* 24: 86–96.

Zuk, M., R. Thornhill, J. D. Ligon, and K. Johnson. 1990. Parasites and mate choice in red jungle fowl. *Amer. Zool.* 30: 235–244.

SECTION III

MORPHOLOGY AND BEHAVIOR

Adaptations discussed in the foregoing chapters imply a fit between organisms and their environment. In evolutionary terms, they imply that observed structures of organisms have superseded others where the fit was not so good, the agency being natural selection.

More is known about the evolution of structures than of functions, in part because structures are often hard, inflexible, and easy to measure reliably, and in part because they fossilize. Investigating the evolution of physiological and behavioral functions lacking those advantages presents some challenges. Yet those challenges need to be met because all structures have functions and functions determine success or failure of organisms, in other words, their fitness. Physiology governs how organisms can operate, and what they can do and cannot do. Behavior can be thought of as the way in which animals perform basic functions within the limits of physiological and biomechanical constraints. It includes inflexible application of rules, such as those applied by spiders to the building of a wondrous web, and the modification of feeding behavior many animals make in the light of experience. In keeping with the theme of this book, the authors of this section of chapters examine some basic questions about the performance of organisms and how performance might evolve. Some look inside organisms and probe genetic mechanisms, others look on the outside at how organisms interact in a geographically varying mosaic of communities. The potential in this field of evolutionary biology is enormous. Its fulfillment is a long way off, but the way itself is clear: it is a chain of connections between the genes that initiate the development of a trait, through the chemical and neurohumoral pathways that mediate its physiological or behavioral expression and integration with other traits, to the ecological interactions that test its performance and determine its success.

The first chapter in this section (chapter 9) examines how the environment impinges physically on morphology. It therefore serves as a link with the developmental chapters in the previous section and with the behavioral and ecological chapters that follow. Species exploiting the environment in different ways have diversified in response to different forces. The chapter shows how a force-response system can be quantified so that it can be interpreted. Mimi Koehl discusses the fundamental relationship between some physical forces and animal form in terms of organism performance using examples from studies of sessile organisms buffeted by water currents and the movement of wingless animals through air. Some findings from biomechanical analyses are surprising, illustrating the fact that a particular morphology is fully understood only when the performance of the organism in its environment is known. An example is given of how an apparently "bad" engineering design of seaweeds makes sense only when considered in terms of the life history of the organisms. The chapter, combined with the preceding ones, is thus a reminder of the chain of functional links between genes, their phenotypic effects, and their fitness consequences in the environment in which they are used.

Behavior is often thought to give flexibility to organisms in changing or fluctuating circumstances, but there are limits to which behavior can accommodate stress and those limits vary among organisms and environments. In chapter 10, Hopi Hoekstra addresses some fundamental questions concerning the evolution of behavior. What are the genes involved, how can they be detected, how do they work, and how are they influenced by the environment? What is the genetic architecture of behavior? Surprisingly little is known about the genetic mechanisms that have given rise to ecologically meaningful behavioral diversity. For behavior to evolve there must be genetic variation. But where is the variation? QTL (quantitative trait loci) mapping is a powerful way to find out. For behavioral traits, which are typically difficult to measure precisely and repeatably, and which are subject to environmental modification, this entails breeding in captivity. Unfortunately, this requirement is restrictive, and is simply infeasible for many behaviors and organisms. But not for all, and the chapter ends at a promising beginning. *Peromyscus polionotus*, the old-field mouse, constructs complex burrows with an escape tunnel; the sister species *P. maniculatus* produces shallow ones without such a tunnel. Crossing experiments show that the behavior producing complex burrows segregates as a dominant trait. The search is on to find the genetic factors, with the ultimate goal of finding allelic differences that cause the difference in behavior between the species, and how they produce their effects. Determining the fitness consequences of the two types of behavior in the natural environment will round out the story.

Behavior mediates diverse interactions between members of an ecological community, from murderous to mutualistic. The next two chapters exemplify some of the evolutionary complexities involved in Darwin's "perfection of structure and coadaptation," the evolution of plant-insect interactions. In the first of them (chapter 11), May Berenbaum stresses what an enormous but unrecognized impact Darwin had on the development of the field of pollination biology. As a corrective, she traces the foundations he laid, starting with observations in Galápagos, and leading, among other things, to the correct prediction of a pollinating moth with an 11-inch proboscis in Madagascar, on being shown a flower with an 11-inch nectar spur! The moth was discovered after he had died. It was given the name *Xanthopan morgani praedicta*. Pollination typically involves advertising, and therein lies a danger to the plant, because visual and chemical signals can attract enemies as well as pollinators. The chapter illustrates the balance between attraction and repulsion with a three-way plant-pollinator-florivore interaction between North American wild parsnips, their dipteran pollinators, and their lepidopteran webworm florivores. Florivores deter pollinator visits, and plants deter florivores, so that when plants are damaged by webworms they release octyl butyrate, a deterrent. Change the community and the balance shifts. For example, in the absence of the webworm in New Zealand, wild parsnips have evolved a much lower production of deterrent and possibly an enhanced attractant to pollinators.

Thus, a species that evolves in response to the local environment may, in turn, induce an ecological and evolutionary response from its community. The last chapter in this section (chapter 12) by John Thompson provides a geographical perspective. It delivers a message that assemblages of interacting species such as plants, their pollinators, and exploiters vary from place to place, hence the evolutionary forces acting on them also vary from place to place. Therefore, it should be no surprise that a mutualism in one area can be a commensalism or even an antagonistic relationship in another. The chapter peels back layers of complexity that are displayed by moths and plants, which have been learned only recently through phylogenetic reconstruction combined with field studies. At least sixteen species of *Greya* moths feed on plants in three families, where only a handful were known thirty years ago, and the number of species of *Tegeticula* yucca moths has jumped from four to twenty. These moths have undergone radiation into different feeding and pollination niches. In both groups, some moths pollinate and oviposit in flowers, others oviposit in other tissues but do not pollinate, and the two ecotypes may coexist with each other as well as with bees, flies, and other pollinators to a geographically varying extent. Yucca plants are extreme, and exceptional, in depending entirely upon moths for pollination in all populations: the mutualism is obligate, whereas with other moths it is facultative. The chapter ends with questions for future research about local, mainly genetic, dynamics: how simple is it genetically to modify floral and moth characteristics in mutualisms, and are some directions of change more constrained than others? These echo a major theme of chapter 5.

Plate 1 *Top*: *Archaeopteryx lithographica* in flight. Note the use of the long feathers behind the legs, as well as the feathers on either side of the long bony tail. *Bottom*: *Hesperonykus elizabethi* from the Late Cretaceous of Canada is one of the many small non-avian theropods that were probably covered with feathers when they were alive. Closely related to *Microraptor*, a feathered, four-winged dinosaur from China, *Hesperonykus* was smaller than most breeds of chickens. (Reconstructions by and courtesy of Nicholas Longrich, Yale University)

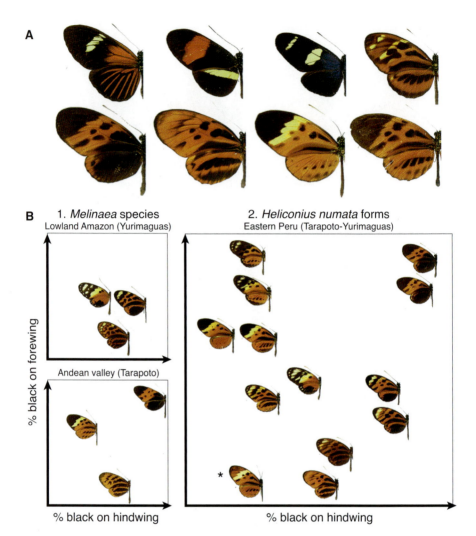

Plate 2 Pattern variation and morphospace occupancy in *Heliconius* butterflies. (a) Mimetic wing pattern themes used by *Heliconius* species in a local community in the Andean valley of the Río Mayo (Tarapoto) in Eastern Peru. These patterns represent largely discrete pattern themes used by at least two different local butterfly species. So-called tiger patterns are somewhat more variable within the genus *Heliconius*, but are clearly mimetic of less variable *Melinaea* species (Nymphalidae: Ithomiinae). The four top butterflies (*H. xanthocles, H. erato, H. sara,* and *H. ethilla*) are monomorphic in the community. The four bottom butterflies are distinct coexisting mimetic morphs of the polymorphic species *H. numata* (*bicoloratus, arcuella, tarapotensis,* and *timaeus*). (b) Diagram of morphospace occupancy in *Heliconius numata* wing patterns and their *Melinaea* models. 1. Wing patterns of different *Melinaea* species representing discrete mimetic patterns which serve as models for *H. numata*. Lowland and Andean communities show different wing pattern themes. *H. numata* populations in each region mimic all three forms through local polymorphism, thereby occupying multiple separate regions in the morphospace. 2. The morphology represented by a star (bottom left corner of the plot) is a non-mimetic variant which crops up in natural populations, albeit very rarely, via recombination between mimetic alleles.

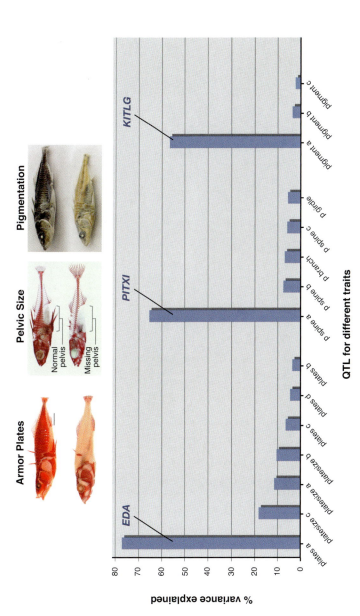

Plate 3 Genetic basis of major morphological differences in sticklebacks. Marine sticklebacks are covered with armor plates, have robust pelvic hindfins, and abundant pigmentation (*upper fish in each panel, from left to right*). Many freshwater fish have evolved greatly reduced plate numbers, complete loss of pelvic hindfins, or lighter skin color (*lower fish in each panel, from left to right*). Genetic crosses show that single major quantitative trait loci (QTL) control each trait, together with multiple independent modifier QTL, with smaller phenotypic effects. Each of the major QTL has been shown to correspond to a key developmental control gene (EDA, PITX1, and KITLG). Modified from data in Colosimo et al. 2004 and 2005; Shapiro et al. 2004; and Miller et al. 2007. (Images courtesy of Pam Colosimo, Mike Shapiro, Frank Chan, Craig Miller, and David Kingsley)

Plate 4 The low (*left*) and complete (*right*) lateral plate morphs in marine stickleback. Low morphs are rare in the ocean, occurring at a frequency of about 0.001 in Oyster Lagoon, the source of the fish used in our experiment. (Photo courtesy of R. Barrett)

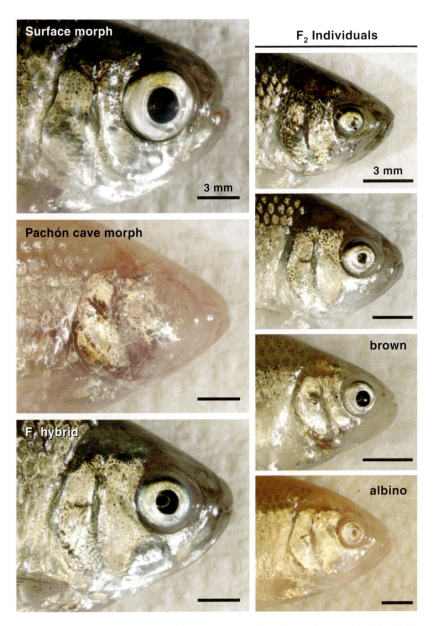

Plate 5 Surface and cave morphs of *Astyanax* can be bred to produce viable F_1 hybrids. Several populations of the surface form of *Astyanax mexicanus* persist in the rivers and streams surrounding the extensive cave network in northeast Mexico inhabited by cave morphs. A surface individual can be successfully mated to a cavefish collected from the Pachón cave. An F_2 pedigree produced from a cross of sibling surface × Pachón F_1 hybrids display variable phenotypes (note, eye size and pigmentation) that range between the two parental phenotypes. Included among these are examples of a brown individual (a non-albino individual carrying two Pachón copies of the gene *Mc1r*) and an albino individual (an individual carrying two Pachón copies of the gene *Oca2*). Scale = 3 mm.

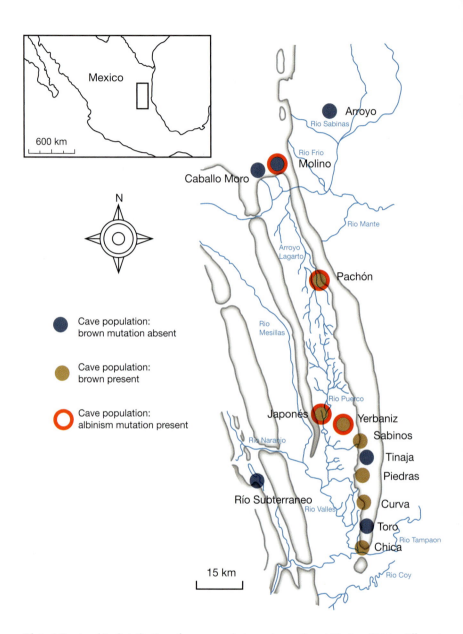

Plate 6 Geographic distribution of cave morphotypes in northeast Mexico. Thirty different cave populations have been described from a network of subterranean limestone caves in the Sierra de El Abra and Guerrero regions of Mexico (*inset*). Those populations in which the light pigmentation brown phenotype has been reported are denoted by a brown dot; those caves in which albinism has been reported are represented by a red outline. Note that the brown phenotype has not been reported in several cave systems (e.g., Caballo Moro, Río Subterraneo). Further, fish derived from the Molino cave are albino, however do not express the brown phenotype. Map of localities adapted from Gross et al. 2009.

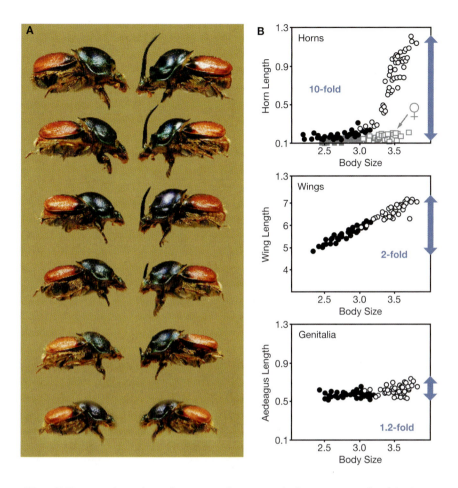

Plate 7 Nutrition-dependent phenotypic plasticity and allometry are related in insects. (a) Female (*left*) and male (*right*) *Proagoderus* (*Onthophagus*) *tersidorsis* showing among-individual variation in body size and, in males, horn size. (b) Scaling relationships (allometries) for three morphological traits in the beetle *O. acuminatus*. Individuals reared with access to large food amounts (high nutrition, open symbols) emerged at larger adult body sizes than full-sibling individuals reared with smaller food amounts (low nutrition, closed symbols). Traits differed in how sensitive (plastic) their growth was to this variation in nutrition. Male horns (*top*) were the most sensitive, and horn lengths were > tenfold longer in the largest individuals than they were in the smallest individuals. (Females of this species do not produce enlarged horns, and the height of the corresponding head region is indicated by the gray squares.) Wing development (*middle*) was also sensitive to nutrition, and wings scale positively with variation in body size. Male genitalia (*bottom*) were almost entirely *in*sensitive to nutrition, and the size of the aedeagus was largely body size invariant. Horns and genitalia are plotted on the same scale to illustrate the relative plasticity (horns) or canalization (male genitalia, female horns) of their development. Wings were much larger, and are shown on their own scale. In all cases, the degree of plasticity/canalization (blue arrows) is reflected in the steepness of the trait size–body size allometries. Redrawn from Emlen et al. 2007.

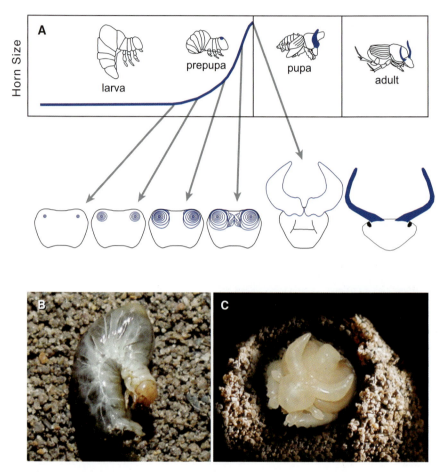

Plate 8 Development of beetle horns. (a) Horns form from clusters of epidermal cells (the horn disc) that undergo a localized burst of growth, ultimately producing a long evagination/outgrowth from the adult body. The cells that will form the horns begin proliferating near the end of the final (third) larval instar as animals stop feeding and purge their guts in preparation for metamorphosis (blue curve); horns have largely completed all growth by the end of the prepupa period. Horns at this stage comprise densely folded tubes of epidermis that unfurl as the animal sheds its larval cuticle. Third instar larva (b) and pupa (c) from *Onthophagus taurus* shown. (Reprinted with permission from Emlen et al. 2006)

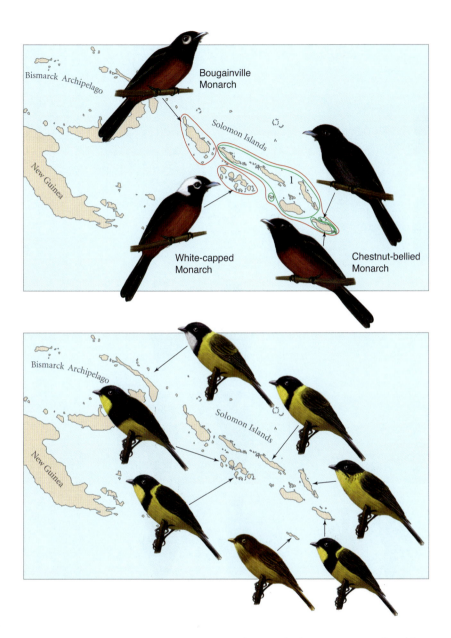

Plate 9 Color pattern variation in northern Melanesian bird species. *Top*: Males of four taxa of chestnut-bellied flycatcher, *Monarcha castaneiventris*, currently all considered to belong to a single species, with a root at about 1 million years (Uy et al. 2009). *Bottom*: Males of subspecies of Golden Whistler *Pachycephala pectoralis*. Note that peripherally isolated Rennell Island has the most distinctive population. (Drawn by Emiko Paul, and reproduced with permission from Price 2008: "Speciation in Birds," Roberts and Co., Boulder, Co.)

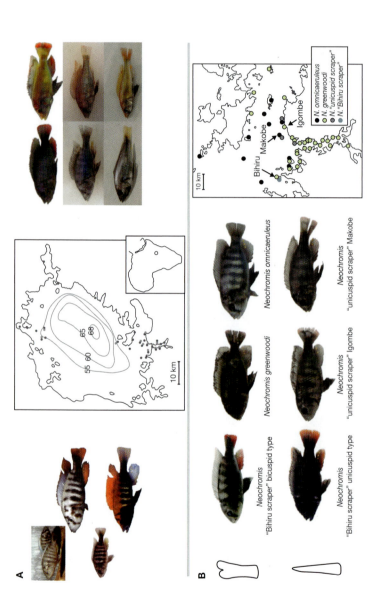

Plate 10 Three major classes of polymorphisms associated with variable stages of geographically sympatric speciation in Lake Victoria cichlids. (a) Blue versus red male nuptial coloration: female and male of *Pundamilia pundamilia* and *P. nyererei* (*left*), the known geographical distributions of both species with depth contours for the lake (center: blue = *P. pundamilia*), and other morph and species pairs with the same male color polymorphism (*right*). *From top to bottom:* color morphs of *Neochromis greenwoodi* (a rock-dwelling species; Terai et al. 2006), the ecologically parapatric sister species *Gaurochromis hiatus* and *G. iris* (soft bottom species), and *Yssichromis piceatus* and *Y. pyrrhocephalus* (pelagic species). (b) Trophic (dental) morphology: symparic bicuspid (*top*) and unicuspid tooth shape ecomorphs and incipient species of *Neochromis* from three islands in Lake Victoria, Bihiru (*left*), Igombe (*middle*) and Makobe. Typical subequally bicuspid (*top*) and unicuspid tooth (*left*). The known geographical distributions.

Plate 11 X-linked (female) color polymorphisms in three species of Lake Victoria cichlids (3 columns), *Neochromis omnicaeruleus* from Makobe Island (*left*), *N. greenwoodi* from Igombe Island (*middle*), *Paralabidochromis chilotes* from Makobe Island (plain female) and Ruti Island (blotched female). Blotched males exist too but are very rare. The known geographical distributions of the color morphs in *Neochromis omnicaeruleus* (circles) and its allopatric sister species *N. greenwoodi* (squares; black = plain morph, red = OB morph, blue = WB morph). The islands chosen for more detailed investigation are indicated by arrows.

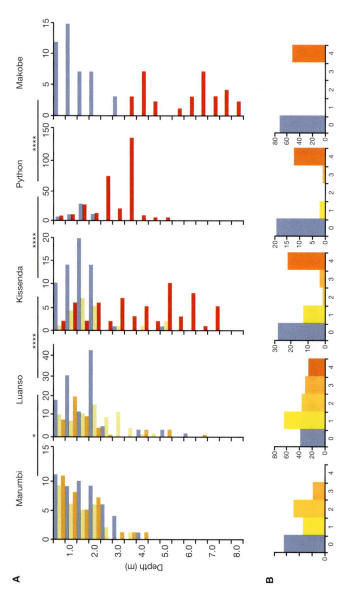

Plate 12 Ecological, phenotypic, and genetic divergence between blue and red *Pundamilia* nuptial phenotypes at five islands (data for the same island are presented in the same column). (a) Depth distributions of male nuptial color phenotypes. Blue bars, blue (class 0 of panel b); pale yellow bars, intermediate (classes 1 and 2); and orange or red bars, red (orange if dominated by class 3; red if dominated by class 4; classes are defined in Seehausen et al. 2008). Significance levels of differences between islands in the divergence between red and blue are reported as P values of G-tests, indicated by asterisks (all tests two-tailed): *P < 0.05, ****P < 0.0001. (b) Frequency distributions of male nuptial color phenotypes in 5 classes, from blue (*left*) to red (*right*). (Part (a) redrawn from Seehausen et al. 2008, with permission)

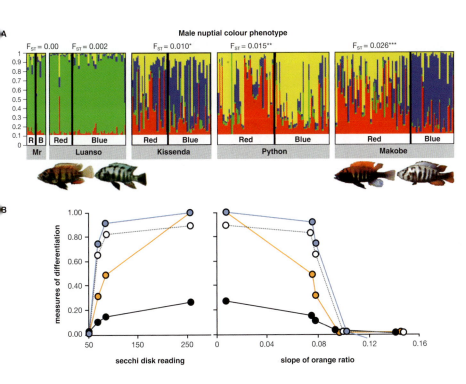

Plate 13 (a) Bayesian assignment probabilities of individuals to the populations defined by the blue or the red male phenotype. Each vertical line represents an individual and colors indicate the proportion of an individual's genotype assigned to a particular population, showing genetic segregation between phenotypes at Kissenda, Python, and Makobe Islands, but not at Luanso and Marumbi Islands. Pairwise multilocus F_{ST} values and their significance (*P < 0.05, **P < 0.01, ***P < 0.001) are reported above the plots. (b) Measures of neutral versus adaptive differentiation between sympatric *Pundamilia* phenotypes plotted against water transparency (*left*) and light slope (*right*). Blue symbols and line: Spearman rank correlations between color and *LWS* genotype. Open symbols and dashed line: *LWS* F_{ST} between red and blue phenotypes. Filled orange symbols and orange line: association between color and water depth. Filled black symbols and black line: microsatellite F_{ST} (multiplied by 10 for display) between red and blue phenotypes. (Reproduced from Seehausen et al. 2008, with permission)

Plate 14 Anoles have an extensible throat fan, termed a dewlap, that is used in intra- and interspecific communication. Sympatric species invariably differ in the size, color, or pattern of the dewlap. A sample of anole dewlaps (*clockwise from top left*): *A.* (Chamaelinorops) *barbouri, A. grahami, A. conspersus, A. lineatopus, A. lividus, A. allogus.*

Plate 15 Sampling an experimental pond with a seine net. (Photo courtesy of A. Paccard)

Plate 16 Immigrant and resident *Geospiza fortis* (medium ground finch) on Daphne Major Island, Galápagos. The upper figure shows a typical adult male of the resident population. The lower figure shows a juvenile (note the yellow mandible) of the sixth generation of a lineage that began with a hybrid male immigrating from the neighboring island of Santa Cruz and breeding with a resident female (also a hybrid). From the fifth generation onward, and possibly earlier, members of the immigrant lineage bred only with each other. They were reproductively isolated from the residents as a result of their distinctive song and morphology, important cues used in mate choice and hence a behavioral barrier to interbreeding (see figure 17.1). (From Grant and Grant 2009)

Chapter Nine

How Does Morphology Affect Performance in Variable Environments?

Mimi A. R. Koehl

The preceding chapters in the section on "Mechanisms, molecules, and evo-devo" deal with the genetic and developmental mechanisms that produce and limit phenotypic diversity and morphological changes as lineages of organisms evolve. To understand the selective consequences of such phenotypic variations, we also need to determine whether or not morphological differences between organisms affect their relative performance in natural environments, as well as how those effects depend on the habitat. Studies of the physiology and biomechanics of organisms are useful tools in assessing the functional consequences of phenotypic differences. The purpose of this chapter is to explore some of the ways in which the performance consequences of differences in morphology are affected by the environment in which an organism lives.

BACKGROUND

The relationship between the form and function of organisms has long been studied (e.g., reviewed in Koehl 1996). One approach to investigating functional morphology is biomechanics, which applies quantitative engineering techniques to study how organisms perform mechanical functions and interact with their physical environments. By elucidating basic physical rules governing how biological structures operate, biomechanical studies can identify which structural characteristics affect the performance of a defined function and can analyze

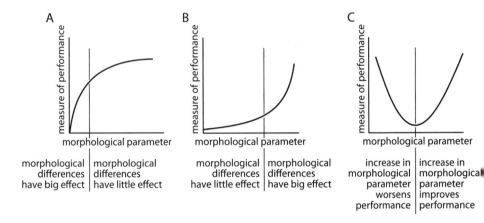

Figure 9.1. Examples of nonlinear ways in which a measured aspect of performance can vary as a function of a quantified morphological parameter. Biological examples of each of these types of curves are reviewed by Koehl (1996).

the mechanisms responsible for the effects of morphological differences on performance.

Biomechanical analyses have revealed that the dependence of measures of performance on quantifiable aspects of morphology is often nonlinear (reviewed by Koehl 1996, 2000). When the effect of morphology on performance is nonlinear, there are ranges of the morphological parameter in which changes in structure have little effect on function, and other ranges where small morphological modifications can have large consequences. For example, if an asymptotic curve describes the dependence of performance on a morphological parameter, then changes in that parameter at low values can have a big effect on performance, whereas changes in that parameter at high values can make little difference to function (fig. 9.1a). Conversely, if an exponential curve describes how function depends on structure, then performance should be insensitive to structural variation at small values of the morphological parameter, but be very sensitive to morphological changes at large values of the parameter (fig. 9.1b). In the range where morphology has little effect on the performance of a particular function, there can be permission for morphological diversity without consequences for that function. Furthermore, selection on that morphological variable based on performance of a different task can occur without jeopardizing the performance of the first function. If the relationship between performance and a morphological variable goes through a maximum or a minimum (fig. 9.1c), then the effect of increasing the morphological variable reverses once it passes a critical value. Passing through such an inflection point represents the acquisition of a novel consequence for a particular type of morphological change. In addition to the examples of nonlinear relationships between morphology and performance illustrated in figure 9.1, some mechanical behaviors (such as buckling under a load) and fluid dynamic processes (such as the transition to turbulence) are unaffected by morphological variations except in a critical range, where a small change in size or shape causes a sudden, drastic switch in performance.

The influence of environmental variables on how morphology affects function can also be nonlinear and can lead to surprising relationships between phenotype and performance. In this chapter, I present some examples from my work and that of my students that illustrate how the consequences of morphological differences can depend on the location of organisms within a habitat, and on the timing of organism behaviors and life history stages relative to temporal changes in the environment.

Consequences of Morphological Differences Can Depend on the Environment
Does Morphology Matter? (Habitat affects whether or not morphology affects performance)

Many bottom-dwelling marine organisms produce microscopic larvae that are dispersed by ocean currents, and then settle back onto the substratum and metamorphose into benthic juveniles. These larvae and juveniles provide an example of how habitat can determine whether or not morphology affects performance. For a larva to recruit to a benthic habitat that is exposed to ambient water flow, it must not wash away during settlement and metamorphosis. Drag is the hydrodynamic force acting in the same direction as the ambient water flow that pushes the larva downstream. Most larvae of benthic marine invertebrates are very small (a few hundred microns). How do the changes in body shape that occur during metamorphosis affect the drag experienced by a larva versus a newly metamorphosed juvenile sitting on a substratum in the ocean?

We are addressing this question using the sea slug, *Phestilla sibogae*, which is an important model organism for studying larval settlement.

Figure 9.2. Diagram of a lateral view of the body shape of a newly settled larva, and of a newly metamorphosed juvenile of the sea slug, *Phestilla sibogae*. The anterior end of each animal is to the left.

A larva of *P. sibogae* settles onto the substratum (fig. 9.2) and then undergoes metamorphosis into a juvenile benthic slug (details in Bonar and Hadfield 1974). *P. sibogae* slugs live on coral reefs dominated by their prey, the branching coral, *Porites compressa*. Coral reefs are porous, so there are surfaces on which larvae can land inside the reef as well as on the top of the reef. *P. compressa* reefs are exposed to turbulent, wave-driven water flow with peak freestream velocities of ~0.4 m/s, and to much slower flow (peak velocities of ~0.04 m/s) through the spaces within the porous reef (Koehl and Hadfield 2004). When fluid flows over a solid surface, a boundary layer of slowed fluid develops along the surface (e.g., Vogel 1994; Koehl 2007). Therefore, to determine the drag that could wash settled larvae and juveniles off a reef, we need to know the water velocities encountered by organisms only 200 μm tall sitting on surfaces of the reef. We used laser-Doppler velocimetry to measure water velocities 200 μm from coral surfaces at the top of a reef (velocity peaks of 0.085 m/s) and at various positions down within a reef (e.g., 0.007 m/s at 200 μm from surfaces 10 cm below the top of the reef) (Reidenbach et al. 2008).

Measuring the drag on larvae and juveniles of *P. sibogae* is challenging because they are so small. We met this challenge by borrowing a technique routinely used by engineers to study fluid dynamic forces on objects of inconvenient sizes: dynamically scaled physical modeling. If a model is scaled dynamically, then the ratios of the forces and of the velocities in the fluid around a model are the same as those for comparable positions around the real larva, and fluid dynamic forces measured on the model can be used to calculate the forces on the real animal (e.g., Koehl 2003). Reynolds number (Re) is the ratio of inertial forces to viscous forces for a particular flow situation (Re = $LU\rho/\mu$, where L is a linear dimension of the body, U is the velocity of the fluid relative to the body, ρ is the density of the fluid, and μm is the dynamic viscosity of the fluid, which is its resistance to being sheared) (e.g., Vogel 1994). If the Re of a model and the Re of its prototype organism are the same, then the model is dynamically similar to the organism. We made big models (L ~ 5 cm) of microscopic larvae and juveniles, but we kept the Re's of the models the same as the Re's of the larvae and juveniles by lowering the velocity of the fluid relative to the models, and by using a fluid (mineral oil) with a higher viscosity than that of water.

We used these dynamically scaled physical models to determine the hydrodynamic forces on larvae and on newly metamorphosed juveniles of *P. sibogae* sitting on solid surfaces at different positions on a coral reef (Kreft, Waldrop, and Koehl unpublished data). Surprisingly, when exposed to flow at Re = 2 (like they experience within the reef), there was no significant difference between the drag on the upright, bulbous larva and the sleek, flat juvenile (drag on both shapes was ~10 nN). Body shape did *not* affect hydrodynamic performance of these small animals in within-reef microhabitats. In contrast, when exposed to flow at Re = 17 (like they experience at the top of a coral reef), body shape had a big effect on hydrodynamic performance: drag on the larva (~240 nN) was nearly three times greater than drag on the juvenile (~90 nN).

How can shape affect drag on organisms in one region of a habitat, but not in another? The mechanisms responsible for drag depend on the Reynolds number, and hence on the fluid velocity relative to an organism. The Re's of the tiny *P. sibogae* larvae and newly metamorphosed juveniles fall in a very interesting transitional Re range where both viscous and inertial forces are important. At low Re (slow flow), drag is due to "skin friction" (fluid is sheared as it flows across a body, and the viscous resistance of the fluid to being sheared drags the body downstream). At high Re (fast flow), in addition to skin friction, bodies also experience "form drag" (a wake forms on the downstream side of a body, and the resulting pressure difference between the upstream and downstream sides of the body push it downstream) (e.g., details in Vogel 1994). Bluff bodies that produce wide wakes experience higher form drag than streamlined shapes that have narrow wakes. By measuring the fluid velocity fields around our models (Kreft, Waldrop, and

Koehl unpublished data), we discovered that wakes formed at the higher Re of larvae and juveniles at the top of the reef (accounting for about 90% of the drag on the bluff larva, but only about 35% of the drag on the streamlined juvenile), hence body shape affected drag. In contrast, at the lower Re they experience within the reef, form drag and body shape were not important.

Comparison of hydrodynamic forces with the attachment strengths of larvae and juveniles of *P. sibogae* (Koehl and Hadfield 2004) indicates that larvae have a high probability of washing away at the top of the reef but not within it (Reidenbach et al. 2008), suggesting that larval settlement occurs within the reef. In contrast, if the juveniles crawl up to the top of the reef where the living coral tissue on which they feed is most abundant, they are unlikely to be swept away.

This example focuses on an ontogenetic transformation in morphology rather than an evolutionary change in form. Nonetheless, it illustrates a general principle for small organisms operating at these intermediate Re's: differences in the water current or wind velocity that occur within a spatially diverse environment can determine whether or not form drag, and hence body shape, affects the fluid dynamic performance of the organisms.

Does a Morphological Change Improve or Worsen Performance?
(Habitat effects on the consequences of a morphological change)

Not only do bottom-dwelling aquatic organisms have to withstand the forces imposed on them by ambient water flow, but they also can use that water motion for the transport of materials. For example, moving water supplies oxygen to benthic organisms, dissolved nutrients to algae, and planktonic food to suspension feeders. Ambient currents also carry away wastes released by organisms and can disperse their gametes, spores, or larvae. We have been investigating ways in which the morphology of benthic organisms affects their performance in utilizing ambient water motion for transport. Several of those studies have revealed examples of how the habitat determines whether a specific change in morphology improves or worsens performance.

Bryozoans, colonial animals that live attached to surfaces in aquatic environments, are suspension feeders that capture planktonic prey from the water flowing past them. For upright arborescent colonies, habitat determines the effect of colony size on the feeding rates of the zooids in the colony (Okamura 1984). An increase in colony size can lead to a decrease in the rate of capture of food particles per zooid in habitats characterized by slow ambient currents because the upstream zooids in a colony deplete the water of planktonic food particles. In contrast, in habitats exposed to rapidly flowing water, colony size has the opposite effect on feeding rate per zooid. If the ambient water current is too fast, zooids cannot hold on to prey particles. As water flows between the branches of a large colony, it is slowed more than it is by a small colony. Therefore, zooids in a large colony are able to catch and retain food particles in ambient water currents that are so fast that zooids in small colonies cannot feed.

Habitat can also determine the effect of morphology on the dispersal of materials (e.g., wastes, gametes) released by bottom-dwelling aquatic organisms. We used physical models of benthic animals of different sizes (1–10 cm in height) to investigate how body size affects the dispersal of materials they shed. The models were affixed to a wave-swept rocky shore, either on bare rock or surrounded by a canopy of models of flexible seaweeds (~50 cm long). Each model animal released dye (an analogue for water-borne substances released by benthic animals) from an opening at its top, and the dispersal of the dye in the ambient water flow was measured as a function of time (techniques described in Koehl et al. 1993). We found that the effect of the height of a model animal on the dispersal of material it released into wave-driven flow depended on its neighbors. Size had no effect on dispersal from solitary model animals on wave-swept rocks. In contrast, height had a surprising effect on dispersal of dye released by the same model animals when surrounded by canopies of flexible model seaweeds: dispersal was faster for short organisms than for tall ones. The mechanism responsible for this effect is that waves whiplash flexible macroalgae back and forth near the substratum (Koehl 1999). The flailing fronds stirred the water down near the rock surface, thereby spreading material released by short organisms more quickly than material released by organisms that stood taller than the whiplashing seaweeds (Koehl and Powell unpublished data).

These examples of mass exchange between benthic organisms and the surrounding water illustrate that both the abiotic and biotic

183

Figure 9.3. Diagram of a dorsal view of a basal non-flying frog and a derived "flying" frog (redrawn from Emerson and Koehl 1990). The hands, feet, and skin flaps are shown in stippled light gray. Both flyers and non-flyers assume the posture shown in these diagrams when they fall through the air (McCay 2001b).

environments of organisms can alter how differences in morphology affect performance.

Which Function Is Important? (Habitat use in spatially complex, temporally varying environments)

We have been using tree frogs in tropical rain forests to study the evolution of a novel mode of locomotion: gliding. In two separate lineages, the Hylidae and the Rhacophoridae, "flying" frogs that glide through the air have evolved. In both clades, the derived "flying" species have enlarged, highly webbed hands and feet and skin flaps on arms and legs, whereas the more basal species do not (Emerson and Koehl 1990; McCay 2001a). We studied aerodynamic behavior of "flyers" and non-flyers in the field (Emerson and Koehl 1990) and in a wind tunnel (McCay 2001a), and used those data to design wind tunnel experiments in which the aerodynamic forces on physical models of the frogs were measured. The models enabled us to vary one or more morphological or postural features at a time to quantify the effects of each; thus we could determine the aerodynamic consequences of the "flying" morphology (Emerson and Koehl 1990; McCay 2001a, 2001b).

The distinctive morphological features of gliding animals traditionally have been evaluated by their effects on glide performance, the horizontal distance traveled per vertical distance fallen, which is equal to the ratio of lift (the force perpendicular to the direction of air flow relative to the body) to drag (the force parallel to the air flow direction relative to the body) (e.g., Vogel 1994). Surprisingly, our experiments revealed that the lift-to-drag ratio was lower (i.e., gliding performance was worse) for "flying" frogs than for non-flyers (Emerson and Koehl 1990; McCay 2001b). However, consideration of how frogs locomote through the air in rain forests suggests that glide performance is not the aspect of aerodynamic function that is most likely to affect fitness.

Airborne frogs maneuver through complex plant canopies to reach breeding pools on the forest floor (e.g., McCay 2001), so we studied how the morphological features of "flying" frogs affected their stability in the face of ambient wind gusts, and their maneuverability in steering their way through the trees. A stable aircraft passively rights itself after being perturbed (for example, by a wind gust), a neutrally stable aircraft does not right itself, and an unstable one continues to tumble passively after the perturbing force has ceased. If a frog falling through the air has an aerodynamically stable shape, then its body passively resists maneuvers that the frog tries to impose on its trajectory by a steering movement. In contrast, a neutrally stable shape does not fight maneuvers, and an unstable shape enhances them. Thus, there is a trade-off between aerodynamic stability and maneuverability. Our model studies showed that the non-flyer tree frogs are aerodynamically stable, but the "flying" frogs are unstable in pitch (rotating nose up or down) and yaw (turning right or left) (Emerson and Koehl 1990; McCay 2001b). McCay's (2001a) wind tunnel studies with living frogs showed that they turn by changing the angles of their feet. Using physical models in the wind tunnel, he measured the moments generated by changing the angle of one foot by frogs of different morphologies and found that "flying" frogs have higher "control effectiveness" than do non-flyers (i.e., they generate bigger moments per foot angle change) (McCay 2001b). Since "flying" frogs are also unstable, they do not offer passive resistance to a moment generated by a foot. Together, these two features make them more "agile" (i.e., they can turn more rapidly for a given change in foot angle) than non-flyer frogs.

Our aerodynamic experiments showed that non-flyer tree frogs are stable and passively right themselves if perturbed by a wind gust, whereas "flying" frogs are unstable, but maneuverable. Do those different attributes affect their parachuting performance under natural wind conditions in a rain forest? McCay (2003) measured wind speeds and turbulent velocity fluctuations at a range of heights in a rainforest canopy at different times during the day and night. He found that daytime winds could be gusty. However, the frogs only glided at night, when the air was generally quite still (average velocities only ~0.01 m/s). Therefore, the danger of an unstable "flying" frog being tumbled by a wind gust is low during the times that the frogs are gliding in the forest.

The example of "flying" frogs illustrates that we can be easily misled about the consequences of morphological differences if the physical structure of the habitat is not considered, and if the temporal patterns of activity by the animals in nature are not known. In complex forest habitats, maneuverability (rather than glide distance) can be the aspect of aerodynamic performance that has the biggest effect on fitness, and the aerodynamic instability that contributes to maneuverability may not pose problems if the animals do not glide during times of day when the wind is gusty.

What Is "Good" Performance? (Ontogenetic changes in the function of a structure in the environment)

Biomechanical investigations sometimes reveal that the morphologies of organisms result in "bad" mechanical performance such as breakage, but field studies of the ways those organisms function in their environments at different stages in their lives can help us understand how such creatures with poor mechanical performance can survive and reproduce. Attached marine organisms (e.g., seaweeds, corals) provide a number of examples of how seemingly "bad" engineering designs can sometimes enhance the ecological performance of organisms (Koehl 1999).

The tropical seaweed *Turbinaria ornata* provides an example of how ontogenetic changes in morphology cause a significant deterioration in mechanical performance (Stewart 2006). An individual *T. ornata* has fleshy blades attached by a stem-like stipe to a holdfast that adheres to the substratum (fig. 9.4). As a *T. ornata* grows larger and ages, gamete-producing reproductive receptacles develop on the blades and the percent of its body mass devoted to reproductive tissue increases (from 0% up to ~25%). As an individual grows, the drag force due to ambient water currents also increases two- to threefold. *T. ornata* blades can develop gas-filled spaces, and the net buoyant force exerted by a plant

Figure 9.4. Diagram of the alga, *Turbinaria ornata*. Each fleshy blade can contain a buoyant gas-filled compartment and can support a gamete-bearing reproductive receptacle. The blades are connected by the stipe to a holdfast, which is attached to the substratum.

due to these floats changes with age as well: young juveniles are negatively buoyant (net buoyant force of about $-0.02N$), and older individuals float (net buoyant force of about $+0.06N$). Since the drag and net buoyant force rise as reproductive effort increases, we might expect the stipe to become wider and stronger as *T. ornata* grow and age. Instead, we find that stipe strength decreases as reproductive effort increases, from ~ 8 MN.m^2 in juveniles to ~ 3 MN/m^2 in the most reproductive individuals (strength is the stress required to break stipe tissue, where stress is force per cross-sectional area of tissue bearing that force).

Ambient forces on sessile organisms (e.g., plants, attached aquatic animals) often vary with season, and behaviors of motile animals (e.g., foraging, predator avoidance, migration, social interactions, and fighting) often change with age; thus we use "environmental stress factor" (ESF) to relate the ability of organisms at their particular stages in ontogeny to resist breakage relative to the maximum loads that they experience in nature at those stages (Johnson and Koehl 1994). The ESF for a *T. ornata* is simply an age- and season-dependent safety factor for the stipe (i.e., the ratio of the stress required to break the stipe to the stress in the stipe due to drag imposed by ambient water currents). If ESF ≤ 1, then an individual breaks and washes away. The ambient current velocity at which ESF = 1 for juvenile *T. ornata* is 3 m/s, whereas for older reproductive individuals, it is only 1 m/s. Thus, from an engineering point of view, reproductive *T. ornata* have a "bad" morphology and are likely to wash away. However, when the weak, buoyant reproductive individuals break, they float to the water surface where they form rafts with many other reproductive *T. ornata*. The potential for sexual reproduction in these rafts is enhanced, as is long-distance dispersal by ocean currents (Stewart 2006), thus a mechanically "bad" mechanical structure leads to "good" ecological performance that can improve fitness.

T. ornata illustrate that the roles organisms play in their environments at different stages in their ontogeny can determine whether the performance consequences of particular differences in morphology enhance or hurt fitness.

Conclusions

To understand the consequences of phenotypic variation, we must determine *if* and *how* morphological differences between organisms affect their relative performance in natural environments. The purpose of this chapter has been to explore ways in which the performance consequences of changes in morphology are affected by the environment in which organisms live. The influence of environmental variables on how differences in phenotype affect function can be nonlinear and can lead to surprising relationships between morphology and performance. The example of settling marine larvae illustrated that the environment can determine *whether or not* particular morphological differences affect performance. The studies of suspension feeding and of waste dispersal by benthic marine organisms showed how the neighborhood of an organism can determine if a specific change in structure *improves or worsens performance*. The investigations of flying frogs illustrated that environmental conditions can determine *which aspects of performance are important* to the success of particular organisms. Research on breakable algae provided an example of how a change in *performance can either enhance or hurt ecological success*, depending on the ontogenetic stage of an organism. In sum, these examples show how the effects of different morphologies on performance depend both on the location of organisms within a habitat, and on the timing of organism behaviors and life history stages relative to temporal changes in the environment.

Environmental conditions can affect the performance consequences of different morphologies in surprising ways. Therefore, it can be all too easy, in the absence of field data, to reach the wrong conclusions about how phenotypic differences between organisms might affect their ecological performance or fitness. Since natural environments vary spatially, it is important to determine what conditions are like in the microhabitats experienced by the organisms in question. Furthermore, because environmental conditions vary with time (e.g., diurnally, seasonally), the timing of specific behaviors as well as ontogenetic changes in the ecological roles of organisms need to be determined relative to the temporal fluctuations in their habitats. Therefore, quantitative field studies of where, when, and how organisms with different phenotypes function in their natural habitats can make important contributions to our understanding of the process of morphological evolution.

The work of the Grants provides many examples of the importance of field work to understanding the process of evolution (e.g., Grant 1999). For instance, they coupled analyses of how beak size and musculature determine the size of seeds that a finch can crack with field data on how climatic variation affects seed supply. Their information about food conditions in the field has led to insights about the composition of finch populations with respect to beak morphology, and also about the role of hybrids with intermediate beak sizes in the evolution of sympatric species of finches (Grant and Grant 2006, 2008).

References

Bonar, D. B., and M. G. Hadfield. 1974. Metamorphosis of the marine gastropod *Phestilla sibogae*. I. Light and electron microscopic analysis of larval and metamorphic stages. *J. Exp. Mar. Biol. Ecol.* 16: 1–29.

Emerson, S. B., and M.A.R. Koehl. 1990. The interaction of behavior and morphology in the evolution of a novel locomotor type: "Flying frogs." *Evolution* 44: 1931–1946.

Grant, B. R., and P. R. Grant. 2008. Fission and fusion of Darwin's finches populations. *Phil Trans. Roy. Soc. B.* 363: 2821–2829.

Grant, P. R. 1999. *Ecology and Evolution of Darwin's Finches.* Princeton, NJ: Princeton University Press.

Grant, P. R., and B. R. Grant. 2006. Evolution of character displacement in Darwin's finches. *Science* 313: 224–226.

Johnson, A. S., and M.A.R. Koehl. 1994. Maintenance of dynamic strain similarity and environmental stress factor in different flow habitats: Thallus allometry and material properties of a giant kelp. *J. Exp. Biol.* 195: 381–410.

Koehl, M.A.R. 1996. When does morphology matter? *Ann. Rev. Ecol. Syst.* 27: 501–542.

———. 1999. Ecological biomechanics: Life history, mechanical design, and temporal patterns of mechanical stress. *J. Exp. Biol.* 202: 3469–3476.

———. 2000. Consequences of size change during ontogeny and evolution. In J. H. Brown and G. B. West, eds., *Scaling in Biology*, 67–86. New York: Oxford University Press.

———. 2003. Physical modeling in biomechanics. *Phil Trans. Roy. Soc. Lond. B* 358: 1589–1596.

———. 2007. Hydrodynamics of larval settlement into fouling communities. *Biofouling* 23: 357–368.

Koehl, M.A.R., and M. G. Hadfield. 2004. Soluble settlement cue in slowly-moving water within coral reefs induces larval adhesion to surfaces. *J. Mar. Systems* 49: 75–88.

Koehl, M.A.R., T. M. Powell, and G. Dairiki. 1993. Measuring the fate of patches in the water: Larval dispersal. In J. Steele, T. M. Powell, and S. A. Levin eds., *Patch Dynamics in Terrestrial, Marine, and Freshwater Ecosystems*, 50–60. Berlin: Springer-Verlag.

McCay, M. G. 2001a. Aerodynamic stability and maneuverability of the gliding frog *Polypedates dennysi*. *J. Exp. Biol.* 204: 2817–2826.

———. 2001b. The evolution of gliding in neotropical tree frogs. Ph.D. dissertation. University of California, Berkeley.

————. 2003. Winds under the rain forest canopy: The aerodynamic environment of gliding tree frogs. *Biotropica* 35: 94–102.

Okamura, B. 1984. The effects of ambient flow velocity, colony size, and upstream colonies on the feeding success of bryozoa. I. *Bugula stolonifera* (Ryland), an arborescent species. *J. Exp. Mar. Biol. Ecol.* 83: 179–193.

Reidenbach, M. A., J. R. Koseff, and M.A.R. Koehl. 2008. Hydrodynamic forces on larvae affect their settlement on coral reefs in turbulent, wave-driven flow. *Limnol. Oceanogr.* 54: 318–330.

Stewart, H. L. 2006. Ontogenetic changes in buoyancy, breaking strength, extensibility, and reproductive investment in a drifting macroalga Turbinaria ornata (Phaeophyta). *J. Phycol.* 42: 43–50.

Vogel, S. 1994. *Life in Moving Fluids*. Princeton, NJ: Princeton University Press.

Chapter Ten

In Search of the Elusive Behavior Gene

Hopi E. Hoekstra

As environmental conditions change or as organisms colonize new habitats, they must adapt to novel conditions. Such adaptive change often involves a suite of traits ranging from morphology to physiology to behavior. Many classical examples of adaptation focus on morphological change, perhaps the most celebrated of which is the variation in bill size and shape of Darwin's finches on the Galápagos Islands (e.g., Grant 1999, 2001; Grant and Grant 2008). Other examples include rapid morphological adaptation in cichlid fish (chapter 14), *Anolis* lizards (chapter 15), or sticklebacks (chapter 16). Such morphological change, however, is often accompanied by behavioral change such as in foraging behavior (in finches), perch-site selection (in lizards), or mating/aggressive display (in fish). In fact, the classical view posits that change in morphology is often preceded by that in behavior—"a shift into a new niche or adaptive zone is, almost without exception, initiated by a change in behavior" (Mayr 1963, p. 604).

Understanding the ultimate causes responsible for generating and maintaining such tremendous variation in both morphology and behavior has been a major theme in both evolutionary biology and behavioral ecology. Many studies attempting to disentangle the relative roles

The work on *Peromyscus* burrowing was done by a team of undergraduates (Dan Brimmer, Alan Chui, Andrew Goldberg, Juani Hoopwood, Adam Weiss, and Stephen Wolff) under the leadership of an enthusiastic and hard-working graduate student, Jesse Weber, who kindly shared preliminary data for inclusion in this chapter. This research was funded by an Arnold and Mabel Beckman Young Investigator Award. Finally, I would like to thank Peter and Rosemary Grant for inspiration, especially in demonstrating how field and laboratory work (both observation and experiment) can be combined to shed new light on evolutionary process.

of different evolutionary forces on diversity are conducted in the field. Such field studies have led to many important insights such as the observation that selection can be very strong (Endler 1986; Kingsolver et al. 2001) and can act over short timescales (Hoekstra et al. 2001; Grant and Grant 2002), leading to rapid evolutionary change. Most recently, ecological studies have reinforced the idea that natural selection plays a major role in speciation and ultimately diversification (Coyne and Orr 2004; Grant and Grant 2008; Schluter 2009).

Recently, a complementary field of study has emerged, one in which elucidating the proximate mechanisms responsible for evolution of diversity is the prime focus. New molecular genetic technologies have enabled genetic architectures of adaptive traits to be characterized in a diverse set of species (Feder and Mitchell-Olds 2003; Abzhanove et al. 2008). These studies are most often performed under controlled conditions in the laboratory. Identifying the genetic basis of underlying adaptive evolution will allow us to address many long-standing questions in evolutionary biology—for example, does adaptation proceed via a few large steps or many small ones? Is genetic dominance important in adaptive change? Where do adaptive alleles come from—preexisting genetic variation or new mutations? While great initial progress has been made in uncovering genes and addressing questions about the genetic basis of adaptative morphological variation, the same tools, techniques, and methods are now being applied with full force to behavioral phenotypes; however, dissecting the molecular basis of behavior comes with a new set of challenges. Yet, it all may be worthwhile because when we eventually have "behavior genes" in hand, we can begin to investigate whether behaviors evolve in a similar way to other traits, such as morphology. And, if not, why not?

In this chapter I will discuss how studies of the ultimate and proximate mechanisms responsible for behavior are not far behind those for morphological evolution. Specifically, in addition to understanding the forces responsible for behavioral evolution, it is now feasible to search for genes that contribute to behavioral differences. Such studies are increasingly being applied to non-model organisms, and it seems likely that we will (soon) be able to uncover alleles that contribute to behavioral variation in the wild.

Linking Genotype to Phenotype

Recently there have been great strides made in identifying developmental pathways, genes and, in some cases, even mutations that contribute to natural variation. While there are many ways to make this connection between gene and trait (Stinchcombe and Hoekstra 2007), arguably one of the most direct and unbiased approaches is quantitative trait locus (QTL) mapping. This method requires that individuals that differ in one or more measurable, heritable traits are interbred, which can be done in controlled laboratory crosses or with pedigree information from field populations (Ellegren and Sheldon 2008). The QTL intervals on a genetic map can then be narrowed by high-resolution mapping, quantitative complementation, or linkage-disequilibrium mapping, all methods that have been described in detail elsewhere (e.g., Mackay 2001). The key advantage of the QTL method, unlike mutant screens, microarray experiments, or candidate-gene approaches (see Nadeau and Frankel 2000), is that it targets the specific genomic regions that contain the mutations which *cause* phenotypic differences and also provide us with the ability to measure how much of the phenotypic variation is explained by allelic variation in that particular genomic region or gene.

In the last decade, this QTL approach has been successful in identifying genes responsible for morphological variation in a wide variety of organisms. Most recently, the precise genes involved in adaptation have been discovered in natural populations. These traits range from pigmentation in animals (Gompel et al. 2005; Hoekstra et al. 2006; Protas et al. 2006; Steiner et al. 2007; Linnen et al. 2009) and plants (Quattrocchio et al. 1999; Zufall and Rausher 2004) to skeletal variation, such as body armor in sticklebacks (Shapiro et al. 2004). Many of these studies have relied on some knowledge of candidate genes, either targeted from the outset or later identified within QTL peaks, which are sometimes plentiful for such morphological traits. But, in at least one case, the *ectodysplasin* locus contributing to lateral plate number sticklebacks, the gene was identified with no prior knowledge of gene function (Colosimo et al. 2005). This elegant study began with a QTL analysis of a large genetic cross maintained in the laboratory, but then narrowed in on the gene of interest using an LD-mapping approach in a field population. The hunt for genes underlying additional ecologically relevant traits in other species is well under way (chapters 6 and 7). Bolstered by this success and a long history of quantitative-genetic analyses of behavior (Boake 1994), the time to take a parallel approach to finding genes that contribute to behavioral variation has come!

EVOLUTIONARY SIGNIFICANCE OF BEHAVIOR

As well as showing exuberant morphological variation, animals also display a stunning variety of behaviors of fundamental importance to their ability to attract mates, find food, avoid predators, and communicate (chapters 5–8). It has been argued that behavior may serve to restrain or limit the course of evolution because an individual can use behavior to avoid the impact of environmental variation (Huey et al. 2003). However, it is also clear that changes in behavior also expose animals to new environments, and thus promote evolutionary divergence. In this way, behavior can ultimately contribute to speciation and diversification. For example, mate choice is an important isolating mechanism that is driven by differences in the behavior of a signaler, receiver, or both whether that signal is visual, auditory, tactile, or olfactory. Thus, behavior clearly plays a leading role in both adaptation and speciation.

GENETICS OF BEHAVIOR

The study of behavioral diversity was pioneered in the field by the classic ethologists—Konrad Lorenz, Niko Tinbergen, and Karl von Frisch—who were interested in understanding both the causes and consequences of behavioral variation across many organisms in their natural environments. One important assumption made by these ethologists was that genes underlie behaviors and that variation in these genes is necessary for behavioral evolution to occur.

Despite the assumption that there is a genetic component to most behaviors, we still know very little about the genetic mechanisms that have given rise to the behavioral diversity found in nature, leaving many fundamental questions unanswered. Here are just a few: What

195

are the relative contributions of genetic and environmental effects (or learning) to behavioral differences? What are the genetic changes that underlie the differences in behavior found both within and between species? Do these genetic changes act early in development to alter neural circuitry or does the circuitry remain constant and changes in gene regulation (e.g., neurotransmitters) underlie behavioral variation? And if so, how? Despite the importance of these questions, the connection between genes, neural circuitry, and the evolution of complex and adaptive behaviors remains a major frontier in biology.

As a first step, researchers have focused on finding genes that affect behavior in laboratory populations, such as psychological traits in mouse strains (e.g., Flint et al. 1995; Yalcin et al. 2004) and a number of behavioral phenotypes that vary between inbred strains of *Drosophila melanogaster* (reviewed in Sokolowski 2002). Technological developments and new resources are making it increasingly possible to complete these connections in genetic model organisms (e.g., in mice, Flint and Mott 2008; and in *Drosophila*, Anholt and Mackay 2004). Importantly, in these model systems, one not only benefits from the tremendous genetic and molecular resources, but one also can carefully control for genetic background, effects of environmental variation, and sexual dimorphism. However, this work often does not address the genetic mechanisms that underlie behavioral *evolution*. For this, we have to turn to natural variation.

In a few cases, genes causing natural variation in behavior have been identified in genetic model organisms, such as *Drosophila*. For example, the *foraging* gene has two widespread alleles, termed rover (*for^R*) and sitter (*for^s*), which affect the movement patterns of *Drosophila* larvae between feeding patches (Osborne et al. 1997; Sokolowski 1980, 2001). It is thought that these alleles are both maintained in natural populations by frequency-dependent selection—the rarer of the two behavioral strategies is at an advantage in poor nutrient conditions (Fitzpatrick et al. 2007). A second case involves amino acid changes in the *period* gene that underlie temporal differences in species-specific courtship song between *Drosophila* species. Functional assays show that transferring the *simulans* allele into the *melanogaster* genome results in a *melanogaster* fly that sings a *simulans* song (Wheeler et al. 1991). Both of these studies demonstrate how even changes in single genes—segregating either within or between species—can have large effects on behavior.

Only a few exceptional studies have identified genes underlying natural behavioral variation in traditionally non-model organisms. Remarkably, amino acid changes in a pheromone binding protein (*General Protein 9*; *Gp9*) are responsible for a social polymorphism—the number of queens tolerated in a colony—in fire ants (Krieger and Ross 2002; Wang et al. 2008). And perhaps, the only example in vertebrates of a genetic change known to contribute to natural behavioral differences is the transition from a polygamous to more monogamous behavior in two species of voles. Changes in the length of a microsatellite in the upstream regulatory region of the *vasopressin-1 receptor* (*v1ar*) cause its differential ViaR expression in specific regions of the brain, which in turn are associated with changes in male affiliative behavior (reviewed in Donaldson and Young 2008). Despite these few compelling examples, genetic studies of natural behavioral variation have been limited, largely due to the challenges associated with dissecting behavior genetically.

THE CHALLENGES

When compared to the study of the genetic basis of morphological traits, the study of behavior shares many of the same obstacles but also introduces a host of new ones. First, at one extreme, some behaviors that vary within or differ between species may not have any heritable basis, and may instead reflect learning or "cultural" inheritance. Is the remarkable tool-use of woodpecker finches (*Camarhynchus pallida*)—the use of a twig or cactus spine to dislodge insect prey from trees—an innate behavior or one that is learned socially (Tebbich et al. 2001)? We must also be concerned with experience or the learning process itself, which may confound the repeatability of some behaviors. And, repeatability of behavior may also change depending on internal conditions, such as motivation. Second, many behaviors are likely to be influenced both by genes and by environment; in fact, many behaviors *require* an environmental stimulus. These environmental effects can act both during development, when neural networks are being laid down, and/or during the real-time performance of behavior. Of course, environmental effects are especially relevant for social behaviors, which by definition involve the interaction among individuals (Robinson et al. 2008). Thus, the importance of the interaction between genes and environment, and their relative roles, remain unclear for many behaviors.

There are also practical concerns to the study of behavior, that is, being able to study an evolutionary interesting behavior in a tractable system. It seems as though many of the most "interesting" behaviors—such as the artful mating displays of a bird of paradise or the deceptive luring of an anglerfish—occur in species with few genetic resources and/or that would be difficult to study in a controlled laboratory environment. And even if such traits could be studied in a laboratory, many behaviors are thought to have a complex genetic architecture that works via an intricate network of gene interactions in the nervous system, thus requiring large sample sizes and an ability to efficiently measure behaviors (although it has been argued that QTL effect sizes for behavioral and morphological traits may be similarly distributed; Flint and Mott 2008). Perhaps one of the largest challenges is just that—measurability; that is, our ability to precisely define and quantify behavior. Many behaviors, like courtship display, are modular—comprised of a series of smaller components, which together make up what we often think of as a single complex behavior. How to quantify these components together or to choose the most biologically relevant one is not always straightforward.

Finally, like the study of morphological traits, we are faced with the challenges of complex architectures, difficulty in identifying genetic regions if they have small effects, and of course actually finding the genes and understanding their function both at a biochemical and organismal level. But, our ultimate ability to find genes underlying behaviors, as for morphologies, will allow us to ask big questions like: how do behaviors evolve at the molecular level?

WHAT IS A "BEHAVIOR GENE"?

The search for genes that affect behavior is now well under way. Yet, it is still worthwhile to pause and ask, what is it that we are searching for? When we talk about a "behavior gene," what do we mean exactly? In most animals, the proper functioning of hundreds, if not thousands, of genes are necessary to perform a behavior, yet we don't often think of *all* these genes as "behavior genes." For example, let us consider a complex behavior like an avian courtship display. A mutation in, let's say, a bone morphogenic protein (BMP) that disrupts proper wing bone development will likely affect courtship display, but a BMP is not what we consider to be a "behavior gene." Thus, while large-scale mutagenesis experiments are useful in producing a list of candidate genes—genes that are necessary to perform a behavior—this approach may be less likely to be relevant to behavioral variation in natural populations. Similarly, there are likely to be changes in the expression of many genes before (even in early development) and during the performance of a behavior. For example, different social situations can radically change the global expression profiles in the brains of swordtail fish (Cummings et al. 2008) and songbirds (Replogle et al. 2008). However, changes in expression of these genes are *correlated* with behavior (see Schadt et al. 2005). Some may consider any gene that affects the expression of a behavioral pattern to be a behavior gene, but to others, the holy grail is to unearth a locus that harbors genetic differences that *causes* a change in behavior between individuals, populations, or species in nature.

Although daunting, one of the most exciting aspects of searching for the causal behavior genes is that we don't have any clear expectations of what we may find—where these causal genes are expressed and when, what their function is, or how they cause changes in behavior. These questions are more mechanistic, but having these genes in hand can also allow us to better understand how behaviors evolve. In any particular system, we may, for example, examine the temporal or spatial distributions of alleles to test for evidence of selectionas well as reconstruct the origin of these alleles. More generally, we also can explore the constraints that may limit (or allow) behaviors to evolve.

Are "Behavior Genes" Specific to Behavior?

The term behavior gene can be misleading because it seems to imply that it functions only to control behavior. However, it is clear from behavioral screens for mutants that most genes that affect behavior also have pleiotropic effects on other traits (Pflugfelder 1998). One notable exception may be *fruitless* in *Drosophila*, whose sex-specific

199

transcripts seem to play a dedicated role in initiating sex-specific behavior (Baker et al. 2001). However, the observation that many behavioral mutants also seem to have developmental abnormalities (see supp. table 1 in Sokolowski 2002) has two interesting implications: (1) that genes that contribute to natural variation in behavior may often act during development, when neural circuits are first being generated, rather than act physiologically during the functioning of behavior; and (2) many genes that could contribute to natural variation in behavior may be constrained by deleterious pleiotropic consequences, suggesting that there may, in fact, be relatively few targets of behavioral evolution.

Do the Same Genes Contribute to Similar Behaviors in Different Taxa?

With the recent ease of sequencing complete genomes in a variety of organisms, it has become clear that even distantly related animals such as fruit flies and humans share a large complement of gene homologues. It has also been argued that neural circuits and mechanisms are also likely to be conserved in animals (Robinson et al. 2008; Donaldson and Young 2008). The next question, however, is: do changes in these genes—either their amino acid sequence or their temporal or spatial expression—have similar effects on behavior? And, are the same genes the repeated target of evolutionary change? While morphological changes have often, but not always, evolved through changes in the same genes (Ardent and Reznick 2008), a few recent studies hint that this also may be the case for behavioral change.

In two cases, the same genes are associated with similar behaviors in diverse animals. First, changes in the expression level of the *foraging* gene, which in *Drosophila* affects foraging activity of larvae, in honey bees (*Apis mellifera*) affects social activity (Ben-Sharar et al. 2002). Specifically, higher mRNA expression of *for* is found in the brains of forager bees compared to nurse bees. Thus, this gene, which causes a polymorphism in behavioral strategy in *Drosophila*, is also associated with an ontogentic transition in honey bees from working in the hive to foraging outside the hive. It has been suggested that socially induced changes in *for* expression may be widespread in social insects (Toth and Robison 2007). In addition, the homolog of *for* in nematodes

(*Caenorhabditis elegans*) influences food-related locomotion (Fujiwara et al. 2002), leading others to suggest that *for* is a strong candidate for any food-related behaviors (Fitzpatrick and Sokolowski 2004). Second, proper functioning of the forkhead box P2 gene (*foxp2*) seems to be essential for animal communication, from human speech (Lai et al. 2001; Enard et al. 2002) to song learning in zebra finch (*Taeniopygia guttata*; Haesler et al. 2004). While these studies suggest that the same genes have similar functions across taxa, these candidate genes are either associated with changes in behavior or necessary to elicit proper behavior, but not likely to directly encode changes in behavior.

In the case of the *vasopressin-1 receptor*, mutations in the locus itself contribute to differences in social behavior between two vole species. Although it has been shown that the precise mutation (i.e., length variation in an upstream microsatellite) may not be responsible for differences in social behavior among all vole species (Fink et al. 2006), there is some suggestion that mutations in *v1ar* may affect human behavior. Recently a study reported an association between *v1ar* allelic variation and partner fidelity in a large Swedish population—men homozygous for the *v1ar* variant allele were twice as likely to experience marital discord (Walum et al. 2008). If this pattern of convergent evolution indeed holds true for many other genes, it suggests that as more genes are identified in our favorite model organisms, they also may serve as strong candidates for natural variation in other species (Fitzpatrick et al. 2005).

Genes Contributing to Natural Variation

Ideally then, evolutionary geneticists are most interested in finding the precise genes that cause variation in behavior between individuals or species that in turn translates to fitness differences in nature. No small task! This goal requires working across disciplines ranging from ecology to molecular genetics in species that are amenable to both field and lab study. In the past, a handful of laboratory-based behaviors in a few genetic model organisms have dominated the fields of behavioral genetics and neuroscience, while behavioral ecology focused on the evolution of naturally occurring behavior with little interest in the underlying genetics. While both these fields have made great progress, a more complete picture of the link between genes and behavior will require the integration of these approaches in novel systems (Boake et al. 2002). This goal is now a possibility in part because of the growing number of developmental and genomic resources available for diverse species (Abzhanov et al. 2008), buttressed by a rich history of detailed field and natural history studies in some species. Although this is a tall order, many recent and ongoing studies are attempting to do just this. To illustrate how natural variation can be dissected, I will use an example from our own research, although there are certainly other examples from which to choose.

Just as the natural history studies of David Lack (1947) on the Galápagos laid the groundwork for the Grants' studies of Darwin's finches, Francis Sumner's studies of natural variation in deer mice set the stage for work on adaptation in rodents. While many of Sumner's classic studies described geographic variation in morphology (Sumner 1926, 1930), he also described some behavioral variation. Most notably, Sumner reported on the unique and stereotyped burrows produced by oldfield mice (*Peromyscus polionotus*) throughout their range in the southeastern United States (Sumner and Karol 1929). Specifically, oldfield mice construct "complex" burrows consisting of a ~60cm entrance tunnel, a large nest chamber, and an escape tunnel that almost reaches the surface (fig. 10.1a). Although not formally tested, it has been hypothesized that this burrow design, and the presence of an escape tunnel in particular, is a behavioral adaptation to escape predators, especially snakes (Hayne 1936). Remarkably, burrowing behavior is quite different in its sister species, *P. maniculatus*, which produces shallow burrows (~10cm) with no escape tunnel (fig. 10.1b). While many field studies have reported interesting behavioral differences either within or between species, it is the ability to measure this behavior in a controlled laboratory environment, the ease of performing large genetic crosses, and the availablility of genomic resources for these species that have made this system so tantalizing.

Nonetheless, there were still major questions to answer. Can these behaviors be recapitulated in the laboratory and reliably measured? Are these behavioral differences heritable? And, if so, is the genetic basis tractable enough that a gene eventually could be found? Finally, is complex burrow a derived trait—so that we may expect to find a

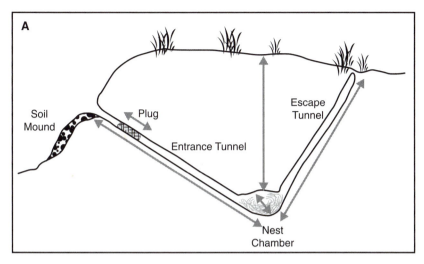

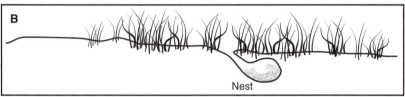

Figure 10.1. Differences in burrow design between two interfertile sister species of *Peromyscus*: (a) Large complex burrows of *P. polionotus* and (b) small simple burrows of *P. maniculatus*. Modified from Dawson et al. (1988).

gene contributing to behavioral novelty rather than behavioral loss? Many of the initial clues came from a simple yet elegant study by Wally Dawson and colleagues. Dawson et al. (1988) demonstrated (1) burrows could be measured in laboratory conditions, (2) the species-specific burrow behaviors of *polionotus* and *maniculatus* were retained after many years of captive breeding, and (3) using a small genetic cross between *polionotus* and *maniculatus,* the complex *polionotus* burrows were genetically dominant over the simple *maniculatus* burrows.

To follow up on these initial studies, we first refined a method to measure burrowing behavior. Specifically, we make polyurethane molds of each burrow (following Felthauser and McInroy 1983), thus visualizing an "extended phenotype" (à la Dawkins 1999). The resulting cast is a physical representation of a behavior, thus ameliorating many of the concerns about measuring artificial components of behavior; casts

203

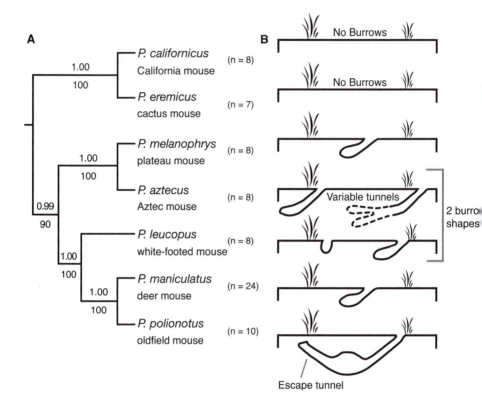

Figure 10.2. Evolution of burrowing in the genus *Peromyscus*. (a) Molecular phylogeny of seven species, representing nine taxa. Numbers above and below the nodes represent posterior probabilities and maximum likelihood bootstrap values, respectively. Precise branch lengths are not shown. (b) The most commonly observed burrow shape of each species. Burrows are not drawn to scale. Sample sizes for each taxa are provided. The two species that regularly produced two burrow shapes are bracketed, and both burrow types are shown. Modified from Weber and Hoekstra (2009).

are measured as any other simple morphological trait. Second, we have shown that burrowing (i.e., burrow size and shape) is highly repeatable at two levels—within a single individual that is allowed to dig on multiple trials and among individuals within species (Weber and Hoekstra 2009). Third, by measuring burrowing in several *Peromyscus* species and examining the results in a phylogenetic context, we have shown that the complex burrowing phenotype is both unique to *polionotus* and derived (fig. 10.2; Weber and Hoekstra 2009). Thus, genetic analyses between *polionotus* and *maniculatus* are likely to reveal genes that contribute to a gain in a complex behavior (rather than its loss). Together, these results lay the groundwork for our QTL studies, in which we are "digging" for the genes important in the evolution of burrow design.

Although preliminary, QTL mapping experiments suggest several interesting results. First and foremost, QTL regions for burrowing can be identified! Some of these initial QTLs explain a large (>10%) percentage of variation in burrow shape. Moreover, different genomic regions contribute to distinct aspects of burrow design (e.g., total length, entrance tunnel length, and presence/absence of an escape tunnel), suggesting that different genes control specific components of this complex behavior (fig. 10.3). For example, the presence of an escape tunnel maps to a genetic region (unlinked marker) that is unique to regions contributing to either total or entrance tunnel length. Although this work is still in its infancy, it is remarkable that we are already unearthing multiple large-effect genetic regions that contribute to differences in burrow size and shape in *Peromyscus*.

With this example, I hope that two complementary points have been highlighted: (1) there are many requirements for a behavior to be successfully dissected at the genetic level, and (2) although these requirements may be both numerous and stringent, if met, there is real promise for finding the elusive behavior gene in nature.

THE FUTURE

Burrowing, of course, represents one (relatively simple) trait in a single species. Understanding the evolution of behavior and its underlying genetic basis will require synthesizing knowledge from a wide variety of behaviors in a diverse set of animals from flies to mice to perhaps even Darwin's finches. In fact, such exciting studies are currently under way in natural populations of *Drosophila*, crickets, songbirds, sticklebacks, cavefish, and cichlids. While elucidating the mechanistic details of how behaviors evolve in nature is still in its infancy, it is exciting to imagine drawing the complete picture of how behaviors evolve—from the evolutionary forces driving differences to their molecular details—and finally to help explain "How and Why Species Multiply."

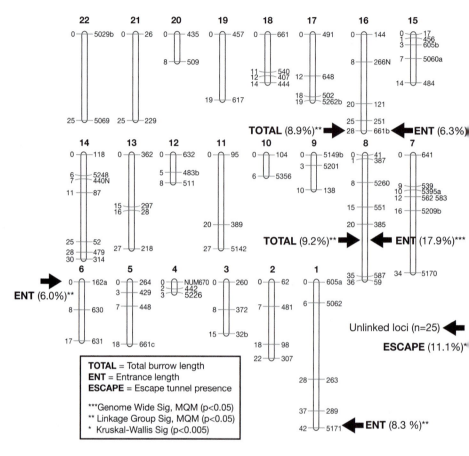

Figure 10.3. The genetic architecture of burrow design in *Peromyscus*. A preliminary genetic linkage map (22 linkage groups covering 447cM) constructed from 110 microsatellite loci in 93 individuals [backcross: *maniculatus* x (F1 *polionotus* x *maniculatus* hybrids)]. Genetic distance in cM is shown on the left of each chromosome, and microsatellite marker number on the right. Twenty-five loci are unlinked. QTL for three burrow traits are shown with the percent variation explained by each locus given in parentheses. Significance values of QTL are provided.

References

Abzhanov, A., et al. 2008. Are we there yet? Tracking the development of new model systems. *Trends in Genetics* 24(7): 353–360.

Anholt, R. H., and T.F.C. Mackay. 2004. Quantitative genetic analyses of complex behaviours in *Drosophila*. *Nature Reviews Genetics* 5: 838–849.

Ardent, J., and D. N. Reznick. 2008. Convergence and parallelism reconsidered: what have we learned about the genetics of adaptation? *Trends in Ecol. Evol.* 23: 26–32.

Baker, B. S., B. J. Taylor, and J. C. Hall. 2001. Are complex behaviors specified by dedicated regulatory genes? Reasoning from *Drosophila. Cell* 105: 13–24.

Ben-Sharar, Y., et al. 2002. Influence of gene action across different time scales on behaviour. *Science* 296: 741–744.

Boake, C.R.B. 1994. *Quantitative Genetic Studies of Behavioural Evolution.* Chicago: University of Chicago Press.

Boake, C.R.B. et al. 2002. Genetic tools for studying adaptation and the evolution of behavior. *Am. Nat.* 160: S143–159.

Colosimo, P. F., K. E. Hoemann, S. Balabhadra, G. Villarreal, M. Dickinson, J. Grimwood, J. Schmutz, R. M. Myers, D. Schluter, and D. M. Kingsley. 2005. Widespread parallel evolution in sticklebacks by repeated fixation of ectodysplasin alleles. *Science* 307: 1928–1933.

Coyne, J. A., and H. A. Orr. 2004. *Speciation.*Sunderland, MA: Sinauer Associates.

Cummings, M. E. et al. 2008. Sexual and social stimuli elicit raid and contrasting genomic responses. *Proc. Natl. Acad. Sci. USA* 275: 393–402.

Dawkins, R. 1999. *The Extended Phenotype: the Long Reach of the Gene.* New York: Oxford University Press.

Dawson, W. D., C. E. Lake, and S. S. Schumpert. 1988. Inheritance of burrow building in *Peromyscus. Behav. Genet.* 18: 371–382.

Donaldson, Z. R., and L. J. Young. 2008. Oxytocin, vasopressin and the neurogenetics of sociality. *Science* 322: 900–904.

Enard, W. et al. 2002. Molecular evolution of FOXP2, a gene involved in speech and language. *Nature* 418: 869–872.

Endler, J. A. 1986. *Natural Selection in the Wild.* Monographs in Population Biology 21. Princeton, NJ: Princeton University Press.

Ellegren, H., and B. C. Sheldon. 2008. Genetic basis of fitness differences in natural populations. *Nature* 452: 169–175.

Feder, M. E., and T. Mitchell-Olds. 2003. Evolutionary and ecological functional genomics. *Nature Reviews Genetics* 4(8): 651–657.

Felthauser, M., and D. McInroy. 1983. Mapping pocket gopher burrowing systems with expanding polyurethane foam. *J. Wildl. Manage.* 47: 555–558.

Fink, S., L. Excoffier, and G. Heckel. 2006. Mammalian monogamy is not controlled by a single gene. *Proc. Natl. Acad. Sci. USA* 103: 10956–10960.

Fitzpatrick, M. J., E. Feder, L. Rowe, and M. B. Sokolowski. 2007. Maintaining a behaviour polymorphism by frequency-dependent selection on a single gene. *Nature* 447: 210–215.

Fitpatrick, M. J., et al. 2005. Candidate genes for behavioural ecology. *Trends in Ecol. Evol.* 20(2): 96–104.

Fitzpatrick, M. J., and M. B. Sokolowski. 2004. In search of food: exploring the evolutionary link between cGMP-dependent protein kinase (PKG) and behavior. *Integrative and Comparative Biology* 44: 28–36.

Flint, J., et al. 1995. A simple genetic basis for a complex psychological trait in laboratory mice. *Science* 269: 1432–1435.

Flint, J., and R. Mott. 2008. Applying mouse complex-trait resources to behavioural genetics. *Nature* 456: 724–727.

Fujiwara, M., et al. 2002. Regulation of body size and behavioural state of *C. elegans* by sensory perception and the EGL-4 cGMP-dependent protein kinase. *Neuron* 36: 1091–1102.

Gompel, N., B. Prud'homme, P. J. Wittkopp, V. A. Kassner, and S. B. Carroll. 2005. Chance caught on the wing: *cis*-regulatory evolution and the origin of pigment patterns in *Drosophila*. *Nature* 433: 481–487.

Grant, P. R. 1999. *Ecology and Evolution of Darwin's Finches*. 2nd ed. Princeton, NJ: Princeton University Press.

———. 2001. Reconstructing the evolution of birds on islands: 100 years of research. *Oikos* 92: 385–403.

Grant, P. R., and B. R. Grant. 2002. Unpredictable evolution in a 30-year study of Darwin's finches. *Science* 296: 707–711.

———. 2008. *How and Why Species Multiply: The Radiation of Darwin's Finches*. Princeton, NJ: Princeton University Press.

Haesler, S., et al. 2004. FoxP2 expression in avian vocal learners and non-learners. *J. Neurosci.* 24: 3164–3175.

Hayne, D.W. 1936. Burrowing habits of *Peromyscus polionotus*. *J. Mammal.* 17: 420–421.

Hoekstra, H. E., R. J. Hirschmann, R. A. Bundey, P. A. Insel, and J. P. Crossland. 2006. A single amino acid mutation contributes to adaptive beach mouse color pattern. *Science* 313: 101–104.

Hoekstra, H. E., J. M. Hoekstra, D. Berrigan, S. N. Vignieri, C. E. Hill, A. Hoang, P. Beerli, and J. G. Kingsolver. 2001. Strength and tempo of directional selection in the wild. *Proc. Natl. Acad. Sci. USA* 98(16): 9157–9160.

Huey, R. B., P. E. Hertz, and B. Sinervo. 2003. Behavioral drive versus behavioral inertia in evolution: a null model approach. *Am. Nat.* 161: 357–366.

Kingsolver, J. G., H. E. Hoekstra, J. M. Hoekstra, D. Berrigan, S. N. Vignieri, C. E. Hill, A. Hoang, P. Gibert, and P. Beerli. 2001. The strength of phenotypic selection in natural populations. *Am. Nat.* 157: 245–261.

Krieger, M.J.B., and K. G. Ross. 2002. Identification of a major gene regulating complex social behavior. *Science* 195: 328–332.

Lack, D. 1947. *Darwin's Finches*. Cambridge: Cambridge University Press.

Lai, C.S.L., et al. 2001. A forkhead-domain gene is mutated in a severe speech and language disorder. *Nature* 413: 519–523.

Linnen, C. R., E. P. Kingsley, J. D. Jensen, and H. E. Hoekstra. 2009. On the origin and spread of an adaptive allele in deer mice. *Science* 325: 1095–1098.

Mackay, T.F.C. 2001. The genetic architecture of quantitative traits. *Annu. Rev. Genet.* 35: 303–339.

Mayr, E. 1963. *Animal Species and Evolution*. Cambridge, MA: Harvard University Press.

Nadeau, J. H., and W. N. Frankel. 2000. The roads from phenotypic variation to gene discovery: mutagenesis versus QTLs. *Nature Genetics* 25: 381–384.

Osborne, K. A., et al. 1997. Natural behavior polymorphism due to a cGMP-dependent protein kinase of *Drosophila*. *Science* 277: 834–836.

Pflugfelder, G. O. 1998. Genetic lesions in *Drosophila* behavioural mutants. *Behav. Brain Res.* 95: 3–15.

Protas, M. E., C. Hersey, D. Kochanek, Y. Zhou, H. Wilkens, W. R. Jeffery, L. I. Zon, R. Borowsky, and C. J. Tabin. 2006. Genetic analysis of cavefish reveals molecular convergence in the evolution of albinism. *Nature Genetics* 38: 107–111.

Quattrocchio F., J. Wing, K. van der Woude, E. Souer, N. de Vetten, J. Mol, and R. Koes. 1999. Molecular analysis of the anothcyanin2 gene of petunia and its role in the evolution of flower color. *Plant Cell* 11: 1433–1444.

Replogle, K., et al. 2008. The songbird neurogenomics (SoNG) initiative: community-based tools and strategies for study of brain gene function and evolution. BMC *Genomics* 9: 131.

Robinson, G. E., R. D. Fernald, and D. F. Clayton. 2008. Genes and social behavior. *Science* 322: 896–900.

Schadt, E. E., et al. 2005. An integrative genomics approach to infer causal associations between gene expression and disease. *Nature Genetics* 37: 710–717.

Schluter, D. 2009. Evidence for ecological speciation and its alternative. *Science* 323: 737–741.

Shapiro, M. D., M. E. Marks, C. L. Peichel, B. K. Blackman, K. S. Nereng, B. Jonsson, D. Schluter, and D. M. Kingsley. 2004. Genetic and developmental basis of evolutionary pelvic reduction in threespine sticklebacks. *Nature* 428: 717–723.

Sokolowski, M. B. 1980. Foraging strategies of *Drosophila melanogaster*—a chromosomal analysis. *Behavior Genetics* 10: 291–302.

———. 2001. *Drosophila*: genetics meets behaviour. *Nature Reviews Genetics* 2: 879–890.

———. 2002. Neurobiology—Social eating for stress. *Nature* 419: 893–894.

Steiner, C. C., J. N. Weber, and H. E. Hoekstra. 2007. Adaptive variation in beach mice produced by two interacting pigmentation genes. *PLoS Biology* 5(9): 1880–1889.

Stinchcombe, J. R., and H. E. Hoekstra. 2007. Combining population genomics and quantitative genetics: finding the genes underlying ecologically important traits. *Heredity* 100: 158–170.

Sumner, F. B. 1926. An analysis of geographic variation in mice of the *Peromyscus polionotus* group from Florida and Alabama. *J. Mammal.* 7: 149–184.

———. 1930. Genetic and distributional studies of three subspecies of *Peromyscus*. *J. Mammal.* 23: 275–376.

Sumner, F. B., and J. J. Karol. 1929. Notes of the burrowing habits of *Peromyscus polionotus*. *J. Mammal* 10: 213–215.

Tebbich, S., M. Taborsky, B. Fessl, and D. Blomqvist. 2001. Do woodpecker finches acquire tool use by social learning? *Proc. R. Soc. B* 268: 2189–2193.

Toth A. L., and G. E. Robison. 2007. Evo-devo and the evolution of social behavior. *Trends in Genetics* 23(7): 334–341.

Walum, H., et al. 2008. Genetic variation in the vasopressin receptor 1a gene (AVPR1A) associates with pair-bonding behavior in humans. *Proc. Natl. Acad. Sci. USA* 105: 14153–14156.

Wang, J., K. G. Ross, and L. Keller. 2008. Genome-wide expression patterns and the genetic architecture of a fundamental social trait. *PLoS Genetics* 4: e1000127.

Weber, J. N., and H. E. Hoekstra. 2009. The evolution of burrowing behavior in deer mice. *Animal Behavior* 77: 603–609.

Wheeler, D. A., et al. 1991. Molecular transfer of a species-specific behavior from *Drosophila simulans* to *Drosophila melanogaster*. *Science* 251: 1082–1085.

Yalcin, B., et al. 2004. Genetic dissection of a behavioral quantitative trait locus shows that Rgs2 modulates anxiety in mice. *Nature Genetics* 36: 1197–1202.

Zufall, R.A., and M. D. Rausher. 2004. Genetic changes associated with floral adaptation restrict future evolutionary potential. *Nature* 428: 847–850.

Chapter Eleven

There Must Be Finches—Charles Darwin, Prickly Pears, and Pollination Biology

May R. Berenbaum

Among the most paradigmatic groups in all of evolutionary biology are the birds collectively known as Darwin's finches—an assemblage of about a dozen species of emberizine passerines restricted to the Galápagos Islands. Although they were first observed and collected almost 170 years ago by Charles Darwin, he actually wrote very little about them. In fact, they did not become known as "Darwin's finches" until about a century after they were first described by Darwin's colleague, John Gould. Truth be told, with regard to eponyms, inasmuch as most of what is known about these birds today is the result of three decades of thorough, dedicated, and insightful investigation by Peter and Rosemary Grant, a compelling argument could be made for calling them "Grants' finches." It is the work of the Grants that ultimately conferred iconic status on these birds.

Darwin first encountered his namesake birds during his five-week sojourn in the Galápagos in September 1835. He was originally uncertain as to their identity and classification and was not sure how many species were represented in the archipelago. Thus, after his return, in 1837, he contacted John Gould, a prominent ornithologist of the era; Gould is the one who realized that Darwin's motley assortment of grosbeaks, blackbirds, and finches in fact consisted entirely of finches. In 1837, Darwin introduced the scientific world to these birds with

Deep appreciation goes to Peter and Rosemary Grant, both for their invitation to contribute this chapter and for their patience in dealing with its deadline-impaired author. Preparation of this chapter, as well as some of the results reported, were supported by NSFDEB0816616.

some "Remarks upon the Habits of the Genera Geospiza, Camarhynchus, Cactornis, and Certhidea of Gould," published in the *Proceedings of the Zoological Society of London*. They next appeared in print in 1839, in his *Narrative of the Surveying Voyages of His Majesty's Ships Adventure and Beagle*, written with Robert Fitzroy. Darwin reported:

> 9th. A group of finches, of which Mr. Gould considers there are thirteen species; and these he has distributed into four new subgenera. These birds are the most singular of any in the archipelago. They all agree in many points; namely, in a peculiar structure of their bill, short tails, general form, and in their plumage. The females are gray or brown, but the old cocks jet-black. All the species, excepting two, feed in flocks on the ground, and have very similar habits. It is very remarkable that a nearly perfect gradation of structure in this one group can be traced in the form of the beak, from one exceeding in dimensions that of the largest grosbeak, to another differing but little from that of a warbler.

The observation of a "nearly perfect gradation of structure" was a crucial element in Darwin's formation of a theory of evolution by common descent. Although Darwin wrote little else about these birds, they have inspired dozens of studies and today remain a compelling example of how the process of speciation proceeds.

But the finches were important in another context as well. Darwin edited a series of books after returning home, including, in 1841, a volume written by Gould and titled *Birds Part 3 No. 5 of The Zoology of the Voyage of* H.M.S. *Beagle*. In this book appeared the first description of *Cactornis scandens* (now *Geospiza scandens*). In addition to the obligatory Latinate accounts of morphology were Darwin's remarkably acute observations of the ecology and behavior of this bird:

1. CACTORNIS SCANDENS. Gould.
PLATE XLII.
C. intensè fuliginosa, crisso albo; rostro et pedibus nigrescenti-brunneis.
Long. tot. 5 unc.; rostri, 3/4; alæ, 2 5/8; caudæ, 1 3/4; tarsi, 3/4.

Fœm. Corpore superiore, gutture pectoreque intensè brunneis, singulis plumis pallidiorè marginatis; abdomine crissoque cinereis, stramineo tinctis; rostro pallidè fusco; pedibus nigrescenti-fuscis. Deep sooty black, with the under tail-coverts white; the bill and feet blackish-brown.

Female: Upper surface of the body, throat and breast intensely brown, with the margins of each feather paler; the abdomen and the under tail coverts cinereous, tinged with straw-colour; the bill pale fuscous, and the feet blackish fuscous.

Habitat, Galápagos Archipelago, (James' Island.)

The species of this sub-genus alone can be distinguished in habits from the several foregoing ones belonging to Geospiza and Camarhynchus. *Their most frequent resort is the Opuntia Gala-pageia, about the fleshy leaves of which they hop and climb, even with their back downwards, whilst feeding with their sharp beaks, both on the fruit and flowers.* Often, however, they alight on the ground, and mingled with the flock of the above mentioned species, they search for seeds in the parched volcanic soil.

Although neither Darwin nor Gould probably recognized it at the time, the cactus finch is the principal and in some places sole pollinator of the Galápagos prickly pear cactus *Opuntia*, and this account is among the first to describe flower-feeding by birds.

Whether observing the flower-feeding behavior of the cactus finch on *Opuntia* piqued Darwin's interest in pollination biology is unknown, but the account of this behavior appeared in press at a critical juncture in his life. In 1841, the same year *Birds Part 3* was published, Darwin's friend Robert Brown, First Keeper of Botany at the British Museum, gave him a copy of a book by Christian Konrad Sprengel written almost a half-century earlier. Sprengel had been a teacher in Berlin with an all-consuming interest in what today would be called floral ecology. In 1793, he published an account of his experiments with flowers in a book titled, *Das entdeckte Geheimnis der Natur im Bau und in der Befruchtung der Blumen.* Sprengel was among the first to recognize that many floral features, including color patterns and nectar production, function primarily as rewards for insects. As he described (in translation):

> The longer I pursued this investigation, the more I perceived that those Flowers which contain nectar are so constructed that insects can get at it very easily, but that the rain cannot spoil it. I therefore conclude that the nectar of these flowers is secreted, at least principally for the sake of insects, and in order that they may enjoy it pure and unspoiled, it is protected against the rain.
>
> In the following summer I investigated the forget me not (*Myosotis palustris*). I found that this flower not only has nectar, but also the nectar is fully protected against the rain. But at the same time I was struck by the yellow ring which surrounds the mouth of the corolla tube, and contrasts so beautifully with the sky-blue limb of the corolla. Might possibly, thought I, this circumstance also relate to insects? Might nature perhaps have colored the ring for the special purpose of showing insects the way to the nectar receptacle? With this hypothesis in mind I examined other flowers and found most of them corroborated it.

His brilliant insights on the nature of the interactions between flowers and insects almost died with him. The book did not sell widely and he died essentially in obscurity, having failed to convince the scientific public of the era that plants are sexual beings that consort with various and sundry insects for sexual gratification, a scandalous notion at the time. This book, however, proved to be Darwin's inspiration to investigate plant pollination and fertilization. According to Darwin's son Francis, "It may be doubted whether Robert Brown ever planted a more fruitful seed than in putting such a book in such hands" (Darwin 1902).

For developing an interest in flowers, the timing couldn't have been better. When William Hooker was appointed as the first director of Kew in 1841, he declared orchids a priority, and by 1848, Kew Gardens could boast of a collection of over 750 species. Darwin's letters reflected a growing interest in flowers; in 1846, for example, he wrote a letter to Hooker about orchids and other plants he had collected, including a Calceolaria he described as a "wonderful production of nature!" (Letter 976—Darwin, C. R. to Hooker, J. D., [19 May 1846]).

After the success of his *Voyage of the* Beagle, Darwin focused his effort on geological work, publishing prolifically on coral reefs and on fossils and living barnacles. But the publication in 1856 of a paper by his friend and colleague, Thomas H. Huxley, provided an intellectual challenge for Darwin. In his discourse on *Natural History as Knowledge Discipline and Power*, Huxley mused, "Who has ever dreamed of finding an utilitarian purpose in the forms and colours of flowers, in the sculpture of pollen-grains." In Darwin's mind, his developing theory of natural selection argued explicitly for a utilitarian purpose for those forms and colors, and a substantial portion of the next twenty years of his life was dedicated to ascertaining those purposes.

Darwin began his investigations into pollination biology not with some "wonderful production of nature" but with the common kidney bean flower. In 1857, Darwin described the fertilization of kidney beans by bees in a brief note in *Gardeners' Chronicle,* pointing out that "the brush on the pistil, its backward and forward curling movement, its protrusion on the left side, and the constant alighting of the bees on the same side, were not accidental coincidences but were connected with, perhaps necessary to, the fertilization of the flower." Darwin wrote one more paper on kidney beans, but his ruminations on pollination profoundly influenced the development of his theory of natural selection, to the point that it was specifically mentioned in the introduction of the *Origin of Species* as an example of "how the innumerable species, inhabiting this world have been modified, so as to acquire that perfection of structure and coadaptation which justly excites our admiration"—specifically, in the case of the mistletoe, "how flowers with separate sexes absolutely requiring the agency of certain insects to bring pollen from one flower to the other."

After the tumult following the publication of *Origin of Species,* Darwin busied himself with studies of pollination, publishing a series of papers in *Gardeners' Chronicle* on orchids, periwinkles, and primroses. A visit to Torquay in the summer of 1861 whetted his interest in orchids, which grew wild in profusion by the sea. Back at home in Down, he sought out orchids in the neighborhood and designated a particularly rich spot "Orchid Bank." He devoted his attention to elucidating their hitherto undescribed pollination mechanisms, finding many remarkable examples of adaptation and natural selection in the process.

In May 1862, this work culminated in the publication of *On the various contrivances by which British and foreign orchids are fertilised by insects and the good effects of intercrossing*; the book presented a compelling argument against natural theology and dramatic vindication of Sprengel's interpretations of the sexual life of flowers and the contributions of animal pollinators to plant reproduction. Together, the *Origin of Species* and *Various Contrivances* provided the foundation for the field of pollination biology.

All told, Darwin ultimately produced close to two dozen publications, including two books, on plant pollination and fertilization. The enormous impact he had on the development of the field of pollination biology often goes unrecognized; rather, his writings on pollination biology are cited in the context of illustrating the tremendous predictive power of the theory of natural selection. One example in particular is a textbook favorite—his speculation on the significance of the structure of a Madagascar orchid:

> I must say a few words on the *Angræcum sesquipedale*, of which the large six-rayed flowers, like stars formed of snow-white wax, have excited the admiration of travellers in Madagascar. A whip-like green nectary of astonishing length hangs down beneath the labellum. In several flowers sent me by Mr. Bateman I found the nectaries eleven and a half inches long, with only the lower inch and a half filled with very sweet nectar. What can be the use, it may be asked, of a nectary of such disproportional length? We shall, I think, see that the fertilisation of the plant depends on this length and on nectar being contained only within the lower and attenuated extremity. It is, however, surprising that any insect should be able to reach the nectar: our English sphinxes have probosces as long as their bodies: but in Madagascar there must be moths with probosces capable of extension to a length of between ten and eleven inches!

Less than 40 years after those words were written, such a moth was found by Rothschild and Jordan (1903) and named *Xanthopan morgani praedicta*, in recognition of Darwin's prediction (although ultimately the subspecies was synonymized out of existence).

Darwin's brilliant insights into pollinator-plant interactions inspired many others to pick up the gauntlet. His *The Origin of Species* was translated into German by paleontologist Heinrich Georg Bronn within a year of its publication in English and stimulated a tremendous amount of interest among German scientists. In 1869, H. Mueller applied evolutionary theory to interactions between flowers and insects in a paper ably translated into English by F. Delpino, who prefaced the paper with a disclaimer indicating that he did not fully endorse all of Mueller's views. In 1876, in a book titled *Schutzmittel der Blüthen gegen unberufene Gäste,* Anton Kerner von Marilaun reexamined aspects of floral morphology meticulously described during the previous twenty-five years by German plant anatomists in the context of evolutionary theory; in particular, as his title suggests, he pointed out the importance of floral morphology in reducing the damage and depredations of non-pollinating floral visitors (or "unbidden guests"). Darwin provided an enthusiastic preface for the English translation, commenting that "the conclusion that flowers are not only delightful from their beauty and fragrance, but display most wonderful adaptations for various purposes."

As Grant and Grant (1981) came to document a century later, pollinators can themselves become unbidden from the plant perspective. As these authors reported (following up on the original descriptions), in the association between *Opuntia helleri* and *Geospiza* species on Isla Genovesa, the finches consume nectar and pollen (along with the aril surrounding the seed and even the seed itself) and in the process pollinate the flowers. Inasmuch as the cactus appears to be obligatorily outcrossed and the birds appear to be the only effective pollinators among the handful of visitors, the interaction would appear to be a mutualism. Paradoxically, however, in the process of seeking out pollen and nectar to eat, particularly during the dry season, the birds pry open flower buds and to gain access to pollen they remove the style and damage the stigmas. Over three-fourths of a total flower crop can be destroyed in this manner, odd behavior for a pollinator because destroying the stigma prevents the process of fertilization—ultimately such behavior could lead to the extinction of the major source of food for the finches on their home islands.

The work on *Geospiza* and *Opuntia* was among the first to focus on the conflict between pollinators and florivores. In the case of *Geospiza*, the mutualist pollinator and antagonist florivore are one and the same; in other systems, the ecological roles are filled by entirely different taxa. This conflict of selection pressures, illustrated so beautifully by Kerner (1876), is now a major focus for research (McCall and Irwin 2006). New tools and approaches are available today that could only be imagined in Kerner's day, and today this area of pollination biology complements ongoing research on the role of defensive chemistry in plant-herbivore interactions. As McCall and Irwin (2006) state, "extant theories of plant chemical defence, including optimal defence theory, growth rate hypothesis and growth differentiation–balance hypothesis, can be used to make testable predictions about when and how plants should defend flowers against florivores. The majority of the predictions remain untested, but they provide a theoretical foundation on which to base future experiments. The approaches to studying florivory that we outline may yield novel insights into floral and defence traits not illuminated by studies of pollination or herbivory alone." It is, as it were, a fruitful interface.

An example of the insights to be gained by examining florivory in two contexts is the seemingly unremarkable interaction between the wild parsnip *Pastinaca sativa*, a noxious roadside weed throughout North America, and its principal (and in many places only) herbivore, the parsnip webworm *Depressaria pastinacella*, the caterpillar stage of a small brown moth. The caterpilllar owes its name to its habit of webbing together and then consuming the buds, flowers, and fruits of the large umbrella-shaped inflorescences. Because the wild parsnip is a monocarpic biennial, flowering and producing seeds only once in its life after at least two years of development, *D. pastinacella*, which consumes the developing reproductive structures, has the potential to act as a major selective agent. Florivory by webworms can reduce fitness both directly, by destroying flowers, and indirectly, by altering the appearance and odor of the inflorescence.

Throughout its native range and Europe and in its introduced range in North America, the wild parsnip is "promiscuously pollinated" by a diversity of insects, particularly dipterans (Lohman et al. 1996) that can access the nectar secreted freely by a uniquely umbelliferous floral structure, the stylopodium, even with their short tongues. Many of the volatile compounds emitted by the flowers of parsnips (Borg-Karlson et al. 1994), including octyl esters (fig. 11.1) and sesquiterpenes, are attractive to flies (Jürgens et al. 2006). Infestation by only a single webworm can decrease visitation by potential pollinators (Lohman et al. 1996). Infested umbels produce fewer seeds, but the impact of the webworm extends beyond simply the consequences of removing flowers. Webworm infestation can reduce attractiveness to pollinators by altering the volatile profile, both by damage-elicited changes in plant chemistry and by the contribution of extraneous odors (notably from the large amounts of frass that accumulate in the webbing).

A parsnip umbel infested by a webworm produces significantly greater quantities of volatile emissions than from intact umbels, most notably octyl butyrate (Zangerl and Berenbaum 2009). Laboratory bioassays with webworms have shown that octyl butyrate is actually repellent to them (Carroll et al. 2006) and the large quantities of octyl butyrate in damaged flowers may represent a response on the part of the plant to minimize damage. Estimates of narrow sense heritability, which reflects the amount of additive genetic variation in a trait available for selection, for octyl butyrate content of mature fruit of wild parsnip, were on the order of 0.63–0.67 (Carroll et al. 2000), suggesting that variation in this trait (at least in mature seeds) may be responsive to selection by the florivore.

A glimpse into the impact of florivory on floral chemistry was afforded in 2004 by the appearance for the first time of *D. pastinacella* in New Zealand, where wild parsnips had been growing essentially free of florivores for over 150 years (Zangerl et al. 2008). With the exception of two monoterpenes (isomers of ocimene), all volatile constituents differed in abundance, with flowers from midwestern U.S. populations producing over twice the amounts of octyl butyrate and myristicin and greater amounts of four sesquiterpenes (caryophellene, α-transbergamotene, cis-β-farnesene, and β-cubebene), but lower amounts of octyl acetate, 1-octanol, palmitolactone, and α-farnesene.

Thus, in comparison with North American parsnips with a long history of association with webworms buds and flowers of New Zealand, parsnips contained lower amounts of at least two compounds, octyl butyrate and myristicin (fig. 11.2), inimical to webworms; octyl butyrate

DARWIN'S PUBLICATIONS ON POLLINATION AND FERTILIZATION

1. Bees and fertilisation of kidney beans. Gard. Chron. 1857: 725.
2. On the agency of bees in the fertilisation of papilionaceous flowers, and on the crossing of kidney beans. Gard. Chron. 1858: 828–9.—Ann. & Mag. of Nat. Hist. iii. 2: 459–465.
3. The origin of species by means of natural selection: or the preservation of favored races in the struggle for life. London, 1859, etc. Index.
4. Fertilisation of British orchids by insect agency. Gard. Chron. 1860: 528.—Entomol. Weekly Intelligencer. 1860.
5. Fertilisation of Vineas. Gard. Chron. 1861: 552.
6. Fertilisation of orchids. Gard. Chron. 1861: 831.
7. Vineas. Gard. Chron. 1861: 831-2.
8. On the two forms, or dimorphic condition, in the species of *Primula*, and on their remarkable sexual relations. Journ. Linn. Soc., Bot. 6: 77–96.—Abst. in Gard. Chron. 1861: 1048–9.
9. On the three remarkable sexual forms of *Catasetum tridentatum*, an orchid in the possession of the Linnean Society. Journ. Linn. Soc., Bot. 6: 151–7.—Abst. in Gard. Chron. 1862: 334–5.—Transl. in Ann. des Sci. Nat., Bot. iv. 19: 204–255.
10. On the various contrivances by which British and foreign orchids are fertilised by insects, and on the good effects of intercrossing. London, 1862, etc.
11. On the existence of two forms, and on their reciprocal sexual relation, in several species of the genus *Linum*. Journ. Linn. Soc., Bot. 7: 69–83.—Read Feb. 5, 1863.
12. On the sexual relations of the three forms of *Lythrum salicaria*. Journ. Linn. Soc., Bot. 8: 169–196.—Read June 16, 1864.
13. On the character and hybrid-like nature of the offspring from the illegitimate unions of dimorphic and trimorphic plants. Journ. Linn. Soc., Bot. 10: 393–437.—Read Feb. 20, 1868.
14. On the specific difference between *Primula veris*, Brit. Fl. (var. *officinalis* of Linn.), *P. vulgaris*, Brit. Fl. (var. *acaulis* Linn.), and *P. elatior* Jacq.; and on the hybrid nature of the common oxlip. With supplementary remarks on naturally produced hybrids in the genus *Verbascum*. Journ. Linn. Soc., Bot. 10: 437–454.—Read Mar. 19, 1868.
15. The variation of animals and plants under domestication. London, 1868, etc. Index.
16. Notes on the fertilization of orchids. Ann. & Mag. of Nat. Hist. iv. 4: 141–159.
17. Fertilisation of *Leschenaultia*. Gard. Chron. Sept. 1871: 1166.
18. Fertilisation of the Fumariaceæ. Nature 9: 460.
19. The effects of cross- and self-fertilisation in the vegetable kingdom. London, 1876, etc.
20. The different forms of flowers on plants of the same species. London, 1877, etc.
21. Fertilisation of plants. Gard. Chron. n. s. 7: 246.
22. Fritz Müller on flowers and insects. Nature 17: 78.

Figure 11.1. Charles Darwin's Pollination Publications (Trelease 1909).

A

Figure 11.2. (a) The parsnip webworm, a major selective agent on the chemistry of wild parsnip reproductive structures, is repelled by octyl butyrate. (b) In New Zealand, where parsnips have grown without webworms for over 140 years, levels of this repellent are significantly lower than in midwestern parsnips, which have been attacked by webworms for over 150 years (Zangerl et al. 2008). (c) Damage to flowers by webworms results in higher levels of octyl butyrate emission (d); (e) this increase cannot be attributed to frass odors inasmuch as very little octyl butyrate is eliminated by webworms in frass. Photo by Art Zangerl.

is, as mentioned, a webworm repellent and myristicin is an inhibitor that interferes with the ability of webworms to metabolize other plant toxins (Mao et al. 2008) (fig. 11.2). Moreover, levels of octyl acetate, which is an olfactory attractant for caterpillars (Carroll et al. 2006), were marginally higher in New Zealand parsnips that had been free of webworms for generations. These results suggest that a potential constraint on floral volatile advertisement for pollinators may be increased

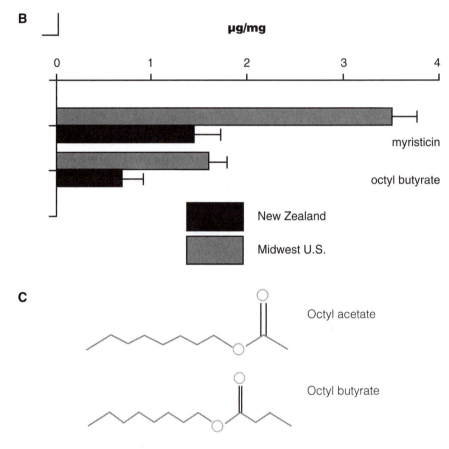

Figure 11.2. (*Continued*)

apparency to florivore enemies. In the absence of florivores, volatile emissions can be released to increase attractiveness to pollinators.

Which of wild parsnip's many floral volatiles attracts which of the plant's many pollinators has yet to be determined, but what is clear is that the ways in which plants and insects interact to bring about pollination and fertilization is no less fascinating than it was in Darwin's day and that the framework of evolution by natural selection continues to inform, guide, and enlighten studies in this area. Observations of natural history, in the spirit of Darwin's work in the Galápagos, continue to reveal new interactions and new variants on themes. Micheneau

rass emissions (dynamic headspace) are similar to those contained in the female flowers
onsumed (scaled to myristicin) (hexane-extracted)

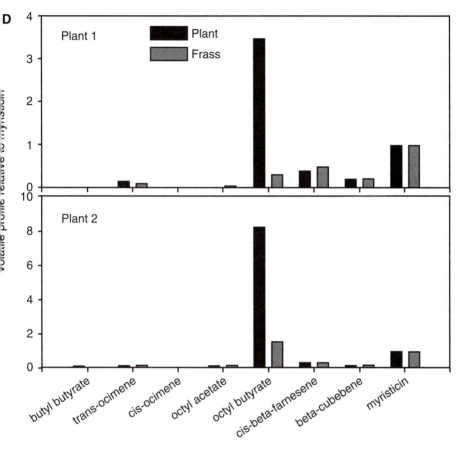

Octyl butyrate is a conspicuous exception

Figure 11.2. (*Continued*)

et al. (2006), for example, reported that the short-spurred *Angraecum
striatum*, from the Mascarene Archipelago, is pollinated not by a long-
tongued moth, as are most of its congeners, but rather is pollinated by
a bird, *Zosterops borbonicus*. Although he may not have predicted it,
Darwin would probably not have been surprised.

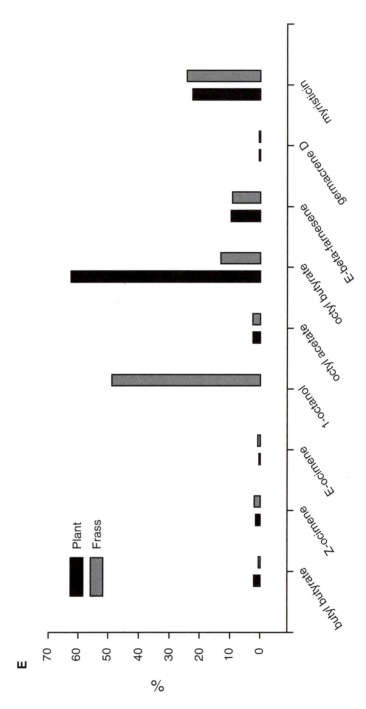

Figure 11.2. (*Continued*)

REFERENCES

Borg-Karlson, A.-K., I. Valterová, and L. A. Nilsson. 1994. Volatile compounds from flowers of six species in the family Apiaceae: bouquets for different pollinators. *Phytochem.* 35: 111–119.

Carroll, M. J., and M. R. Berenbaum, 2001. Behavioral responses of the parsnip webworm to hostplant volatiles. *J. Chem. Ecol.* 28: 1365–1375.

Carroll, M. J., A. R. Zangerl, and M. R. Berenbaum, 2000. Octyl acetate and octyl butyrate in the mature fruits of the wild parsnip, *Pastinaca sativa* (Apiaceae). *J. Heredity* 91: 68–71.

Darwin, C. 1837. Remarks upon the Habits of the Genera Geospiza, Camarhynchus, Cactornis, and Certhidea of Gould. Proceedings of the Zoological Society of London. Part V. London.

———. ed. 1841. Birds Part 3 No. 5 of *The Zoology of the Voyage of* H.M.S. Beagle, by John Gould. Edited and superintended by Charles Darwin. London: Smith Elder and Co.

———. 1859. *On the Origin of Species by Means of Natural Selection.* London: Murray.

———. 1862. *On the various contrivances by which British and foreign orchids are fertilised by insects, and on the good effects of intercrossing.* London: John Murray.

Darwin C., and R. Fitzroy. 1839. *Narrative of the Surveying Voyages of His Majesty's Ships* Adventure *and* Beagle, *between the Years 1826 and 1836, describing their Examination of the Southern Shores of South America, and the* Beagle's *Circumnavigation of the Globe.* London: Henry Colburn.

Darwin, E. 1789. *The Loves of the Plants.* London: J. Johnson.

Darwin, F. 1902. *Charles Darwin: His Life Told In an Autobiographical Chapter.* London: Murray.

Delpino F. 1868–1875. *Ulteriori osservazione sulla dicogamia nel regno vegetale.* Atti della Societa Italiana di Scienze Naturali Milano, Vols. 1 and 2.

Fenster C. B., and S. Marten-Rodriguez. 2007. Reproductive assurance and the evolution of pollination specialization. *Int. J. Plant Sci.* 168: 215–228.

Gould J. 1841. *Birds Part 3 No. 5 of The Zoology of the Voyage of* H.M.S. Beagle. Edited and superintended by Charles Darwin. London: Smith Elder and Co.

Grant, B. R. 1996. Pollen digestion by Darwin's finches and its importance for early breeding. *Ecology* 77: 489–499.

Grant, B. R., and P. R. Grant. 1981. Exploitation of *Opuntia* cactus by birds on the Galápagos. *Oecologia* 49: 179–187.

Huxley, T. H. 1856. On natural history, as knowledge, discipline, and power. *Proceedings of the Royal Institution* (1856), Scientific Memoirs I.

Jürgens, A., S. Dötterl, and U. Meve. 2006. The chemical nature of fetid floral odours in stapeliads (Apocynaceae-Asclepiadoideae-Ceropegieae). *New Phytologist* 172: 452–468.

Kerner, A. 1876. *Schutzmittel der Blüthen gegen unberufene Gäste* ('Protective means of flowers against unbidden guests'). Innsbruck, Wagner'sche Universitäts-Buchhandlung (English translation 1878).

Knuth, P. 1906. *Handbook of Flower Pollination*. Vol. I. Transl. J. R. Ainsworth Davis. Oxford: Clarendon.

———. 1908. *Handbook of Flower Pollination*. Vol. II. Transl. J. R. Ainsworth Davis. Oxford: Clarendon.

Kritsky, G. 1991. Darwin's Madagascan moth prediction. *American Entomologist* X: 207–211.

Lohman, D., A. R. Zangerl, and M. R. Berenbaum, 1996. Impact of floral herbivory by parsnip webworm (Oecophoridae: *Depressaria* pastinacella Duponchel) on pollination and fitness of wild parsnip (*Apiaceae: Pastinaca sativa L.*). *Am. Midl. Nat.* 136: 407–412.

Mao, W., A. R. Zangerl, M. R. Berenbaum, and M. A. Schuler. 2008. Metabolism of myristicin by *Depressaria pastinacella* CYP6AB3v2 and inhibition by its metabolite. *Insect Biochem Mol. Biol.* 38: 645–651.

Micheneau, C., J. Fournel, and T. Pailler. 2006. Bird pollination in an angraecoid orchid on Reunion Island (Mascarene Archipelago, Indian Ocean). *Annals of Botany* 97(6): 965–974.

Müller, H. 1883. *The Fertilization of Flowers*. Transl. D'Arcy W. Thompson. London: Macmillan.

Müller, H., and F. Delpino. 1869. Application of the Darwinian theory to flowers and the insects which visit them. Transl. R. L. Packard. 1871, in *Am. Nat.* 5: 27197.

Rothschild, W., and K. Jordan. 1903. A revision of the Lepidopterous family Sphingidae. *Novit. Zool.* 9 (Suppl.): 1–972 (p. 30).

Sprengel, C. K. 1793. *Das entdeckte Geheimnis der Natur im Bau und in der Befruchtung der Blumen* (Berlin).

Trelease, W. 1909. Darwin as a naturalist: Darwin's work on cross pollination in plants. *Amer. Nat.* 43: 131–.

Wallace, A. R. 1889. *Darwinism, an Exposition of the Theory of Natural Selection*. London: Macmillan.

Zangerl, A. R., and M. R. Berenbaum. 2009. Effects of florivory on floral volatile emissions and pollination success in the wild parsnip. *Arthropod-Plant Interactions*.

Zangerl, A. R., M. C. Stanley, and M. R. Berenbaum, 2008. Selection for chemical trait remixing in an invasive weed after reassociation with a coevolved specialist. *Proc. Natl. Acad. Sci.* 105: 4547–4552.

Chapter Twelve

The Adaptive Radiation of Coevolving Prodoxid Moths and Their Host Plants: *Greya* Moths and Yucca Moths

John N. Thompson

Coevolutionary research in recent decades has shown us that the coevolutionary process is much more ecologically dynamic than we previously suspected. A pair or group of interacting species may co-evolve in fundamentally different ways in different ecosystems. The traits favored by natural selection, the number of coevolving species, and even the ecological basis of the interaction may differ among environments, resulting in a complex mosaic of natural selection across landscapes (Thompson and Cunningham 2002; Berenbaum and Zangerl 2006; Laine 2006; Toju and Sota 2006; Hanifin et al. 2008). At the extreme, an interaction may be under mutualistic selection in some environments but under antagonistic selection in some others (Hoeksema and Thompson 2007; Piculell et al. 2008). Elsewhere, it may be commensalistic, creating coevolutionary coldspots in which selection is not reciprocal on the interacting species. Gene flow, random genetic drift, and metapopulation dynamics further fuel the coevolutionary process by remixing the geographic distribution of coevolving traits.

The geographic mosaic created by geographic selection mosaics, co-evolutionary hotspots and coldspots, and trait remixing among populations results in constantly changing coevolutionary experiments among

I am very grateful to Peter and Rosemary Grant for inspiring so many of us to study the ecological underpinnings of adaptive radiations. I thank Olle Pellmyr and Kari Segraves for helpful discussions. This work was supported by NSF grants DEB-0344147 and DEB-0839853.

interacting species (Thompson 1994, 2005, 2009). These experiments, in turn, become the raw material for the adaptive radiation of evolving interactions. Such radiations often involve diverging interactions with other species (Grant 1994; Schluter 2000; Losos et al. 2003; Grant and Grant 2008), but we are only now beginning to understand how the geographic mosaic of coevolution may contribute directly to these adaptive radiations.

Studies of the same interspecific interactions in multiple ecological settings, however, are starting to show us why the co-radiation of co-evolving lineages often shows complex, rather than simple, patterns in who interacts with whom as populations diverge and speciate (Thompson and Cunningham 2002; Barrett et al. 2007; Benkman et al. 2008; Nash et al. 2008). These studies have shown that the geographic mosaic of coevolution sometimes leads to differences among ecosystems in the combinations of interacting and coevolving species. From these geographic differences a very novel coevolutionary interaction may occasionally take hold in one part of the geographic range of an interaction and become the basis for a new radiation of coevolving species. The wide range of evolutionary experiments generated by the geographic mosaic of coevolution therefore almost guarantees that parallel diversification of coevolving taxa will be the rare exception rather than the rule. In fact, the only coevolving interactions likely to result in sustained parallel cladogenesis at the species level are those between vertically transmitted symbionts and their hosts (Thompson 2005). In these interactions, gene flow in the symbiont population is completely governed by mating patterns in the host populations.

In this chapter I summarize what we have learned in recent decades about one adaptive radiation involving coevolving lineages: the prodoxid moths and their host plants. These moths include the yucca moths and the *Greya* moths that have formed coevolved pollination mutualisms with their host plants, as well as other moths that feed parasitically on some of the same host plants. The plant families used by these moths have similarly diversified in their interactions. The result is both a geographic mosaic and a phylogenetic mosaic of coevolving interacting insects and plants.

229

Diversification of Prodoxid Moths and Their Interactions

The moth family Prodoxidae includes a group of about a hundred described species distributed throughout the Northern Hemisphere, and one basal species, placed in the genus *Prodoxoides*, known from South America (fig. 12.1). It is an ancient moth family estimated to have arisen about 95 million years ago (Pellmyr 2003), and has therefore been present through much of the 130 million-year history of the angiosperms (Soltis et al. 2005).

Thirty years ago, the coevolved mutualisms between prodoxid moths and their host plants seemed fairly simple. About four pollinating yucca moth species were thought to specialize on different combinations of yucca species, a few additional prodoxid species were known to feed parasitically on yuccas or on plants in a related plant family, and some remaining, poorly known, moth species fit into the genera *Greya, Tetragma*, or *Lampronia* (Riley 1892; Davis 1967; Powell 1992). A few *Lampronia* species had been recorded feeding on roses or currants *(Ribes)*, but little more was known about the ecological relationships of these or any other close relatives of yucca moths.

The first host plants of *Greya* species were discovered in 1980 (Thompson 1986, 1987), and the first pollinating species of *Greya* was discovered a few years later (Thompson and Pellmyr 1992). These studies indicated that *Greya* moths and yucca moths were closely related but used different mechanisms to pollinate their host plants (Pellmyr and Thompson 1992). Additional discoveries made it clear that the diversification and coevolution of prodoxid moths and their host plants were much richer than previously suspected. After another two decades of study, we now know that there are at least sixteen *Greya* species (Davis et al. 1992; Brown et al. 1994; Kozlov 1996), which feed on plants in three plant families (fig. 12.2). Recent molecular studies have indicated that some additional cryptic species may yet be undescribed (Rich et al. 2008). Similarly, the number of yucca moth species has risen considerably. *Tegeticula* is now known to include twenty species rather than four, and the remaining monocot-feeding genera also continue to grow in the number of known species (Pellmyr et al. 2008). *Lampronia*, including a species currently in the monotypic genus

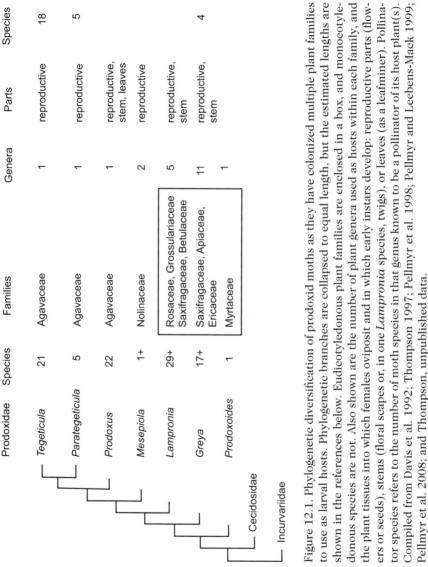

Figure 12.1. Phylogenetic diversification of prodoxid moths as they have colonized multiple plant families to use as larval hosts. Phylogenetic branches are collapsed to equal length, but the estimated lengths are shown in the references below. Eudicotyledonous plant families are enclosed in a box, and monocotyledonous species are not. Also shown are the number of plant genera used as hosts within each family, and the plant tissues into which females oviposit and in which early instars develop: reproductive parts (flowers or seeds), stems (floral scapes or, in one *Lampronia* species, twigs), or leaves (as a leafminer). Pollinator species refers to the number of moth species in that genus known to be a pollinator of its host plant(s). Compiled from Davis et al. 1992; Thompson 1997; Pellmyr et al. 1998; Pellmyr and Leebens-Mack 1999; Pellmyr et al. 2008; and Thompson, unpublished data.

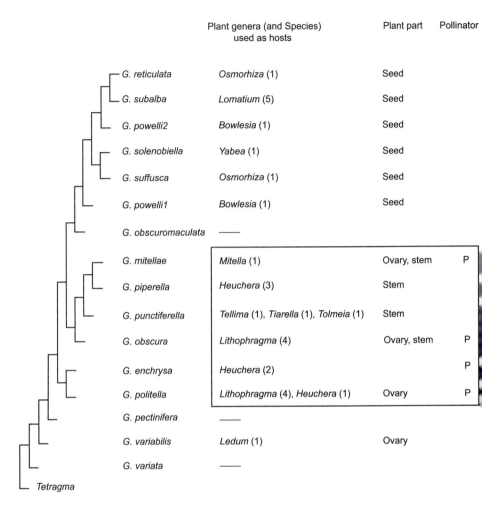

Figure 12.2. Phylogenetic diversification of the prodoxid moth genus *Greya* onto multiple plant families, showing the occurrence of pollinator species (P) within the genus. Phylogenetic branches are collapsed here to equal length, but the estimated lengths are shown in Brown et al. (1994) and Pellmyr et al. (1998). Plant genera in the Saxifragaceae are enclosed in a box. Plant genera above the box are in the Apiaceae. *Ledum* is in the Ericaceae. Two species are missing from the phylogeny: *G. sparsipunctella*, which is known only from a museum specimen and is thought now to be extinct, and *G. marginimaculata* described by Kozlov (1996) from Japan. Some remaining complexities are not shown (e.g., *G. mitellae* is embedded within *G. piperella* in analyses that include more populations of *G. piperella*). A dash indicates that the host of that *Greya* species is not yet known. Compiled from Thompson 1987; Davis et al. 1992; Brown et al. 1994; Pellmyr et al. 1996; Pellmyr et al. 1998; Thompson and Merg 2008; and references therein.

Tetragma, is a complex genus that is under reevaluation (Olle Pellmyr, personal communication).

We now know that the complex radiation of prodoxid species has been driven to a large extent by shifts onto novel host plant species (figures 12.1 and 12.2). In a few instances, these host shifts have involved jumps onto plants that are only remotely related in angiosperm phylogeny. All basal prodoxid genera use eudicotylendous plants, but the most derived prodoxid genera use monocotyledonous plants. Among the eudicot-feeders, there have been rare shifts onto plant families as different as the Ericaceae, Saxifragaceae, and Apiaceae.

These studies of host associations have shown that pollination mutualisms between plants and prodoxids have arisen at least several times from closely related non-pollinating species (figures 12.1 and 12.2). Many prodoxid species lay their eggs in the reproductive tissues of their host plants, but some oviposit into plants that lack a floral architecture conducive to pollination by floral parasites. For example, in the Apiaceae, which is used by multiple *Greya* species, each flower contains only a pair of seeds. Larvae developing in a flower kill one or both of those developing seeds (Thompson 1987). In contrast, plants in the Saxifragaceae, which are hosts of some *Greya* moths, and in the Agavaceae, which are the hosts of yucca moths, make flowers that each contains multiple ovules. Pollination mutualism with floral parasites is possible in these families, because a subset of the developing seeds often survives in each flower after larvae complete their feeding in the ovary. The seeds eaten by prodoxid larvae are part of the cost to the plant of reliable pollination.

Two other features of prodoxids increase the odds for the evolution of pollination mutualisms. Most prodoxid species take nectar as adults only from the local host plant species in which they oviposit, thereby increasing their potential as reliable pollinators. Moreover, because some prodoxids oviposit through the corolla, pollination becomes directly associated with oviposition. In *Greya* moths, pollen that is held at the tip of the abdomen rubs off onto the stigma as a female oviposits through the corolla. In yucca moths, females actively take pollen that they carry on specialized tentacles on their mouthparts and place it directly on the stigma of flowers in which they oviposit.

COMPARATIVE DIVERSIFICATION OF *GREYA* MOTHS
AND YUCCA MOTHS

As prodoxid moths have radiated onto new hosts, they have also diversified in their use of plant parts. *Greya* encompasses about as much life history diversity as the three yucca-feeding genera (*Tegeticula, Parategeticula,* and *Prodoxus*), and these two groups show multiple similarities in their adaptive radiations. Both groups include species that oviposit into flowers and other species that oviposit into other plant tissues. In both groups, the major pollinator species occur in the clade that oviposits into flowers. The only partial exception among the pollinating species is *G. mitellae*, which oviposits into floral pedicels and sometimes into leaf petioles.

The radiations of *Greya* moths and yucca-feeding moths have also often resulted in coexistence on the same host plant of floral-feeding species that are efficient pollinators and non-pollinating or inefficiently pollinating congeners that oviposit into floral stalks, flowers, or capsules. In *Greya*, the efficient pollinator *G. politella* co-occurs in part of its range with the inefficient pollinator *G. obscura* (Thompson, Laine, and Thompson, unpublished data) or the non-pollinator *G. piperella* (Nuismer and Thompson 2001). Similarly, non-pollinating yucca moths lay their eggs into the ovaries of their host plants (Althoff et al. 2006). In both groups, the non-pollinating or inefficiently pollinating species occur in only part of the geographic range of the major pollinating species, creating the potential for a multispecific geographic mosaic of coevolution in these interactions.

As the interactions between prodoxids and their host plants have continued to diversify within both moth lineages, they have, then, repeatedly evolved into small and geographically varying multispecific networks. These local networks often include one or a few host species, a pollinating moth species, and a non-pollinating or inefficiently pollinating moth species. For interactions involving *Greya*, the local network also sometimes includes yet other co-pollinators and, in a few cases, one or more species of insect herbivore (Nuismer and Thompson 2001; Thompson and Merg 2008). For interactions involving yucca moths, the local interaction sometimes involves other insect herbivores (Althoff et al. 2005). Our understanding of the ecology, phylogeny, and geography of these interactions is now reaching the point where we can begin to assess how these geographically varying multispecific networks shape the further diversification of these interactions.

THE ORIGINS OF DIVERSIFICATION AS SEEN IN THE GEOGRAPHIC MOSAIC

The links between the geographic mosaic of coevolution and the diversification of the prodoxids and their host plants are becoming clearer as we have studied, in ecosystem after ecosystem, how one of these moths, *Greya politella*, interacts with its host plants. This moth is the most widely distributed *Greya* species and one of the most widely distributed prodoxids.

Throughout most of its geographic range, *G. politella* pollinates and lays its eggs into *Lithophragma* (fig. 12.3), which is a small genus of less than ten species distributed across the western United States and southwestern Canada. The moths vary geographically in the *Lithophragma* species they use locally as hosts (fig. 12.4).

In some communities, bombyliid flies and solitary bees are effective co-pollinators of *Lithophragma* and can swamp the mutualism between *Greya politella* and its local host plant. In effect, these co-pollinators create a selection mosaic—that is, a genotype by genotype by environment interaction—such that the presence or absence of the co-pollinators constitutes different biotic environments for the pairwise interaction between *G. politella* and its local host. This selection mosaic results in interactions between *Greya* and *Lithophragma* that vary from mutualism to commensalism to antagonism among ecosystems, depending upon the relative abundance of effective co-pollinators (Thompson and Cunningham 2002; Thompson and Fernandez 2006). Geographic diversification of the interaction between *G. politella* and *Lithophragma* appears also to be shaped by the inefficient pollinator *G. obscura*, which occurs in some communities in southern Oregon and California but not in other communities (fig. 12.4).

The geographic mosaic of coadaptation between *Greya* and *Lithophragma* also may have undergone speciation that was too cryptic

Figure 12.3. *Greya politella* ovipositing into, and simultaneously pollinating, *Lithophragma bolanderi* (Saxifragaceae) in Sequoia National Park, USA.

to detect prior to recent molecular studies. Some populations of *G. politella* do not share any haplotypes with populations from other regions (Rich et al. 2008). Other populations, although molecularly similar to those that feed on *Lithophragma*, have shifted onto a closely related plant genus, *Heuchera*, and are, at least in some ecosystems, temporally separated from sympatric *Lithophragma*-feeding populations (fig. 12.4). These *Heuchera*-feeding populations may have diverged from the *Lithophragma*-feeding populations during a period of isolation in the Pleistocene (Rich et al. 2008). Most *Heuchera*-feeding populations of *G. politella* use only *H. grossulariifolia*, although at least

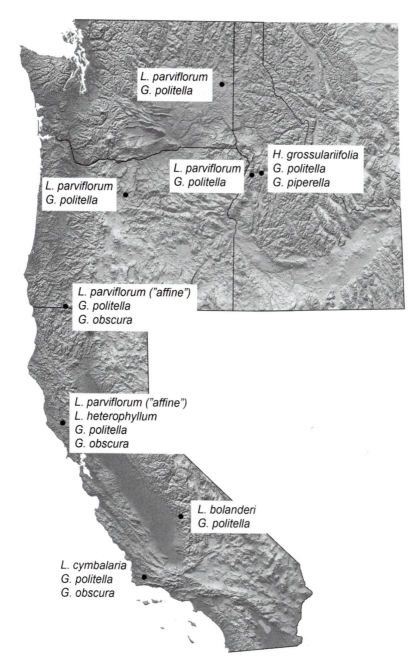

Figure 12.4. Geographic pattern of the number of interacting *Greya* moths and *Lithophragma* plants in various ecosystems throughout far western North America. The map shows eight representative localities over the entire latitudinal range of the interaction.

one population in a deep canyon in northern Idaho has shifted onto a disjunct population of *Heuchera micrantha* (personal observation). Experimental trials on *Lithophragma*-feeding populations and *Heuchera*-feeding populations have shown that the moths distinguish between their normal plants and plants in the other genus (Janz and Thompson 2002).

The evolution of autopolyploid populations of *Heuchera* has added another layer to the geographic mosaic of these interactions. The *Heuchera*-feeding moths preferentially lay their eggs in tetraploid individuals of *H. grossulariifolia* (Thompson et al. 1997; Nuismer and Thompson 2001), but they are only a minor pollinator of these host plants (Thompson and Merg 2008). A non-pollinating species, *G. piperella*, also attacks *H. grossulariifolia*, but preferentially attacks diploid individuals (Nuismer and Thompson 2001). Sympatric tetraploid and diploid populations of *H. grossulariifolia* differ in multiple other ways, including the other herbivores and pollinators they attract (Nuismer and Thompson 2001; Thompson and Merg 2008). Some river drainages in the Rocky Mountains include along their slopes populations of *L. parviflorum* that interact with *G. politella*, populations of tetraploid *H. grossulariifolia* that interact commonly with *G. politella* and less so with *G. piperella*, and populations of diploid *H. grossulariifolia* that interact commonly with *G. piperella* and less so with *G. politella*.

These geographically variable interactions between *G. politella*, *Lithophragma*, *Heuchera*, and other *Greya* species illustrate why we should not expect any simple co-diversification between *Greya* moths and their host plants. As the moth and plant species have locally adapted and diversified, so have the opportunities for reassembling in different combinations in different geographic regions. Various combinations of these moths and their host plants have engaged in mini-coevolutionary experiments in different community contexts, and some of these experiments must inevitably have failed even as others have succeeded over longer periods of time. The complex larger-scale phylogenetic patterns that we see in the ancestral-descendent sequence of divergence of these interactions must surely have arisen from this ongoing geographic mosaic of coevolution.

Diversification of Pollination in the Agavacaeae and Saxifragaceae

Diversification on the plant side of these interactions has been as geographically and phylogenetically varied as on the moth side. Pollination mutualisms with prodoxid moths are restricted to the Agavaceae and the Saxifragaceae. Both plant families exhibit a wide range of floral morphologies adapted to different pollinator taxa. The Agavacaeae, which has nine genera and about three hundred species, includes species pollinated exclusively by prodoxid moths (i.e., the yucca moths) and others pollinated primarily by bats, hawkmoths, bees, hummingbirds, or other birds (Good-Avila et al. 2006; Rocha et al. 2006). Similarly, the Heucherina group of the Saxifragaceae, which has nine genera and about eighty species, includes species pollinated by *Greya* moths and others pollinated primarily by bumblebees, fungus gnats, or other bees and flies (Okuyama et al. 2008). The great divergence in pollination mutualisms within these two relatively small plant lineages suggests that natural selection mediated by pollinators likely has been responsible for some of their diversification.

Pollination mutualisms with prodoxids form a continuum from those in which the moths are relatively unimportant pollinators to those in which the moths are the sole pollinators. At the one extreme, the *Heuchera* species that have been studied in detail are sometimes pollinated by *Greya* moths, but these plants are pollinated primarily by bumblebees (Pellmyr et al. 1996; Segraves and Thompson 1999; Thompson and Merg 2008). One species of *Mitella* is pollinated primarily by *Greya*, at least in part of its geographic range, but this species is an exception within *Mitella*, many of which are pollinated by fungus gnats (Goldblatt et al. 2004; Okuyama et al. 2008). The *Lithophragma* species pollinated by *Greya* show a continuum from almost complete dependence on the moths in some populations to complete swamping of the mutualism by co-pollinators in other populations (Thompson and Cunningham 2002; Thompson and Fernandez 2006). At the extreme of obligate mutualism are the yucca moths in which the plants depend entirely upon the moths for pollination in all populations (Pellmyr and Krenn 2002).

239

Even more broadly, we now know from a growing number of ecological studies that prodoxid moths interact with their host plants as antagonists, commensals, minor mutualists, geographically varying mutualists, and, rarely, obligate mutualists with their host plants. Seen in this context, the yucca moths are unusual only in the sense that the most extreme conditions along a continuum are always unusual. Through the lens of the geographic mosaic of coevolution, we can sometimes find almost the full range of these ecological outcomes within a single moth species and its interactions with its various hosts. As shown by the studies of *Greya*, evolving traits are only one of the drivers of coevolutionary change. Ecological context is equally important.

FROM PATTERNS AND PROCESSES TO MECHANISMS

We still need to understand more deeply the genetic mechanisms that shape the ecological outcomes of these interactions and mold their diversification. Four questions currently stand out.

How has plant hybridization shaped the diversification of these interactions? For example, *Lithophragma bolanderi* is a complex hybrid species that forms a geographic "bridge" in the Sierra Nevada of California, connecting the ranges of other *Lithophragma* species and allowing *Lithophragma* and *G. politella* to form a "ring interaction" around the Central Valley. Similarly, hybrid zones occur in yuccas and in yucca moths (Leebens-Mack et al. 1998; Segraves et al. 2005). We do not yet know whether hybridization has constrained or augmented the diversification of these interactions.

Why is polyploidy so important in shaping the use of *Heuchera* by *Greya* moths? Polyploidy is common in many plant lineages, and we therefore need to understand if and how this genetic mechanism of evolutionary change differs from allelic changes at individual loci in shaping the dynamics of coevolutionary change.

How simple is it genetically to alter floral characters in the Saxifragaceae and Agavaceae and the morphological traits associated with pollination in prodoxids? Floral ovary position, for example, is clearly important to the evolution of interactions between prodoxid moths and their host plants, and it has diverged greatly from that found in related plant taxa. We need to know if alteration of floral and moth characters, and hence the potential for mutualism, is more constrained in some directions than in others as selection acts on these interactions.

How is oviposition preference for different plant taxa determined genetically in prodoxid moths, and do major jumps to novel plant families involve different genetic mechanisms than shifts onto closely related plant species? Some host shifts in lepidopterans are known to be associated with genes on the X-chromosome (Thompson 1988; Janz 2003), but variation in host use within populations of some species appears not to be sex-linked (Nylin et al. 2005). The limited number of studies so far available for Lepidoptera indicates that the genetics of host shifts may involve a complex interplay of sex-linked and autosomal genes.

We still have few coevolved interactions in which the phylogeny, phylogeography, geographic mosaic, ecology, and genetics of a set of diverging interactions have been studied from the perspectives of both (or all) participants. As more interactions are studied in this way, we are likely to find that cases of extreme reciprocal specialization in coevolving interactions are one end of a continuum that has been explored repeatedly by interacting lineages as they have diversified through the geographic mosaic of coevolution.

REFERENCES

Althoff, D. M., K. A. Segraves, J. Leebens-Mack, and O. Pellmyr. 2006. Patterns of speciation in the yucca moths: parallel species radiations within the *Tegeticula yuccasella* species complex. *Syst. Biol.* 55: 398–410.

Althoff, D. M., K. A. Segraves, and O. Pellmyr. 2005. Community context of an obligate mutualism: pollinator and florivore effects on *Yucca filamentosa*. *Ecology* 86: 905–913.

Barrett, L. G., P. H. Thrall, and J. J. Burdon. 2007. Evolutionary diversification through hybridization in a wild host-pathogen interaction. *Evolution* 61: 1613–1621.

Benkman, C. W., A. M. Siepelski, and T. L. Parchman. 2008. The local introduction of strongly interacting species and the loss of geographic variation in species and species interactions. *Mol. Ecol.* 17: 395–404.

Berenbaum, M. R., and A. R. Zangerl. 2006. Parsnip webworms and host plants at home and abroad: trophic complexity in a geographic mosaic. *Ecology* 87: 3070–3081.

Brown, J. M., O. Pellmyr, J. N. Thompson, and R. G. Harrison. 1994. Phylogeny of *Greya* (Lepidoptera: Prodoxidae) based on nucleotide sequence variation in mitochondrial cytochrome oxidase I and II: congruence with morphological data. *Mol. Biol. Evol.* 11: 128–141.

Davis, D. R. 1967. A revision of the moths of the subfamily Prodoxinae (Lepidoptera: Incurvariidae). *Bull. U.S. Nat. Hist. Museum* 255: 1–170.

Davis, D. R., O. Pellmyr, and J. N. Thompson. 1992. Biology and systematics of *Greya* Busck and *Tetragma,* new genus (Lepidoptera: Prodoxidae). *Smithsonian Contrib. Zool.* 524: 1–88.

Goldblatt, P., P. Bernhardt, P. Vogan, and J. C. Manning. 2004. Pollination by fungus gnats (Diptera: Mycetophilidae) and self-recognition sites in *Tolmeia menziesii* (Saxifragaceae). *Plant Syst. Evol.* 244: 55–67.

Good-Avila, S. V., V. Souza, B. S. Gaut, and L. E. Eguiarte. 2006. Timing and rate of speciation in *Agave* (Agavaceae). *Proc. Natl. Acad. Sci. USA* 103: 9124–9129.

Grant, P. R. 1994. Ecological character displacement. *Science* 266:746–747.

Grant, P. R., and B. R. Grant. 2008. *How and Why Species Multiply: The Radiation of Darwin's Finches.* Princeton, NJ: Princeton University Press.

Hanifin, C. T., E. D. J. Brodie, and E.D.I. Brodie. 2008. Phenotypic mismatches reveal escape from arms-race coevolution. *PLOS Biol.* 6: e60.

Hoeksema, J. D., and J. N. Thompson. 2007. Geographic structure in a widespread plant-mycorrhizal interaction: pines and false truffles. *J. Evol. Biol.* 20: 1148–1163.

Janz, N. 2003. Sex linkage of host plant use in butterflies, In C. L. Boggs, W. B. Watt, and P. R. Ehrlich, eds. *Butterflies: Ecology and Evolution Taking Flight*, 229–239. Chicago: University of Chicago Press.

Janz, N., and J. N. Thompson. 2002. Plant polyploidy and host expansion in an insect herbivore. *Oecologia* 130: 570–575.

Kozlov, M. V. 1996. Incurvariidae and Prodoxidae (Lepidoptera) from Siberia and the Russian Far East, with descriptions of two new species. *Entomol. Fennica* 7: 55–62.

Laine, A.-L. 2006. Evolution of host resistance: looking for coevolutionary hotspots at small spatial scales. *Proc. Roy. Soc. London B* 273: 267–273.

Leebens-Mack, J., O. Pellmyr, and M. Brock. 1998. Host specificity and the genetic structure of two yucca moth species in a yucca hybrid zone. *Evolution* 52: 1376–1382.

Losos, J. B., M. Leal, R. E. Glor, K. de Queiroz, P. E. Hertz, L. R. Schettino, A. C. Lara, et al. 2003. Niche lability in the evolution of a Caribbean lizard community. *Nature* 424: 542–545.

Nash, D. R., T. D. Als, R. Maile, G. R. Jones, and J. J. Boomsma. 2008. A mosaic of chemical coevolution in a large blue butterfly. *Science* 319: 88–90.

Nuismer, S. L., and J. N. Thompson. 2001. Plant polyploidy and non-uniform effects on insect herbivores. *Proc. Roy. Soc. London B* 268: 1937–1940.

Nylin, S., G. H. Nygren, J. J. Windig, N. Janz, and A. Bergstrom. 2005. Genetics of host-plant preference in the comma butterfly *Polygonia c-album* (Nymphalidae), and evolutionary implications. *Biol. J. Linnean Soc.* 84: 755–765.

Okuyama, Y., O. Pellmyr, and M. Kato. 2008. Parallel floral adaptations to pollination by fungus gnats within the genus *Mitella* (Saxifragaceae). *Mol. Phylo. Evol.* 46: 560–575.

Pellmyr, O. 2003. Yuccas, yucca moths, and coevolution: a review. *Ann. Missouri Bot. Gard* 90: 35–55.

Pellmyr, O., M. Balcázar-Lara, K. A. Segraves, D. M. Althoff, and R. J. Littlefield. 2008. Phylogeny of the pollinating yucca moths, with revision of Mexican species (*Tegeticula* and *Parategeticula*; Lepidoptera, Prodoxidae). *Zool. J. Linnean Soc.* 12: 297–314.

Pellmyr, O., and H. W. Krenn. 2002. Origin of a complex key innovation in an obligate insect-plant mutualism. *Proc. Natl Acad. Soc. USA* 99: 5498–5502.

Pellmyr, O., and J. Leebens-Mack. 1999. Forty million years of mutualism: evidence for Eocene origin of the yucca–yucca moth association. *Proc. Natl. Acad. Sci. USA* 96: 9178–9183.

Pellmyr, O., J. H. Leebens-Mack, and J. N. Thompson. 1998. Herbivores and molecular clocks as tools in plant biogeography. *Biol. J. Linnean Soc.* 63: 367–378.

Pellmyr, O., and J. N. Thompson. 1992. Multiple occurrences of mutualism in the yucca moth lineage. *Proc. Natl. Acad. Sci. USA* 89: 2927–2929.

Pellmyr, O., J. N. Thompson, J. M. Brown, and R. G. Harrison. 1996. Evolution of pollination and mutualism in the yucca moth lineage. *Amer. Natur.* 148: 827–847.

Piculell, B. J., J. D. Hoeksema, and J. N. Thompson. 2008. Interactions of biotic and abiotic environmental factors on an ectomycorrhizal symbiosis, and the potential for selection mosaics. *BMC Biol.* 6: 23.

Powell, J. A. 1992. Interrelationships of yuccas and yucca moths. *Trends Ecol. Evol.* 7: 10–15.

Rich, K. H., J. N. Thompson, and C. C. Fernandez. 2008. Diverse historical processes shape deep phylogeographic divergence in a pollinating seed parasite. *Mol. Ecol.* 17: 2430–2448.

Riley, C. V. 1892. The yucca moths and *Yucca* pollination. *Annu. Rep. Missouri Bot. Gard.* 3: 99–158.

Rocha, M., S. V. Good-Ávila, F. Molina-Freaner, H. T. Arita, A. Castillo, A. García-Mendoza, A. Silva-Montellano, et al. 2006. Pollination biology and adaptive radiation of Agavaceae, with special emphasis on the genus *Agave*. *Aliso* 22: 3329–3344.

Schluter, D. 2000. *The Ecology of Adaptive Radiation*. Oxford: Oxford University Press.

Segraves, K. A., D. M. Althoff, and O. Pellmyr. 2005. Limiting cheaters in mutualism: evidence from hybridization between mutualist and cheater yucca moths. *Proc. Roy. Soc. London B* 272: 2195–2201.

Segraves, K. A., and J. N. Thompson. 1999. Plant polyploidy and pollination: floral traits and insect visits to diploid and tetraploid *Heuchera grossulariifolia*. *Evolution* 53: 1114–1127.

Soltis, D. E., P. S. Soltis, P. K. Endress, and M. W. Chase. 2005. *Phylogeny and Evolution of Angiosperms*. Sunderland, MA: Sinuer Associates.

Thompson, J. N. 1986. Oviposition behaviour and searching efficiency in a natural population of a braconid parasitoid. *J. Anim. Ecol.* 55: 351–360.

———. 1987. Variance in number of eggs per patch: oviposition behaviour and population dispersion in a seed parasitic moth. *Ecol. Entomol.* 12: 311–320.

———. 1988. Variation in preference and specificity in monophagous and oligophagous swallowtail butterflies. *Evolution* 42: 118–128.

———. 1994. *The Coevolutionary Process*. Chicago: University of Chicago Press.

———. 1997. Evaluating the dynamics of coevolution among geographically structured populations. *Ecology* 78: 1619–1623.

———. 2005. *The Geographic Mosaic of Coevolution*. Chicago: University of Chicago Press.

———. 2009. The coevolving web of life. *Amer. Natur.* 173: 125–140.

Thompson, J. N., and B. M. Cunningham. 2002. Geographic structure and dynamics of coevolutionary selection. *Nature* 417: 735–738.

Thompson, J. N., B. M. Cunningham, K. A. Segraves, D. M. Althoff, and D. Wagner. 1997. Plant polyploidy and insect/plant interactions. *Amer. Natur.* 150: 730–743.

Thompson, J. N., and C. C. Fernandez. 2006. Temporal dynamics of antagonism and mutualism in a geographically variable plant-insect interaction. *Ecology* 87: 103–112.

Thompson, J. N., and K. Merg. 2008. Evolution of polyploidy and diversification of plant-pollinator interactions. *Ecology* 89: 2197–2206.

Thompson, J. N., and O. Pellmyr. 1992. Mutualism with pollinating seed parasites amid co-pollinators: constraints on specialization. *Ecology* 73: 1780–1791.

Toju, H., and T. Sota. 2006. Imbalance of predator and prey armament: geographic clines in phenotypic interface and natural selection. *Amer. Natur.* 167: 105–117.

SECTION IV

ECOLOGICAL DIVERSITY

The chapters in this section are concerned with explaining biological diversity in an ecological context. They are the ecological equivalent in contemporary time of the paleontological chapters in historical time. And they confront many of the same issues, for example what are the environmental factors that permit or actually drive the diversification of major lineages, discussed in chapters 1 and 2, and why do certain ones diversify and others do not, discussed in chapter 3? The answers obtained by ecologists can inform the field of paleontology, just as the findings from the latter provide perspective as well as missing information to those studying present-day organisms. The principal advantages of working with living organisms are the potential for genetic analysis, experimentation and, often, very large sample sizes.

The first chapter (chapter 13) adopts a population genetic approach to the question of how speciation occurs. Populations of different size are likely to have different population genetic structures and dynamics. Starting from this premise, Trevor Price and colleagues explore the implications for evolutionary radiations of birds on islands. The chapter brings out the importance of time, for mutations to arise and for endemics to form. Mutations arise more frequently in large populations than small ones, and, all other things being equal, large populations have a higher mutational potential for evolutionary divergence than do small ones. Moreover, populations persist for longer on large islands. As a result, large and old islands have a high proportion of endemic species and subspecies. Endemics are usually classified as such by plumage traits, which diverge under sexual selection, or possibly by drift. Small islands, in contrast to large islands, differ ecologically from each other to a large extent. Thus, the relative importance of natural and sexual selection in contributing to speciation varies according to geography. Paraphrasing Herbert Spencer, the population proposes, by genetic variation, and the environment disposes, by selection. The chapter concludes appropriately with the suggestion that the rate at which divergence takes place is set by the genetic proposal, specifically mutation producing novel alleles, especially on large islands, and not by environmental disposal.

Many taxa are extremely diverse, and so they are especially suitable for examining questions of speciation. Foremost among them are the cichlid fish of the African Great lakes, because not only do they display a great variety of color, pattern, and form, and can be bred in captivity, but in the case of the inhabitants of Lake Victoria they have evolved remarkably rapidly. From a small number of colonists, five hundred species evolved in the last 15,000 years (chapter 14). Is this due to exceptional properties of the genome, for example unusually high mutation rates at key regulatory loci, or to their environment? The answer is unknown; most research is directed toward the environment. To many investigators, application of the allopatric model of speciation to account for the enormous number of cichlids has seemed like a Procrustean act of accommodation, because topographical relief and hence ecological heterogeneity in the lakes is not pronounced. An alternative model of sympatric speciation has recently gained increasing acceptance. With this as background, Ole Seehausen and Isabel Magalhaes investigate the middle ground between the two models with a small number of well-investigated species (chapter 14). They find evidence for speciation along a depth gradient; adaptation to depth-related microenvironments results in diminished gene exchange between fish at different depths. Therefore, assuming the ancestral species split into two where they are now observed, speciation is geographically sympatric but ecologically parapatric, driven by selection and aided by some level of spatial environmental heterogeneity. As they suggest, this could be typical of the rapid radiation of Lake Victoria cichlid fish. It would be interesting to know if shallow water species give rise to deep water ones, but not vice versa, as David Jablonski has found for marine invertebrates in the fossil record (chapter 2).

As adaptive radiations such as the cichlids become better understood, it is possible to compare radiations of different organisms in different environments. The goal, as in the case of the cichlid fish study (chapter 14), is to seek both similarities and differences, as both can help us to understand how species multiply. And what is found in one radiation could stimulate a search in another.

In the following chapter (chapter 15), Jonathan Losos compares the radiation of fourteen species of Darwin's finches on the Galápagos islands in the last two or three million years with the radiation of 120 species of *Anolis* lizards in the greater Antilles in the last 40 million years. Interspecific competition is the driving force leading to resource partitioning in anoles and finches (possibly also in cichlids), and reproductive isolation may be the incidental by-product of adaptation to different environments. The differences between organisms in their dispersal potential, and the age and degree of isolation of their environments, go a long way toward explaining why, for example, anoles have radiated within islands, repeatedly and in parallel on different islands, and Darwin's finches have not. But then, why have birds not radiated in the Caribbean and lizards have, whereas birds have radiated in Galápagos and the *Microlophus* lizards there have not? The anoles may be more similar to the cichlid fish than to the finches in diverging in ecological parapatry. Exploitation by anoles of a gradient of height and substrates in trees can be thought of as being equivalent to exploitation by cichlids of a depth gradient in lakes. Cichlids and finches are phylogenetically much younger than the anoles, and for this reason they hybridize, with little or no loss of fitness, whereas anoles apparently do not. The potential to exchange genes could be an important factor contributing to the early stages of evolutionary radiations.

The final chapter (chapter 16) demonstrates the power of field experiments to help close the gap between evolution of phenotypes and the genomic base-pair differences that create them: between internal events governing morphological traits during development and external determinants of fitness of those same traits. It describes an experimental test of natural selection on a trait governed by a single locus. There is a natural link to chapters in all sections of the book that discuss evolution of traits by natural selection.

Several species of temperate zone freshwater fish show interesting variation in body size and proportions in relation to lake depth. Benthic and limnetic fish display predictable and repeatable differences.

249

One such group, North American sticklebacks, has been intensively studied, yielding insights into a wide range of biological phenomena, from the genetic control of ecologically important traits to the evolution of reproductive isolation in speciation. Freshwater sticklebacks are likely to experience a lower risk of predation from other species of fish than their marine relatives. Correspondingly, they are protected by far fewer lateral bony plates. A single genetic locus, *Eda*, is responsible for the majority of the marine-freshwater differences (also chapter 6). The homozygous low allele is found in fish with very few or no plates, and in many isolated bodies of freshwater. Recurrent fixation of the low allele implies that it undergoes positive selection in freshwater (see also chapter 15). Why? One possibility is that it enhances growth rate, which in turn reduces fish vulnerability to insect predation. A combination of lab and field experiments carried out by Rowan Barrett and Dolph Schluter shows that the *Eda* low allele does indeed confer a growth advantage allowing improved overwinter survival. However, unexpectedly, it also results in poor juvenile performance before the allele is expressed! The eyes of the investigators have now been directed back to the field where the initial observations were made on body armor for new insights into additional factors that need to be investigated to fully understand how the trait has been fixed by selection.

Chapter Thirteen

Ecological and Geographical Influences on the Allopatric Phase of Island Speciation

Trevor Price, Albert B. Phillimore, Myra Awodey, and Richard Hudson

T he founding of a new island population is followed by genetic divergence from the ancestor, perhaps eventually leading to taxonomic distinctness. Evolution results from both selection and genetic drift, and it depends on variation brought in by the founders, on new mutations, and on any variants obtained from immigrants or from hybridization events. Focusing on new mutations, we review ecological and geographical factors that facilitate divergence under selection. Ecological factors include environmental differences that precipitate strong divergent selection pressures ("ecological speciation"), population persistence (i.e., factors that reduce extinction propensity), and population size, all of which interact. Geographical factors include island size (as it affects population size) and isolation. Large, isolated islands tend to contain the most endemics and large, isolated islands contain old populations. This implies that age is critical to the persistence, and possibly formation, of endemics. Divergent selection pressures are less important; indirect evidence suggests they are stronger on small islands than large islands. We relate these findings to two examples in which small populations have been invoked as a cause of rapid evolutionary change: (1) the distinctness of peripherally isolated populations, and (2) the pattern of punctuation and stasis in the fossil record. We argue

We thank Sonya Clegg, Ian Owens, and Jason Weir for help or advice, and Peter and Rosemary Grant for the invitation to the meeting at which this chapter was presented, as well for comments on the manuscript.

that both these patterns can be understood in the context of long persistence times of isolated populations, which need not be small.

Colonization of a new island from a source sets in process a period of differentiation, which may eventually lead to a population so distinct as to be labeled a new species (e.g., plate 9). Evolution in the population results from sorting of genetic variation carried by the founders and from the accumulation of new mutations, possibly augmented by the introduction of new genetic material through immigration and hybridization (Grant and Grant 1996; B. R. Grant and P. R. Grant 2008). Here we review the way ecological factors affect differentiation of island populations. Among birds, endemism, as often assessed using color differentiation (plate 9), is positively correlated with island area and island isolation. We argue that this is due to the overwhelming importance of the old age of populations on large, isolated islands, and that differences in ecological conditions between source and founding population are less important. In our review, we consider the role of fixation of new mutations, and in the discussion briefly consider the roles of sorting genetic variation in the founders and ongoing immigration and hybridization.

We emphasize the role of selection in driving evolution. However, genetic drift continues to be raised as a possibly important factor in the evolution of phenotypic traits, including sexually selected traits such as coloration. Most recently, following Lande (1981), Uyeda et al. (2009) have used simulations to show how sexually selected traits could rapidly diversify, as a result of drift in female preferences for such traits. These models assume the female preferences are subject to no (or very weak) direct selection. By weak direct selection, we mean that the average fitness of females with strong preferences for a certain kind of male is similar to the average fitness of females who mate with the first male they encounter. Because of both search costs and variance in male quality, this seems unrealistic (cf. Uyeda et al. 2009). Additional factors lead us to believe that the role of drift in the evolution of phenotypic traits is minor. First, the measured intensity of selection in nature is often high (Hoekstra et al. 2001). Second, comparative studies are increasingly demonstrating associations of not only morphology but also colors, vocalizations, and perceptual systems with environmental factors (e.g., Cummings 2007; Price 2008, chapter 12; see

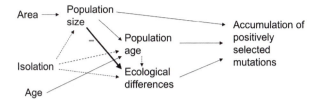

Figure 13.1. Factors affecting accumulation of genetic differences between island populations, assuming effects of gene flow are negligible. Only one path (in bold) is negative. Dashed lines are effects that have been hypothesized in the literature, but remain to be quantified. New data are presented in this chapter regarding the paths from isolation to population age, and population size to ecological differences. Isolation may affect age indirectly through population size, because fewer species are present on more isolated islands, so each has higher population size (Diamond 1980). Isolation may also affect age directly because fewer enemies invade isolated islands (Price 2008, chapter 8). Island age also affects population age directly: young islands have young populations (Heaney 2007; Weir and Schluter 2008; Whittaker et al. 2008).

also chapter 14, this volume). Third, the widespread importance of pleiotropy implies that even if a character does not affect fitness directly, it is likely to be genetically correlated with one that does (Wright 1964). These and other theoretical and empirical considerations (Coyne and Orr 2004) lead us to emphasize selection as the dominant mode of evolution of conspicuous phenotypic traits, but we consider drift models further in the discussion.

We begin by considering three ecological factors and one geographical factor that have been invoked to explain why some populations have diverged from the source more than others (fig. 13.1). These are (1) differences between environments experienced by each population, (2) population persistence, (3) population size, and (4) population isolation. Population size could be considered both an ecological and geographical factor, for it is correlated with island size (MacArthur and Wilson 1967; Mayr and Diamond 2001).

Environmental Differences

Recent emphasis has been placed on the role of differences between the ancestral and new environment (in both biotic and abiotic features) that precipitate novel selection pressures when a population colonizes a new environment. This forms the basis of the research program into "ecological speciation" (Schluter 2001, 2009; Rundle and Nosil 2005). For example, extensive work on Darwin's ground finches has highlighted the role of differences in food resources and competitors in driving differentiation between populations in size and shape, which has also produced some correlated responses in song, and both morphological and song differences contribute to reproductive isolation (reviewed in Grant 1999; Grant and Grant 2008b). The more different the environments, the more likely it is that a mutation which was not favored in the ancestral environment will be favored in the novel one.

The slowest rate of evolution occurs when all favored mutations are subject to similar selection pressures in both the ancestral and novel environment, because environments are very similar. In this case, divergence is attributable solely to the randomness in the mutation process, in which different mutations, by chance, arise and are fixed in different locations ("mutational-order," Mani and Clarke 1990; Schluter 2009; see also chapter 7, this volume). This may especially apply to sexually selected traits (Grant 2001; Price 2008) but can also operate *via* natural selection. A novel mutation that makes an individual less conspicuous to a predator, or better able to exploit a resource, may not have become established in the ancestral population prior to colonization of the new location, but after colonization arises in one or other population. Perhaps this is unlikely: most mutations may already have been tested in the ancestral location but once genetic backgrounds differ between populations, a different suite of naturally selected mutations is favored. For example, if a new trait becomes established in one population under sexual selection, mutations that reduce conspicuousness of the trait to predators will now be selected for. Mutational-order processes are expected whether or not environments differ, so any environmental differences will always accelerate the rate of divergence (Price 2008, chapter 2).

Time

It seems reasonable that the longer the time an isolated population persists, the more distinct it is likely to become (Mayr 1965; Diamond 1984). Age has been commonly invoked to explain differentiation of island populations both within archipelagoes (e.g., Grant et al. 2000), and between archipelagoes (Grant 1994). Clearly, the older the population, the greater is the opportunity for mutations to arise and become fixed. Because many mutations are rare, differences continue to accumulate over long timescales (millions of years). Age can also affect divergent selection pressures: occasional environmental changes (e.g., as a result of climate change, or a biotic invader) result in populations evolving in new directions, and when environmental conditions change again, the populations do not simply evolve back to the ancestral condition, but rather in additional new directions (Price et al. 1993; Grant 1998). For example, the establishment of a predator on an island may cause prey species to exploit resources in different ways, which are further refined even when the predator disappears.

Population Size

Both ecological and geographical factors (island size) affect population size. Small populations have been given a special place in the modern synthesis (i.e., the integration of genetics into evolution, Simpson 1953; Mayr 1954). For example, Mayr (in Haffer 2007, p. 220) stated, "Fisher was completely and utterly wrong, when he said 'the larger the population the faster it will evolve.' We now know the truth is exactly the opposite." Debate on the relative importance of small and large populations in adaptive evolution continues (Gavrilets and Gibson 2002; Gavrilets 2004; Coyne and Orr 2004).

In the simplest selection-based model, larger populations diverge faster than smaller ones. First, note that the majority of favorable mutations are lost during the first few generations. Thus, an individual that carries an attractive new mutation may die, or fail to reproduce, for reasons entirely unrelated to its phenotype. This accidental loss in the early generations is a result of random birth and death, and hence could be called drift (Gillespie 2004), but it is essentially independent of population size. Consider a population fixed for allele a at a given locus, in other words, all individuals are homozygous aa. A new mutation in this population, A, is assumed to have a constant fractional selective advantage (independent of density, frequency, or genetic background), s, in the Aa heterozygote and $2s$ in the AA homozygote. The probability that the new mutation will spread to fixation is $\sim 2s$ (Haldane 1939). Thus, if the mutation has an advantage of 1 percent in the heterozygote, the chances of it becoming fixed are only 2 percent: most mutations are lost soon after they appear. This is an approximation, and at very small population sizes the probability of fixation increases to more than $2s$, because of the effects of drift (the probability of fixation of a neutral allele is $1/2N$, so, for example, in a population of 20 it would be 2.5%).

The number of new mutations arising each generation is proportional to population size, and is given by $2N\mu$, where N is the population size, and μ is the mutation rate (new mutations per gamete). Thus, the number of new mutations that are destined to be fixed (i.e., the fixation rate per generation at equilibrium) is $\sim 4Ns\mu$, which increases monotonically with population size: two isolated populations of 10,000 individuals will diverge from each other ten times faster than two populations of 1,000 individuals, if s remains constant and mutations are unique.

These results are illustrated in figure 13.2, which show simulations of selectively favored mutations arising in populations differing in size, assuming selective coefficients are constant. Many more mutations are fixed in the large population. An important point to note is that the transit times (the time during which favored mutations are spreading to fixation) make up a small proportion of the total time. If favorable mutations arise infrequently, most populations, most of the time, are not segregating for such mutations.

Several factors might reduce the disparity between large and small populations in their rate of evolution. For example, Gillespie (1999) found that if favored mutations are spreading at the same time, they interfere with each other's spread. Because more mutations are spreading in a large population, this interference is greater in a larger population; however, it is possible that favorable mutations arise sufficiently infrequently that this problem is minimized (as illustrated in fig. 13.2). Mayr's verbal objection to faster evolution in large populations was that the selective value of genes varies with the genetic background ("epistasis"), and a novel mutation encounters a greater diversity of genetic backgrounds in a larger population (Mayr 1959). Ohta (1972) suggested an environmental version of this argument. She argued that small populations are likely to experience a less diverse range of environments than large ones. She also suggested that many mutations that have a selective advantage in one environment may be neutral or deleterious in other environments. Thus, if a few mutations in each

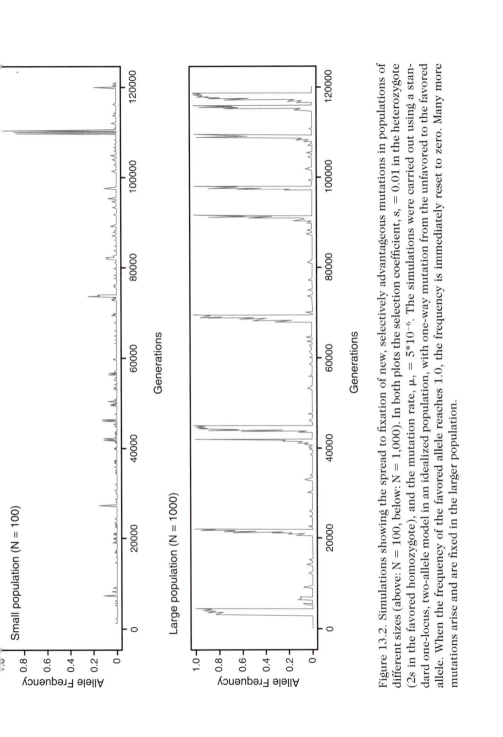

Figure 13.2. Simulations showing the spread to fixation of new, selectively advantageous mutations in populations of different sizes (above: N = 100, below: N = 1,000). In both plots the selection coefficient, s, = 0.01 in the heterozygote (2s in the favored homozygote), and the mutation rate, μ, = $5*10^{-6}$. The simulations were carried out using a standard one-locus, two-allele model in an idealized population, with one-way mutation from the unfavored to the favored allele. When the frequency of the favored allele reaches 1.0, the frequency is immediately reset to zero. Many more mutations arise and are fixed in the larger population.

TABLE 13.1

Expected (Average) Time to Fixation (in Discrete Generations) of Selectively Favored Mutations as a Function of Population Size

	Selection Coefficient, s			
Population Size, N	0.001	0.005	0.01	0.05
100	398	393	378	219
1,000	3,795	2,212	1,406	412
10,000	14,078	4,136	2,346	597
100,000	23,480	5,985	3,269	781

Notes: Based on summing expression (5.53) of Ewens (2004) with $x = i/2N$, and $i = 1, \ldots, 2N-1$. s is the selective advantage in the heterozygote, $2s$ is the advantage in the favored homozygote. Here N is the effective population size (cf. in the text, where it is the census population size).

single environment are favored, when they are averaged across the entire range of environments virtually all mutations might be deleterious. The implication is that adaptive changes are more likely in smaller rather than larger populations. Variants on this theme have been expressed in discussions of morphological stasis (Eldredge et al. 2005). Although in general, experimental studies have found larger populations to evolve more rapidly (Weber and Diggins 1990), it is possible that in nature the increased variability of the genetic and environmental backgrounds experienced by large populations is sufficient to lower their rate of evolution below that of small populations.

A possible objection to the importance of large population size is that populations are unlikely to be in steady state, and any new mutation that is destined to be fixed takes much longer to spread through a large population than a small population. Surprisingly, the difference in fixation times is small, as shown in table 13.1. Thus, when s = 0.01, a mutation takes just over twice as long to spread throughout a population of 100,000 than it does through a population two orders of magnitude smaller. This is because those mutations that do fix are the ones that during the early stochastic events are pushed to high frequency. Once an appreciable number of copies are present in the population, fixation is rapid, whether the population is small or large.

Although definitive tests are needed, if populations diverge from each other due to the accumulation of new, selectively favored, mutations, it is probable that larger populations will diverge from each other faster than smaller ones. On the other hand, if population divergence is due to the fixation of deleterious mutations (i.e., those that are selected against, and can only increase via genetic drift), fixation probability is higher in smaller populations (Ohta 1973). Woolfit and Bromham (2005) used seventy island-mainland comparisons to show that on islands the rate of non-synonymous substitutions is elevated relative to synonymous substitutions, which they attribute to this effect. Whether such a process applies to traits such as morphology, vocalizations, and colors that characterize taxonomic differences is debatable.

Population size, Environmental Differences, and Time

Large populations accumulate favorable mutations more quickly than small ones. Population size also affects differentiation in other ways. First, the extent to which a colonizing population experiences a different environment from the source is likely to depend on the size of the island the population becomes established on. Second, small populations are more extinction prone.

Novel features in the biotic composition of an island—predators, parasites, resources, and competitors—likely provide an important contributor to any directional selection pressures experienced by a colonist population. As an expected result of random sampling, small islands should vary more in the biotic composition than large ones: drawing a few species from a large pool will inevitably result in greater variance among samples than if many species comprise each sample. In turn, this should lead to a population on a small island experiencing particularly strong divergent selection from the source ("biotic drift," Turner and Mallet 1996). In a test of this idea, Price (2008, chapter 3) found that bird communities across small islands differed more from each other than bird communities across large islands, suggesting the competitive environment does vary more across small than across large islands. We extended the test to consider groups ranging from plants to mammals (i.e., with respect to birds, resources and predators as well as competitors), at several different taxonomic levels, across true archipelagoes, mountaintops, and relatively recently isolated reserves, and across multiple spatial scales, as documented in the appendix. In 81 percent of all cases ($N = 81$ comparisons), variance in species composition among small islands (as estimated following Lande 1996; see appendix and Price 2008, chapter 3) was higher than variance among large islands. The figure was essentially the same when we restricted the analysis to true archipelagoes (84%, $N = 50$ comparisons). These results imply that colonists of small islands are likely to experience stronger directional selection away from a source population than are colonists of large islands. At any rate, there is no clear prediction that colonists of large islands should experience stronger divergent selection pressures than colonists of small islands.

Unusual environmental conditions on small islands might favor population divergence, but smaller populations should go extinct more frequently than larger ones (MacArthur and Wilson 1967). Molecular data are now available to test extinction risk across pre-historical timescales. To date, this has only been done for the Lesser Antilles: Ricklefs and Bermingham (1999) and Cadena et al. (2005) showed that old populations are confined to the large islands. Such populations have been present on an island for perhaps 0.5–5 million years (Cadena et al. 2005), assuming mitochondrial sequence divergence between populations proceeds at the rate of ~2%/million years (Weir and Schluter 2008).

Population Isolation

Population isolation is largely a result of distance from other islands. It may, however, be influenced by ecological conditions on the island, because well-adapted residents affect the chances of immigrants becoming established, raising effective isolation.

Unlike potentially conflicting effects of population size on differentiation, greater isolation appears to always favor greater differentiation, for at least three reasons. First, the traditional interpretation is that the more isolated an island is, the less gene flow on to the island, so the population can respond to selection pressures unique to that island (e.g., Johnson et al. 2000; Mayr and Diamond 2001, p. 284; Whittaker et al. 2008). Second, isolated islands have very unusual biotic environments, because only highly dispersive groups are able to reach such islands. We have not quantified this, but it is clear it could be important in driving divergent selection pressures. For example, the endemic moa-nalos of Hawaii apparently evolved from an ancestor resembling the mallard duck, exploiting the grazing niche in the absence of grazing mammals or reptiles (Sorenson et al. 1999).

The third reason why isolated islands might have more differentiated populations is that these populations are older (Diamond 1980, 1984). MacArthur and Wilson (1967) and Diamond (1980, 1984) note that if islands contain relatively few species for their size because of low colonization rates, the species on these islands should have relatively high population sizes, hence low extinction rates (i.e., for a given island area, the probability of extinction increases with species numbers). Isolated islands might also contain older populations because these islands receive fewer invasions of enemies, including predators, parasites, and competitors, reducing extinction rates (Ricklefs and Cox 1972; Ricklefs and Bermingham 2002; Price 2008, chapter 8). We tested for correlates of age with isolation by studying bird superspecies across archipelagoes for which genetic data are available. Older populations do indeed occur on more isolated islands (table 13.2). We will argue that this third effect is likely the most important for the patterns we describe.

In figure 13.1 we summarized these effects of population size and isolation on differentiation of an island population. Only one path is negative: smaller rather than large populations may experience greater divergent selection pressures, because they occupy unusual habitats. We now turn to consider patterns.

PATTERNS OF ISLAND ENDEMISM

In birds, subspecies and species have generally been given endemic status based on unique colors and/or feather ornaments, and to a lesser extent on unusual morphology (Watson 2005). Across northern Melanesia, as well as the Indian and Pacific oceans, endemism is correlated

TABLE 13.2

Associations of Population Age with Island Isolation

Superspecies	Archipelago	Correlation (signed)	Regression ± s.e. (my/100 km)		Number of Populations	Age of Oldest Population (my)	Age of Root (my)	Diversification Interval (my)
Acrocephalus	Northern Marquesas	0.22	0.12	0.19	9	1.04	1.29	0.86
Acrocephalus	Southern Marquesas	0.61	0.15	0.09	6	0.59	0.59	0.54
Amazona	West Indies	0.87	0.91	0.22*	8	1.43	1.88	1.36
Certhidea	Galápagos	−0.14	−0.28	0.53	15	1.38	1.68	0.83
Cinclocerthia ruficauda	West Indies	0.68	1.29	0.79	5	0.68	0.86	0.94
Coereba flaveola	West Indies	0.60	0.29	0.12*	12	0.77	0.77	0.43
Hypsipetes	Indian Ocean	0.30	0.20	0.26	8	2.04	2.04	1.47

Myadestes	West Indies	0.66	0.46	0.30		5	0.99	0.99	1.08
Nectarinia sovimanga	Indian Ocean	−0.28	−0.22	0.29		9	2.47	2.99	1.99
Parus caeruleus	Canaries	0.47	1.39	1.15		7	1.53	1.53	1.22
Zosterops flavifrons	South Pacific	0.50	1.54	0.81†		13	2.58	3.19	1.70

Notes: Not all groups are monophyletic. Isolation is distance from nearest island obtained from the UN database (http://islands.unep.ch/Tiarea .htm). Ages are from an analysis of mitochondrial DNA sequences assuming a relaxed molecular clock in BEAST (Drummond and Rambaut 2007), setting a rate of divergence of 2%/million under a GTR + I + γ model of substitution. The 'age' of each population is based on the estimated time to most recent common ancestor between a population and its closest relative (not necessarily the nearest island). In a limited number of cases an archipelago (e.g., Kiribati and Cayman Islands) is treated as a single island. Significance of individual regressions is: †P < 0. 1, *P < 0.05. Overall, the association is highly significant (two-tailed one sample t-test, comparing signed correlation coefficients, with the null hypothesis that the mean = 0, P = 0.004). The final three columns give the age of the oldest population, the distance to the root of all populations (t), and the diversification interval [average times between branching along a lineage, assuming exponential growth, calculated as t/(ln(N/2), where N is the number of populations (Magallón and Sanderson 2001)].

with both island area and island isolation (Adler 1992, 1994; Diamond 1980, 1984; Mayr and Diamond 2001; Price 2008; see fig. 13.3).

The correlation of endemism with area has been widely recognized (e.g., Mayr 1965). Mayr argued that it is only on large islands that populations persist long enough to differentiate from the source. In addition, population size per se directly promotes differentiation because they accumulate more favorable mutations. Because the evidence we have presented in the appendix suggests that, if anything, small islands differ more from each other ecologically than large ones differ from each other, the positive association of endemism with island size argues against divergent selection between source and colonist as a primary factor affecting endemism.

Endemism is correlated with isolation, as has also been noted previously (Hamilton and Rubinoff 1967; Diamond 1980, 1984). Correlates of endemism and isolation are present over such large (e.g., oceanic) spatial scales that differential gene flow is unlikely to be the major cause of this variation. Many island populations, even within archipelagoes, are separated by very long divergence times, on the order of hundreds of thousands to millions of years (table 13.2, last three columns), implying significant gene flow has been absent for a long time (age of population may often correspond to age of appearance of the island; Heaney 2007; Weir and Schluter 2008). The correlation of endemism with isolation is also present when the analysis is restricted to strongly differentiated subspecies and higher taxa, which are unlikely to receive much, or above the species level, any gene flow (Diamond 1984).

Assuming gene flow is an unlikely explanation for the correlation of isolation with endemism across large scales, we are left with two other explanations: age and ecological distinctness of isolated islands. Based on our finding that age and isolation are correlated, we suggest that a large part of the effect of isolation on endemism is population age, as in the explanation for the correlation with area. Population age is itself expected to partly result from the ecological distinctness of isolated islands, because the more distinct the island, the more difficult it should be for biological enemies to become established and usurp residents. Presently, we are unable to directly assess an independent role for strong divergent selection pressures associated with ecological uniqueness of isolated islands.

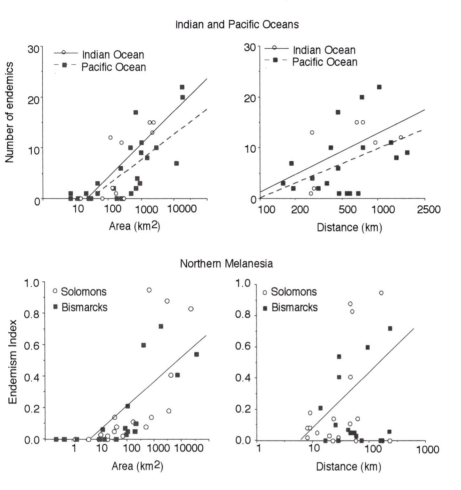

Figure 13.3. Associations of endemism with area (*left*) and isolation (*right*) for the Indian and Pacific Oceans and for Northern Melanesia (with the more remote Solomon islands separated from the Bismarcks). Least squares regression lines are fit to the data (for the Melanesian distance plot islands with zero endemics are omitted in calculating the regression: these islands are small and confound the pattern). Redrawn from Price (2008), based on original data in Mayr and Diamond (2001) and Adler (1992, 1994). Island isolation is a measure of distance from a potential source. For Northern Melanesia, the endemism index is a measure of the degree to which an island contains unique populations (Mayr and Diamond 2001). For both oceans, and across Northern Melanesia, in multiple regression analyses, both area and isolation are significant predictors of endemism (see Price 2008, pp. 61, 143, for further details).

Discussion

Under theories of mutation accumulation, large, old, populations are expected to be most distinct. This leads us to expect positive correlates of island endemism with both isolation and area, as observed here. The importance of age seems especially strong: old populations are much more likely to differentiate to a level where they are recognized as taxonomically distinct than are young populations.

On the other hand, we have little evidence that (time-averaged) ecological differences between islands, and associated divergent selection pressures, are of major importance in accounting for differences in the level of endemism. It is clear that strong selection associated with entry into ecologically unusual environments does drive the formation of some young species (Schluter 2001), but this process does not seem to be prevalent in driving divergence among endemic island species, many of which are morphologically (e.g., plate 9) and apparently ecologically similar to each other. Most sister species within archipelagoes fall into this class. Sympatric, morphologically and ecologically divergent species produced as a result of *in situ* radiations are rare. We know of only five examples of *in situ* island radiations that have led to more than two sympatric species. These are one radiation each within the Galápagos Islands and Hawaii and three in the West Indies (Ricklefs and Bermingham 2007; appendix in Weir et al. 2009).

The larger the differences between environments experienced by the source and colonizing population, the higher the fraction of new mutations that are likely to be favored in the colonists. This is because those mutations favored in both environments should already have arisen and been fixed in the source. Although some novel mutations may fall into the class where they would be favored in either location, it may be that most mutations that are favored in the new location are favored in this location alone, responding to small environmental differences (no two islands are identical, e.g., see the appendix). In either case, when islands are similar, the relatively few mutations favored in the novel environment result in time becoming the limiting factor to divergence. For example, a melanic form of the bananaquit, *Coereba flaveola,* occurs in the higher altitudes of the islands of Grenada and San Vincent.

Behavioral differences plus an analysis of the cline between the yellow and melanic forms suggest that climatic differences are a selective agent favoring the melanic form in these locations (Wunderle 1981). Melanism is known to result from a single base substitution in the MC1R gene (Theron et al. 2001) and phylogenetic analysis suggests that the mutation arose late in the history of colonization of the West Indies (Price and Bontrager 2001). In other words, this seems likely to be an adaptation to a specific environment, but the adaptation has arisen only after some time, as expected if mutation limitation is the main cause of the correlates of age and area with endemism.

Two other genetic factors besides mutation limitation affect the rate of divergence. The first is hybridization and immigration. In this regard, it is notable that in classic examples where ecological speciation has happened rapidly, hybridization and/or migration has been invoked as an important feature that inputs genetic variation, thereby facilitating rapid evolution, in lieu of new mutations. This applies to Darwin's ground finches (B. R. Grant and P. R. Grant 2008), cichlid fish (Terai et al. 2006; see also chapter 14 this volume), and sticklebacks (Colosimo et al. 2005) and implies that a limited amount of gene flow actually promotes, rather than retards, divergence in a new environment. If this is the case, the fact that more isolated islands are more differentiated again suggests that time for mutations to accumulate is a more critical factor affecting differentiation than is divergent selection: more isolated islands are less likely to receive input of new genetic material through immigration.

A second factor affecting differentiation is the genetic variation brought into the population by the founders. As in mutation accumulation, we expect standing genetic variation, and hence response to novel selection pressures, to correlate with population size (both the size of the founder population, and the rate at which it grows). Of additional interest is the possibility that founders carry a nonrandom sample of genes from the ancestral population, causing rapid change in the founder (reviewed by Barton 1996; Rundle et al. 1998; Templeton 2008). The idea is that rare alleles at two or more loci can be raised to such a high frequency that they regularly occur in combination; if this combination is especially favored, it can spread. The reasoning stems from a paper by Mayr (1954), written partly to explain his observation

267

that peripherally isolated populations of birds sometimes differ strikingly in color patterns from other populations on nearby larger land masses (e.g., plate 9, lower panel). In reviews of both the empirical evidence and theory, Coyne and Orr (2004, pp. 388–402) and Templeton (2008) conclude that such events are likely to be very rare. With respect to color patterns in particular, it remains unclear how much hidden variation of the sort that could be revealed by a founder effect is present in ancestral populations.

Thus, "while the hypothesis of founder effects as the cause of change in geographical isolates is perhaps the most novel and influential contribution of the last century to ideas about how evolution of organisms occurs on islands" (Grant 2001), it appears to have little support empirically (Grant 2001; Clegg et al. 2002; P. R. Grant and B. R. Grant 2008). On the basis of the analysis presented here, we suggest that peripheral isolates are divergent primarily because they have persisted for a relatively long time, and only secondarily because populations experience unusual environments and hence divergent selection pressures. However, ecologically unusual conditions in the peripheral isolate may contribute to population persistence, by causing biotic enemies to be relatively maladapted in the face of competition from increasingly well-adapted residents. On this argument, across less isolated areas, extinction-colonization dynamics promote homogeneity, so any particular population remains relatively young.

An important role for small populations in evolution and speciation has been raised in connection with theories of punctuated equilibrium and stasis (reviewed by Eldredge et al. 2005). Morphological stasis seems to be the predominant mode in the fossil record, persisting for millions of years, with transitions between types rarely documented. Because rapid evolution in small populations is unlikely to be recorded in the fossil record, Eldredge et al. (2005), following a model of Ohta (1972), emphasize the idea that small, peripherally isolated populations experience an unusual environment, whereas large ones experience an "average" environment. In this chapter, however, we have noted that small islands are generally not associated with much differentiation, and we have argued that this is because they are usually young. It may be that our results are not relevant to the debate on punctuation, because of matters of scale: even a large island in our data set, with populations persisting for millions of years, is small on a continental scale. Findings highlighted here add to this debate by emphasizing that differentiation in peripheral isolates takes time, and this is the primary factor, rather than the unusual selection pressures isolates may experience.

In conclusion, our analysis indicates population age as a major factor affecting population differentiation across islands. Age may contribute because fluctuating environmental conditions on islands drive populations on different trajectories over long periods of time (Price et al. 1993; Grant 1998). However, we suggest that a major cause of the association of island differentiation with age is mutation limitation. This applies both if the mutations are fixed in a colonizing population due to novel selection pressures, or if the new mutations are generally advantageous and just happen to arise in the colonist, but not other, populations. Studies that isolate those mutations affecting traits, followed by ecological assessments of their significance and phylogenetic reconstructions of their history, could be used to test the mutation-limitation explanation.

References

Adler, G. H. 1992. Endemism in birds of tropical Pacific islands. *Evol. Ecol.* 6: 296–306.

———. 1994. Avifaunal diversity and endemism on tropical Indian Ocean islands. *J. Biogeog.* 21: 85–95.

Barton, N. H. 1996. Natural selection and random genetic drift as causes of evolution on islands. *Phil. Trans. Roy. Soc. Lond.* B 351: 785–794.

Cadena, C. D., R. E. Ricklefs, I. Jimenez, and E. Bermingham. 2005. Ecology—Is speciation driven by species diversity? *Nature* 438: E1–E2.

Clegg, S. M., S. M. Degnan, J. Kikkawa, C. Moritz, A. Estoup, and I. P. F. Owens. 2002. Genetic consequences of sequential founder events by an island-colonizing bird. *Proc. Natl. Acad. Sci. USA* 99: 8127–8132.

Colosimo, P. F., K. E. Hosemann, S. Balabhadra, G. Villarreal, M. Dickson, J. Grimwood, J. Schmutz, R. M. Myers, D. Schluter, and D. M. Kingsley. 2005. Widespread parallel evolution in sticklebacks by repeated fixation of ectodysplasin alleles. *Science* 307: 1928–1933.

Coyne, J. A., and H. A. Orr. 2004. *Speciation.* Sunderland, MA: Sinauer.

Cummings, M. E. 2007. Sensory trade-offs predict signal divergence in surfperch. *Evolution* 61: 530–545.

Diamond, J. M. 1980. Species turnover in island bird communities. *Proc. 17th Int. Ornithological Congress* 2: 777–782.

———. 1984. "Normal" extinctions of isolated populations. In M. H. Nitecki, ed., *Extinctions*, 191–246. Chicago: University of Chicago Press.

Drummond, A. J., and A. Rambaut. 2007. BEAST: Bayesian evolutionary analysis by sampling trees. *BMC Evolutionary Biology* 7: 214.

Eldredge, N., J. N. Thompson, P. M. Brakefield, S. Gavrilets, D. Jablonski, J.B.C. Jackson, R. E. Lenski, B. S. Lieberman, M. A. McPeek, and W. Miller. 2005. The dynamics of evolutionary stasis. *Paleobiology* 31: 133–145.

Ewens, W. J. 2004. *Mathematical Population Genetics*, 2nd ed. New York: Springer-Verlag.

Gavrilets, S. 2004. *Fitness Landscapes and the Origin of Species.* Princeton, NJ: Princeton University Press.

Gavrilets, S., and N. Gibson. 2002. Fixation probabilities in a spatially heterogeneous environment. *Population Ecology* 44: 51–58.

Gillespie, J. H. 1999. The role of population size in molecular evolution. *Theor. Pop. Biol.* 55: 145–156.

———. 2004. *Population Genetics: A Concise Guide.* Baltimore: Johns Hopkins University Press.

Grant, B. R., and P. R. Grant. 2008. Fission and fusion of Darwin's finches populations. *Phil. Trans. Roy. Soc. Lond. B* 363: 2821–2829.

Grant, P. R. 1994. Population variation and hybridization—comparison of finches from two archipelagos. *Evol. Ecol.* 8: 598–617.

———. 1998. Speciation. In P. R. Grant, ed., *Evolution on Islands*, 83–101. Oxford: Oxford University Press.

———. 1999. *Ecology and Evolution of Darwin's Finches*, 2nd ed. Princeton, NJ: Princeton University Press.

———. 2001. Reconstructing the evolution of birds on islands: 100 years of research. *Oikos* 92: 385–403.

Grant, P. R., and B. R. Grant. 1996. Speciation and hybridization in island birds. *Phil. Trans. Roy. Soc. Lond. B* 351: 765–772.

———. 2008. *How and Why Species Multiply.* Princeton, NJ: Princeton University Press.

Grant, P. R., B. R. Grant, and K. Petren. 2000. The allopatric phase of speciation: the sharp-beaked ground finch *(Geospiza difficilis)* on the Galápagos Islands. *Biol. J. Linn. Soc.* 69: 287–317.

Haffer, J. 2007. *Ornithology, Evolution and Philosophy: The Life and Times of Ernst Mayr 1904–2005.* Berlin, Heidelberg, and New York: Springer.

Haldane, J.B.S. 1939. The equilibrium between mutation and random extinction. *Annals of Eugenics* 9: 400–405.

Hamilton, T. H., and I. Rubinoff. 1967. On predicting insular variation in endemism and sympatry for the Darwin finches in the Galápagos archipelago. *Am. Nat.* 101: 161–171.

Heaney, L. R. 2007. Is a new paradigm emerging for oceanic island biogeography? *J. Biogeog.* 34: 753–757.

Hoekstra, H. E., J. M. Hoekstra, D. Berrigan, S. N. Vignieri, A. Hoang, C. E. Hill, P. Beerli, and J. G. Kingsolver. 2001. Strength and tempo of directional selection in the wild. *Proc. Natl. Acad. Sci. USA* 98: 9157–9160.

Johnson, K. P., F. R. Adler, and J. L. Cherry. 2000. Genetic and phylogenetic consequences of island biogeography. *Evolution* 54: 387–396.

Lande, R. 1981. Models of speciation by sexual selection on polygenic traits. *Proc. Natl. Acad. Sci. USA* 78: 3721–3725.

———. 1996. Statistics and partitioning of species diversity, and similarity among multiple communities. *Oikos* 76: 5–13.

MacArthur, R. H., and E. O. Wilson. 1967. *The Theory of Island Biogeography.* Princeton, NJ: Princeton University Press.

Magallón, S., and M. J. Sanderson. 2001. Absolute diversification rates in angiosperm clades. *Evolution* 55: 1762–1780.

Mani, G. S., and B. C. Clarke. 1990. Mutational order—a major stochastic process in evolution. *Proc. Roy. Soc. Lond. B* 240: 29–37.

Mayr, E. 1954. Change of genetic environment and evolution. In J. Huxley, A. C. Hardy, and E. B. Ford, eds., *Evolution as a Process,* 157–180. London: Allen and Unwin.

———. 1959. Where are we? *Cold Spring Harbor Symposia on Quantitative Biology* 24: 1–24.

———. 1965. Avifauna: turnover on islands. *Science* 150: 1587–1588.

Mayr, E., and J. M. Diamond. 2001. *The Birds of Northern Melanesia: Speciation, Ecology, and Biogeography.* New York: Oxford University Press.

Ohta, T. 1972. Population size and rate of evolution. *J. Mol. Evol.* 1: 305–314.

———. 1973. Slightly deleterious mutant substitutions in evolution. *Nature* 246: 96–98.

Price, T., and A. Bontrager. 2001. Evolutionary genetics: The evolution of plumage patterns. *Current Biology* 11: R405–R408.

Price, T., M. Turelli, and M. Slatkin. 1993. Peak shifts produced by correlated response to selection. *Evolution* 47: 280–290.

Price, T. D. 2008. *Speciation in birds.* Boulder, CO: Roberts and Co.

Ricklefs, R. E., and E. Bermingham. 1999. Taxon cycles in the Lesser Antillean avifauna. *Ostrich* 70: 49–59.

———. 2002. The concept of the taxon cycle in biogeography. *Global Ecology and Biogeography* 11: 353–361.

271

———. 2007. The causes of evolutionary radiations in archipelagoes: passerine birds in the Lesser Antilles. *Am. Nat.* 169: 285–297.

Ricklefs, R. E., and G. W. Cox. 1972. Taxon cycles in the West Indian Avifauna. *Am. Nat.* 106: 195–219.

Rundle, H. D., A. O. Mooers, and M. C. Whitlock. 1998. Single founder-flush events and the evolution of reproductive isolation. *Evolution* 52: 1850–1855.

Rundle, H. D., and P. Nosil. 2005. Ecological speciation. *Ecol. Letters* 8: 336–352.

Schluter, D. 2001. Ecology and the origin of species. *Trends Ecol. Evol.* 16: 372–380.

———. 2008. Evidence for ecological speciation and its alternative. *Science* 266: 798–801.

Schluter, D., and T. Price. 1993. Honesty, perception and population divergence in sexually selected traits. *Proc. Roy. Soc. Lond. B* 253: 117–122.

Seehausen, O., Y. Terai, I. S. Magalhaes, K. L. Carleton, H. D. Mrosso, J. R. Miyagi, I. van der Sluijs, et al. 2008. Speciation through sensory drive in cichlid fish. *Nature* 455: 620–623.

Simpson, G. G. 1953. *The Major Features of Evolution.* New York: Columbia University Press.

Sorenson, M. D., A. Cooper, E. E. Paxinos, T. W. Quinn, H. F. James, S. L. Olson, and R. C. Fleischer. 1999. Relationships of the extinct moa-nalos, flightless Hawaiian waterfowl, based on ancient DNA. *Proc. Roy. Soc. Lond. B* 266: 2187–2193.

Templeton, A. R. 2008. The reality and importance of founder speciation in evolution. *Bioessays* 30: 470–479.

Terai, Y., O. Seehausen, T. Sasaki, K. Takahashi, S. Mizoiri, T. Sugawara, T. Sato, et al. 2006. Divergent selection on opsins drives incipient speciation in Lake Victoria cichlids. *Plos Biology* 4: 2244–2251.

Theron, E., K. Hawkins, E. Bermingham, R. E. Ricklefs, and N. I. Mundy. 2001. The molecular basis of an avian plumage polymorphism in the wild: A melanocortin-1-receptor point mutation is perfectly associated with the melanic plumage morph of the bananaquit, *Coereba flaveola. Current Biology* 11: 550–557.

Turner, J.R.G. and J.L.B. Mallet. 1996. Did forest islands drive the diversity of warningly coloured butterflies? Biotic drift and the shifting balance. *Phil. Tran. Roy. Soc. Lond. B* 351: 835–845.

Uy, J.A.C., R. G. Moyle, and C. E. Filardi. 2009. Changes in call and plumage color drives premating isolation between two populations of endemic flycatchers in the Solomon islands. *Evolution* 63: 153–164.

Uyeda, J. C., S. J. Arnold, P. A. Hohenlohe, and L. S. Mead. 2009. Drift promotes speciation by sexual selection. *Evolution* 63: 583–594.

Watson, D. M. 2005. Diagnosable versus distinct: evaluating species limits in birds. *Bioscience* 55: 60–68.

Weber, K. E., and L. T. Diggins 1990. Increased selection response in larger populations .2. selection for ethanol vapor resistance in *Drosophila melanogaster* at two population sizes. *Genetics* 125: 585–597.

Weir, J. T., E. Bermingham, and D. Schluter. 2009. The Great American Biotic Interchange in birds. *Proc. Natl. Acad. Sci. USA* 51: 21737–21742.

Weir, J. T., and D. Schluter. 2008. Calibrating the avian molecular clock. *Mol. Ecol.* 17: 2321–2328.

Whittaker, R. J., K. A. Triantis, and R. J. Ladle. 2008. A general dynamic theory of oceanic island biogeography. *J. Biogeog.* 35: 977–994.

Woolfit, M., and L. Bromham. 2005. Population size and molecular evolution on islands. *Proc. Roy. Soc. Lond. B* 272: 2277–2282.

Wright, S. 1964. Pleiotropy in the evolution of structural reduction and of dominance. *Am. Nat.* 98: 65–69.

Wunderle, J. M. 1981. An analysis of a morph ratio cline in the bananaquit (*Coereba flaveola*) on Grenada, West Indies. *Evolution* 35: 333–344.

APPENDIX

STATISTICS ON SPECIES TURNOVER ACROSS SMALL AND SEPARATELY ACROSS LARGE ISLANDS WITHIN ARCHIPELAGOES

Taxon	Climate	Location*	Island Sizes			Small > Large?	Partitions (km²)†	Average Number of Species on an Island (Number of Islands)			Reference
			Small	Medium	Large			Small	Medium	Large	
Eucalyptus	Tropical	Australia	53	38		yes	40	11.75 (4)	13.75 (4)		Margules and Stein 1989
Plants–Vascular	Temperate	Boreal Forest	50	55		no	0.02	57.25 (4)	51.33 (6)		Ohlson et al. 1996
Plants–Mosses	Temperate	Boreal Forest	55	63		no	0.02	41 (4)	43 (6)		Ohlson et al. 1996
Plants–Hepatics	Temperate	Boreal Forest	55	57		no	0.02	19.25 (8)	23.5 (2)		Ohlson et al. 1996
Plants–Lichens	Temperate	Boreal Forest	52	60		no	0.02	26.13 (8)	28 (2)		Ohlson et al. 1996
Plants	Temperate	Boreal Forest	69	43	28	yes	0.18	6.72 (33)	17.13 (8)	20.25 (4)	Berglund and Jonsson 2003
Plants	Tropical	Australian Cays	48	44		yes	0.19	29.8 (5)	42.8 (5)		Chaloupka and Domm 1986
Plants	Temperate	Chicago Lots	53	71	58	no	.0003, .0015	17.25 (4)	30.93 (15)	43.83 (7)	Crowe 1979
Plants	Temperate	California	35	27		yes	0.01	7.83 (6)	13.14 (7)		Riebesell 1982
Plants	Tropical	Hawaii Frigate*	75	40		yes	0.01	1 (5)	4.2 (5)		Amerson 1975
Plants	Tropical	Hawaii Reefs*	75	47		yes	0.04	1 (4)	8.5 (4)		Amerson 1975
Plants	Tropical	Hawaii*	25	4		yes	1295	9 (3)	12.5 (4)		Carlquist 1974

Fungi	Temperate	Boreal Forest	52	61		no	0.02	7.25 (4)	12 (6)		Ohlson et al. 1996
Nematodes	Temperate	W. Mediterranean	57	61		no	300	8.6 (5)	18.4 (6)		De Bellocq et al. 2002
Landsnails	Temperate	Aegean Islands*	90	88	73	yes	80, 1000	1 (34)	1.47 (19)	2.75 (4)	Heller 1976
Millipedes	Temperate	Canary Islands*	75	67		yes	1000	3.75 (4)	10.67 (3)		Enghoff and Baez 1993
Ants	Temperate	Thimble Islands*	67	63	21	yes	900	9.56 (9)	8.4 (5)	26 (2)	Goldstein 1975
Ants	Temperate	Frisian Islands*	63	28	18	yes	10, 30	6.29	14.17	15.6	Boomsma et al. 1987
Ants	Temperate	Florida*	44	39		yes	0.015[1]	5 (4)	5.5 (2)		Simberloff and Wilson 1969
Insects	Temperate	Japanese Arch.*	60	71	48	yes	700, 1400	2 (4)	2 (4)	6.75 (2)	Millien-Parra and Jaeger 1999
Insects	Tropical	Hawaii*	40	0		yes	1295	3 (3)	5 (4)		Carlquist 1974
Beetles	Temperate	Stockholm Ils*	62	50		yes	0.3	10.33 (6)	19 (4)		As 1984
Spiders	Tropical	Hawaii*	57	57	36	no	1000	4.33 (3)	4.5 (2)		Garb 1999
Bats	Tropical	West Indies*	40	70	64	no	100, 1000	5.4 (5)	7.6 9 (10)	16.67 (6)	Griffiths and Klingener 1988
Bats	Tropical	South America*		45	46	no	1000	11 (4)	28.5 (3)		Koopman 1958
Bats	Tropical	Krakatau*	50	25		yes	10	4 (2)	3 (2)		Thornton et al. 2002
Carnivora	Temperate	Japanese Arch*	90	75	48	yes	700, 1400	.5 (4)	1.75 (4)	6.75 (4)	Millien-Parra and Jaeger 1999

(Continued)

APPENDIX
(Continued)

Taxon	Climate	Location*	Island Sizes			Small > Large?	Partitions (km²)†	Average Number of Species on an Island (Number of Islands)			Reference
			Small	Medium	Large			Small	Medium	Large	
Insectivora	Tropical	Mindanao*	75	73		yes	1000	.5 (4)	1.6 (4)		Heaney 1986
Rodentia	Temperate	Japanese Arch*	54	54	42	yes	700, 1400	3.25 (4)	2.75 (4)	9.25 (4)	Millien-Parra and Jaeger 1999
Rodentia	Tropical	Mindanao*	43	53		no	1000	6.25 (4)	9 (4)		Heaney 1986
Mammals	Temperate	Virginia Barrier*	66	56		yes	10	2.75 (7)	4 (4)		Dueser et al. 1979
Mammals	Temperate	Japanese Arch*	50	71	38	yes	500, 5000	3 (4)	3.7 (4)	11.25 (4)	Millien-Parra and Jaeger 1999
Mammals	Arctic	Alaska*	44	57	45	no	200, 1000	4.5 (8)	9.56 (9)	11 (7)	Conroy et al. 1999
Mammals	Temperate	Great Basin	59	19		yes	70	5.36 (11)	9.75 (8)		McDonald and Brown 1992
Mammals	Tropical	Australia	67	55		yes	5	5.92	9.4		Kitchener et al. 1980
Mammals	Tropical	Australia	53	24		yes	0.2	2.33 (6)	3.8 (5)		Pahl et al. 1988
Mammals	Temperate	Southwest	78	58	42	yes	200, 800	3.38 (16)	7.5 (6)	11 (5)	Lomolino et al. 1989

Mammals	Temperate	Southwest	51	50		yes	1000	3.4 (5)	6 (5)		Davis et al. 1988
Mammals	Temperate	Wisconsin Forest	45	32		yes	0.04	6 (16)	8.83 (6)		Matthaie and Stearns 1981
Mammals	Arctic	Alexander Arch.*	58	46	48	yes	400, 2000	5 (10)	10.75 (8)	10.33 (6)	Conroy et al. 1999
Mammals	Temperate	Southwest	65	48		yes	1000	2.43 (7)	6.2 (5)		Davis et al. 1988
Mammals	Temperate	Australia	82	69		yes	30	1.06 (16)	2.17 (6)		Main and Yadov 1971
Mammals	Temperate	Maine*	55	43	36	yes	.06, 1	2.25 (8)	3.4 (5)	5.75 (4)	Crowell 1986
Mammals	Temperate	Maine*	45	34		yes	50	12.6 (4)	21.7 (3)		Crowell 1986
Mammals	Temperate	Lake Michigan*	54	33		yes	3	0.56 (9)	4 (5)		Lomolino 1986
Mammals	Temperate	Thousand Isles*	84	71	53	yes	0.02, 0.08	0.8 (5)	1.71 (7)	4.25 (7)	Lomolino 1986
Mammals	Temperate	Massachusetts*	52	60	24	yes	69, 20	2.88 (25)	3.2 (5)	5.33 (3)	Adler and Wilson 1985
Mammals	Temperate	Oxford lots–wood	81	33		yes	0.08	7.83 (6)	11. 33 (6)		Dickman 1987
Mammals	Temperate	Oxford lots–scrub	51	34		yes	0.01	8.4 (5)	9.25 (4)		Dickman 1987
Mammals	Temperate	Oxford lots–grass	57	30		yes	0.01	5.55 (9)	9.75 (4)		Dickman 1987

Notes: Islands within archipelagoes were divided into small, medium, and large (or small and medium) groups based on area, with the †partitions as indicated in km^2 (except [1]km diameter, and [2]log number of bushes). Variance among islands in species composition within each group was calculated following the formula of Lande (1996, eq. 8, ignoring population densities; see also Price 2008, chapter 3 for a numerical example). If all islands within an island group have the same species, the variance is 0, and if all islands within a size contain different species, variance is a maximum. In the majority of cases, variance among small islands is larger than variance among large islands, as indicated in the column (Small > Large?), indicating species turnover is larger across small islands. *True islands (i.e., land surrounded by water, others are unusual land masses surrounding by other land, e.g., mountaintops).

REFERENCES TO APPENDIX ENTRIES

Adler, G. H., and M. L. Wilson. 1985. Small mammals on Massachusetts Islands—the use of probability functions in clarifying biogeographic relationships. *Oecologia* 66: 178–186.

Ahlen, I. 1983. The bat fauna of some isolated islands in Scandinavia. *Oikos* 41: 352–358.

Alberto, J.A.N., and B.F.J. Manly. 2006. The generation of diversity in systems of patches and ranked dominance. *J. Biogeog.* 33: 609–621.

Amerson, A. B. 1975. Species richness on nondisturbed northwestern Hawaiian islands. *Ecology* 56: 435–444.

As, S. 1984. To fly or not to fly—colonization of Baltic islands by winged and wingless carabid beetles. *J. Biogeog.* 11: 413–426.

Bengston, S. A., and D. Bloch. 1983. Island land bird population densities in relation to island size and habitat quality on the Faroe Islands. *Oikos* 41: 507–522.

Berglund, H., and B. G. Jonsson. 2003. Nested plant and fungal communities; the importance of area and habitat quality in maximizing species capture in boreal old-growth forests. *Biol. Conserv.* 112: 319–328.

Boomsma, J. J., A. A. Mabelis, M.G.M. Verbeek, and E. C. Los. 1987. Insular biogeography and distribution ecology of ants on the Frisian Islands. *J. Biogeogr.* 14: 21–37.

Carlquist, S. 1974. *Island Biology*. New York: Columbia University Press.

Chaloupka, M. Y., and S. B. Domm. 1986. Role of anthropochory in the invasion of coral cays by alien flora. *Ecology* 67: 1536–1547.

Conroy, C. J., J. R. Demboski, and J. A. Cook. 1999. Mammalian biogeography of the Alexander Archipelago of Alaska: A north temperate nested fauna. *J. Biogeogr.* 26: 343–352.

Crowe, T. M. 1979. Lots of weeds: insular phytogeography of vacant urban lots. *J. Biogeogr.* 6:169–181.

Crowell, K. L. 1986. A comparison of relict versus equilibrium models for insular mammals of the Gulf of Maine. *Biol. J. Linn. Soc.* 28: 37–64.

Davis, R., C. Dunford, and M. V. Lomolino. 1988. Montane mammals of the American southwest: The possible influence of post-Pleistocene colonization. *J. Biogeogr.* 15: 841–848.

de Bellocq, J. G., S. Morand, and C. Feliu, C. 2002. Patterns of parasite species richness of Western Palaeartic micro-mammals: island effects. *Ecography* 25: 173–183.

Diamond, J. M., and A. G. Marshall. 1976. Origins of the New Hebridean avifauna. *Emu* 76: 187–200.

Dickman, C. R. 1987. Habitat fragmentation and vertebrate species richness in an urban-environment. *J. Appl. Ecology* 24: 337–351.

Dueser, R. D., W. C. Brown, G. S. Hogue, C. McCaffrey, S. A. McCuskey, and G. J. Hennessey. 1979. Mammals on the Virginia Barrier Islands. *J. Mammal.* 60: 425–429.

Enghoff, H., and M. Baez. 1993. Evolution of distribution and habitat patterns in endemic millipedes of the genus *Dolichoiulus* (Diplopoda, Julidae) on the Canary Islands, with notes on distribution patterns of other Canarian species swarms. *Biol. J. Linn. Soc.* 49: 277–301.

Garb, J. E. 1999. An adaptive radiation of Hawaiian Thomisidae: Biogeographic and genetic evidence. *J. Arachnology* 27: 71–78.

Gardner, A. S. 1986. The biogeography of the lizards of the Seychelles Islands. *J. Biogeogr.* 13: 237–253.

Gill, A. E. 1981. Morphological features and reproduction of *Perognathus* and *Peromyscus* on northern islands in the Gulf of California. *Amer. Midl. Nat.* 106: 192–196.

Goldstein, E. L. 1975. Island biogeography of ants. *Evolution* 29: 750–762.

Gotelli, N. J., and L. G. Abele. 1982. Statistical distributions of West Indian land bird families. *J. Biogeogr.* 9: 421–435.

Griffiths, T. A., and D. Klingener. 1988. On the distribution of Greater Antillean bats. *Biotropica* 20: 240–251.

Haila, Y., O. Jarvinen, amnd S. Kuusela. 1983. Colonization of islands by land birds: prevalence functions in a Finnish Archipelago. *J. Biogeogr.* 10: 499–531.

Harris, M. P. 1973. The Galápagos avifauna. *Condor* 75: 265–278.

Heaney, L. R. 1986. Biogeography of mammals in SE Asia: estimates of rates of colonization, extinction and speciation. *Biol. J. Linn. Soc.* 28: 127–165.

Heatwole, H. 1975. Biogeography of reptiles on some of the islands and cays of eastern Papua-New Guinea. *Atoll Research Bulletin* 180.

Heller, J. 1976. The biogeography of Enid Landsnails on the Aegean Islands. *J. Biogeogr.* 3: 281–292.

Johnson, N. 1975. Controls of number of bird species on montane islands in the Great Basin. *Evolution* 29: 545–567.

Jones, K. B., L. P. Kepner, and T. E. Martin. 1985. Species of reptiles occupying habitat islands in western Arizona: a deterministic assemblage. *Oecologia* 66: 595–601.

King, R. B. 1987. Reptile distributions on Islands in Lake Erie. *J. Herpetol.* 21: 65–67.

Kitchener, D. J., A. Chapman, J. Dell, B. G. Muir, and M. Palmer. 1980. Lizard assemblage and reserve size and structure in the Western Australian wheat-belt—some implications for conservation. *Biol. Conserv.* 17: 25–62.

Koopman, K. F. 1958. Land bridges and ecology in bat distribution on islands off the northern coast of South America. *Evolution* 12: 429–439.

Kratter, A. W. 1992. Montane avian biogeography in souther California and Baja California. *J. Biogeogr.* 19: 269–283.

Lomolino, M. V. 1986. Mammalian community structure on islands—the importance of immigration, extinction and interactive effects. *Biol. J. Linn. Soc.* 28: 1–21.

Lomolino, M. V., J. H. Brown, and R. Davis. 1989. Island biogeography of montane forest mammals in the American Southwest. *Ecology* 70: 180–194.

Main, A. R., and M. Yadov. 1971. Conservation of macropods in reserves in western Australia. *Biol. Conserv.* 3: 123–133.

Margules, C. R., and J. L. Stein. 1989. Patterns in the distributions of species and the selection of nature reserves—an example from *Eucalyptus* forests in southeastern New South Wales. *Biol. Conserv.* 50: 219–238.

Matthiae, P. E., and F. Stearns. 1981. Mammals in forest islands in southeastern Wisconsin. *Ecological Studies* 41: 55–66.

McDonald, K. A., and J. H. Brown. 1992. Using montane mammals to model extinctions due to global change. *Conserv. Biol.* 6: 409–415.

Millien-Parra, V., and J. J. Jaeger. 1999. Island biogeography of the Japanese terrestrial mammal assemblages: an example of a relict fauna. *J. Biogeogr.* 26: 959–972.

Murphy, R. W., and G. Aguirre-Leon. 2002. Nonavian reptiles: origins and evolution. In T. J. Case, M. L. Cody, and E. Ezcurra, eds., *A New Island Biogeography in the Sea of Cortés*, 181–220. New York: Oxford University Press.

Nevarro Alberto, J. A., and B.F.J. Manly. 2006. The generation of diversity in systems of patches and ranked dominance. *J. Biogeogr.* 33: 609–621.

Ohlson, M., L. Söderström, G Hörnberg, O. Zackrisson, and J. Hermansson. 1997. Habitat qualities versus long-term continuity as determinants of biodiversity in boreal old-growth swamp forests. *Biol. Conserv.* 81: 221–231.

Pahl, L. I., J. W. Winter, and G. Heinsohn. 1988. Variation in responses of arboreal marsupials to fragmentation of tropical rainforest in north eastern Australia. *Biol. Conserv.* 46: 71–82.

Power, D. M. 1972. Numbers of bird species on California Islands. *Evolution* 26: 451–463.

Raven, H. C. 1935. Wallace's line and the distribution of Indo-Australian mammals. *Bull. Amer. Mus. Nat. Hist.* 68: 179–293.

Riebesell, J. F. 1982. Arctic Alpine plants on mountain tops—agreement with island biogeography theory. *Am. Nat.* 119: 657–674.

Rusterholz, K. A., and R. W. Howe. 1979. Species-area relations of birds on small islands in a Minnesota lake. *Evolution* 33: 468–477.

Silva, M., L. A. Hartling, S. A. Field, and K. Teather. 2003. The effects of habitat fragmentation on amphibian species richness of Prince Edward Island. *Canadian Journal of Zoology—Revue Canadienne De Zoologie* 81: 563–573.

Simberloff, D. 1976. Experimental zoogeography of islands—effects of island size. *Ecology* 57: 629–648.

Simberloff, D. S. and E. O. Wilson. 1969. Experimental zoogeography of islands—colonization of empty islands. *Ecology* 50: 278–296.

Simberloff, D. S., and F. Vuilleumier. 1980. Ecology versus history as determinants of patchy and insular distributions in high Andean birds. *Evol. Biol.* 12: 235–379.

Thornton, I.W.B., D. Runciman, S. Cook, L. F. Lumsden, T. Partomihardjo, N. K. Schedvin, J. Yukawa, and S. A. Ward. 2002. How important were stepping stones in the colonization of Krakatau? *Biol. J. Linn. Soc.* 77: 275–317.

Ward, L. K., and K. H. Lakhani. 1977. Conservation of juniper fauna of food plant island sites in southern England. *J. Appl. Ecology* 14: 121–135.

Wright, S. J. 1985. How isolation affects rates of turnover of species on islands. *Oikos* 44: 331–340.

Chapter Fourteen

Geographical Mode and Evolutionary Mechanism of Ecological Speciation in Cichlid Fish

Ole Seehausen and Isabel Santos Magalhaes

The spatial context of speciation has been a major issue in evolutionary biology and systematics for nearly seventy years (Abbott et al. 2008; Bolnick and Fitzpatrick 2007; Bush 1969; Coyne and Orr 2004; Gavrilets 2004; Mayr 1942, 1963). Specifically, the theoretical possibility, empirical reality, and relevance of sympatric speciation have been, and continue to be, a significant battleground (Bush 1998; Mayr 1982). Whereas a handful of compelling empirical cases has lately led to fairly broad agreement that sympatric speciation has indeed happened in nature (Bolnick and Fitzpatrick 2007; Coyne and Orr 2004), disagreement prevails over two consequential questions. The first is the question, what in fact constitutes sympatric speciation (Fitzpatrick et al. 2008)? The other one is the question whether sympatric speciation is common, and hence of relevance to understanding the origins and structure of species diversity (Bolnick and Fitzpatrick 2007). Several authors have suggested that instead of discussing the geographical mode of speciation, a more fruitful approach would be to focus on the mechanisms that play major roles in speciation, such as the sources of divergent selection (Grant and Grant 2006; Rundle et al. 2000) and the balance between gene flow, selection, and mate choice (Butlin et al. 2008; Grant and Grant 2008; Nosil and Crespi 2004; Rundle et al. 2000; Thibert-Plante and Hendry 2009). Whereas we strongly advocate the mechanistic approach, we wish to stress that the geography of speciation remains equally important for understanding the structure of biodiversity.

We agree with others that discrete classification of speciation modes into allopatric, sympatric, and parapatric, using criteria of theoretical population genetics (i.e., m = 0, m = 0.5, and 0 < m < 0.5), is not likely to produce major new insights into the importance of modes of speciation. Most cases of speciation are likely to be parapatric by this definition (Bolnick and Fitzpatrick 2007; Fitzpatrick et al. 2008). We further agree with Bolnick and Fitzpatrick (2007) that counting compelling cases of sympatric and allopatric speciation is suffering from ascertainment bias. We believe that the way forward necessitates development of methods for quantitative comparative investigations of the role of spatial structure in as many as possible cases of speciation that cannot be categorized as either allopatric or sympatric. In this chapter we investigate several examples of speciation in an unusually rapid large animal radiation, the cichlid fish of Lake Victoria, East Africa.

We suggest ways to extract quantitative rules about the role of space in speciation. The idea requires investigating many evolutionary replicates of divergence of essentially similar phenotypes that vary in the amount of spatial structure and in the levels of completion of speciation. We apply this idea to twelve divergent phenotype pairs of Lake Victoria cichlid fish, representing 5, 4, and 3 replicates of parallel divergence along water depth gradients in three different trait complexes. We proceed in three steps: first we present a short overview of current thinking on modes of speciation in the large cichlid fish radiations. Then we review the replicates of parallel divergence that our group investigates. Finally, we subject these data to an integrated comparative analysis for testing effects of variation in spatial structure on the progress in speciation.

Geographical Modes of Speciation in African Great Lake Haplochromines

The very high ratio of endemic species to river-dwelling species in and around the large lakes, Lake Victoria and Malawi, has sometimes been taken as support for sympatric speciation at coarse geographical grain (Meyer et al. 1990). On the other hand, in recent years allopatric speciation has been suggested to be the predominant mechanism of speciation in the African Great Lake cichlids (Kocher 2004). In these large lakes, two models of speciation involving geographical isolation may account for a significant portion of the species origination. The first involves geographical speciation in isolated lake basins or satellite lakes during lake level low stands, followed by secondary sympatry during subsequent lake level high stands (Sturmbauer 1998). Although proposed dozens of times, compelling evidence is rare. A well-supported case comes from Lake Malawi (Genner et al. 2007). Indirect evidence exists for Lake Tanganyika haplochromines (Sturmbauer et al. 2003), whereas for the classical case of the Lake Victoria satellite Lake Nabugabo (Greenwood 1965), evidence remains anecdotal. More generally, the explanatory power of this mechanism appears limited in lakes where many species evolved during periods of relatively insignificant water level fluctuations. The most extreme such case is Lake Victoria after the latest refilling 15,000 years ago (Stager and Johnson 2008).

The second model involves intralacustrine allopatric speciation when populations diverge in isolation on patches of the same type of habitat separated from each other by other, unsuitable habitat. Such speciation has often erroneously been called microallopatric in the cichlid literature, but the latter concept more correctly applies to speciation that appears at coarse grain sympatric (geographical ranges fully overlapping) but allopatric at fine grain (i.e., host-race formation) (Fitzpatrick et al. 2008). Support for intralacustrine allopatric speciation mostly comes from patterns of species distribution and intraspecific genetic differentiation and is mostly confined to cichlid groups that are restricted to patchily distributed habitats like rocky shores (Kocher 2004). However, given that such habitat specialization does not exist in typical riverine haplochromines, and by inference did probably have to evolve from habitat generalist ancestors after these colonized the lakes, it is perhaps unlikely that geographical isolation alone would have been sufficient to initiate the radiations.

On the other hand, there is compelling evidence for sympatric speciation in cichlid fish of other evolutionary lineages that occupy small isolated crater lakes in West Africa and Nicaragua. While these lineages have made small radiations in isolated crater lakes, they have not made larger species flocks despite their occurrence in large lakes too. Mitochondrial and nuclear genomic evidence taken together suggest that all eleven endemic species of crater lake Barombi Mbo in Cameroon have arisen within the lake, which is thought to be too small to allow within-lake geographical isolation (Schliewen and Klee 2004).While allopatric speciation seems very unlikely indeed in this lake, and also in the nearby lakes Ejagham and Bermin that have smaller radiations too, detailed population-biological analyses required to reveal the fine grain of spatial structure are yet to be conducted. The latter concern also applies to the other compelling case of sympatric speciation in cichlid fish, the heroine cichlids of crater lakes in the vicinity of Lake Nicaragua (Barluenga et al. 2006). The question really is, how much or how little spatial differentiation is required for speciation in cichlid fish and beyond. In this chapter, we attempt to provide an answer to this question for Lake Victoria cichlids.

Mode and Mechanism of Speciation in Replicated Cases of Incipient Speciation
General Approach

We identify replicate pairs of phenotypically similar cichlid morphs and species that occur at multiple islands and vary between islands in the extent of their genetic and phenotypic differentiation, that is, in their "progress in speciation" (Nosil and Sandoval 2008). We investigate such replicate pairs for each of the three major classes of phenotypic polymorphisms that have been associated with incipient speciation in Lake Victoria cichlids; male nuptial coloration, trophic (dental) morphology, and X-linked (female) coloration. Using these, we explore the conditions that are associated with variation in the extent of differentiation (i.e., the "progress" in speciation). We ask:

1. Does ecological habitat structure matter to speciation?
2. Which trait complexes are most strongly affected by divergent selection?

Throughout, our definition of "species" is one close to the genotypic cluster definition (Mallet 1995). We consider two groups species if they differ significantly in neutral multilocus genotypes and maintain these differences in sustained sympatry. We measure such differentiation using at least nine microsatellite loci.

Replicated Polymorphisms

MALE NUPTIAL COLORATION

The most frequent intra- and interspecific polymorphism in male nuptial coloration involves blue versus yellow-red flank coloration of males (Seehausen and van Alphen 1999). We investigated replicate populations of one taxon pair that epitomizes this type of polymorphism. *Pundamilia pundamilia* and *P. nyererei* are a geographically completely sympatric pair of sibling species endemic to the lake proper (plate 10a). The distribution of *P. nyererei* is nested within that of *P. pundamilia.* Independent evidence for speciation in this phenotype pair comes from neutral marker differentiation among sympatric populations (plate 13). Where the phenotypes are significantly differentiated incipient species, many differences between them have been documented (reviewed in Seehausen 2009). These include differences in depth distribution (Seehausen 1997; Seehausen et al. 2008), stomach contents (Bouton et al. 1997), stable isotopes (Mrosso et al. unpublished), parasites (Maan, Van Rooijen, et al. 2008), color vision (Carleton et al. 2005; Maan et al. 2006; Seehausen et al. 2008), female mate choice (Haesler and Seehausen 2005; Seehausen and van Alphen 1998; Stelkens et al. 2008; van der Sluijs, van Dooren, et al. 2008), male aggression bias (Dijkstra, Hemelrijk, et al. 2008; Dijkstra et al. 2006; Dijkstra et al. 2007; Verzijden et al. 2009), and male behavioral dominance (red > blue) (Dijkstra et al. 2005). To test for evidence of selection, we calculated and compared F_{ST} in opsin genes with F_{ST} in unlinked microsatellites.

Trophic Morphology

The genus *Neochromis* is composed of epilithic algae scrapers with various degrees of specialization. At sites in turbid waters, *Neochromis greenwoodi* is a rather unspecialized omnivorous scraper. At islands with clearer water and hence better algae growth, the species assumes a more specialized scraper morphology with a strongly decurved dorsal head profile, subequally bicuspid oral teeth in the outer tooth rows in both oral jaws, and many bands of oral teeth. At clear water offshore islands it is generally replaced by its allopatric sister species *N. omni-caeruleus* with similar but even more specialized algae scraper morphology. At Bihiru Island the species *N.* "Bihiru scraper" takes their position (Bouton et al. 1999; Seehausen 1996). At some of the islands with a relatively specialized scraper population of either species, a second sympatric phenotype occurs. This morph is distinguished by unicuspid teeth in the outer tooth rows, teeth arranged in fewer bands, and usually a less strongly decurved dorsal head profile (plate 10b). The distribution of the ratio unicuspid/bicuspid teeth is at all three islands consistent with disruptive selection on the dental morphology (fig. 14.1c). We found that the replicate pairs of these sympatric morphs show within-island phenotype-environment correlations between dentition type and water depth: the unicuspid morphs tend to occupy deeper waters (Magalhaes et al. unpubl. data). At one island (Makobe) we collected stable isotope data, and these suggest long-term differences in diet: the unicuspid morph tends to be more "limnetic" (H.D.J. Mrosso et al. unpublished). Subsequently, we refer to these sympatric phenotypes as ecomorphs. We studied the sympatric polymorphism at three different islands, using phenotypic information and neutral genetic markers.

To test for evidence of selection, we estimated P_{ST} values for morphology and for dentition. P_{ST} values are equivalent to Q_{ST} values (Spitze 1993) but may be influenced by environmental and non-additive genetic effects. P_{ST} values for morphology were estimated from Principal Components. Between- and within-ecomorph components of phenotypic variances were estimated by performing an analysis of variance using SPSS version 14.0 (SPSS Inc.). Bootstrapping over individuals, using R (http://www.r-project.org/), was performed to calculate confidence intervals. P_{ST} values were quantified as the proportion of variance in quantitative traits attributable to differences among ecomorphs (P_{ST} = $\sigma^2 gb/\sigma^2 gb + 2(h^2\sigma^2 gw)$, where $\sigma^2 gb$ and $\sigma^2 gw$ are the among-ecomorph and within-ecomorph variance components respectively (Spitze 1993) and h^2 is the heritability. We have assumed a heritability of 0.5 (Merilä 1997; Ostbye et al. 2005; Leinonen et al. 2006). P_{ST} values were considered significantly different from multilocus F_{ST} values when their 95 percent confidence intervals did not overlap.

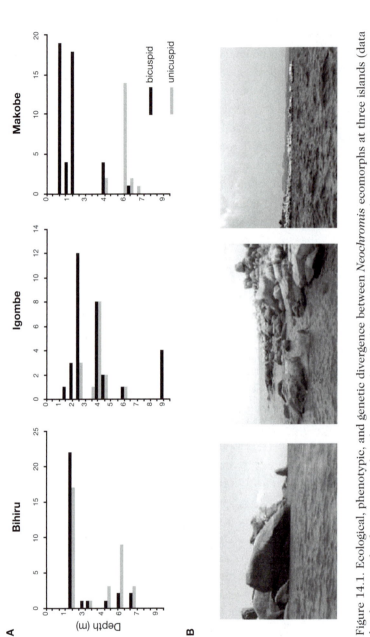

Figure 14.1. Ecological, phenotypic, and genetic divergence between *Neochromis* ecomorphs at three islands (data for the same island are presented in the same column). (a) Depth distributions of ecomorphs. (b) The shore types (*from left to right*): steep with large boulders, intermediate slope and boulder size, gentle with small boulders. (c) Frequency distributions of dentition phenotypes, quantified as the ratio unicuspid/bicuspid teeth. (d) Measures of neutral versus adaptive differentiation between sympatric ecomorphs. P_{ST} values (with their 95% confidence intervals) versus multilocus F_{ST} values (with their 95% confidence intervals; 9 µsat loci). Data from left to right refer to morphometric PC1, PC2, and PC3, number of tooth rows and percentage of unicuspid teeth in the outer tooth row. The solid lines indicate the multilocus F_{ST} value. The dotted lines indicate the 95% confidence intervals of the F_{ST} value.

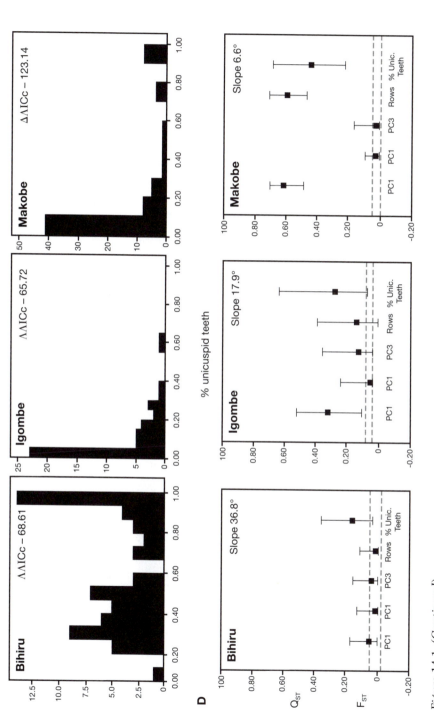

Figure 14.1. (*Continued*)

X-linked (Female) Color

Very similar X-linked female color polymorphisms are found in several genera of Lake Victoria cichlids, and in multiple species within some of the genera. This type of polymorphism involves an orange blotched (OB; black blotches on orange), a white blotched (WB; black blotches on white), and a normal "plain" phenotype (plate 11). OB and WB color is sex linked in all populations for which data are available (Lande et al. 2001). Sex linkage of both OB and WB has been attributed to weak physical linkage between two different major effect genes for color and (dominant) female determination, which together are linked to the original (recessive) female determining X chromosome (Seehausen et al. 1999). Blotched individuals are female, unless additional autosomal male determining genes are present that more than compensate for the dominant female effect of the OB and WB genes. OB and WB males are extremely rare (<1%) in nature in most studied populations (Lande et al. 2001; Maan, Eshuis, et al. 2008; Seehausen et al. 1999). Species possessing OB or WB morphs usually, but not always, also possess a plain (P) morph in both sexes, which is considered the ancestral condition. Theoretical studies modeling this polymorphism suggested that sympatric speciation may be possible through the interaction of selection on sex reversal and sexual selection (Lande et al. 2001).

Most of the knowledge on this type of polymorphism derives from studies on *N. omnicaeruleus*. Some populations of this species possess all three female color morphs: WB, OB, and P, others have just two (generally OB and P), and again others have just P morphs. Geographically the three morphs are fully sympatric with WB nested in OB and P, and OB nested in P (plate 11). There is no evidence of microhabitat differentiation (fig. 14.4a) (Maan, Eshuis, et al. 2008; Seehausen and Bouton 1997) and only weak evidence for morphological differentiation between color morphs (Magalhaes et al. 2010; Seehausen et al. 1999). Yet, behavioral investigations had found partially assortative mating preferences based on this color variation (Seehausen et al. 1999; Pierotti et al. 2009), and assortative aggression biases among the females (Dijkstra, Seehausen, et al. 2008). We have reported frequencies of the three *N. omnicaeruleus* color morphs in a natural population for several years and found a large numerical deficiency of intermediate phenotypes suggesting nonrandom mating and/or strong disruptive selection on the phenotype (fig. 14.2b) (Maan, Eshuis, et al. 2008; Magalhaes et al. 2010; Seehausen et al. 1999). We investigated the differentiation between the color morphs at this island and in *N. greenwoodi* at Igombe Island using phenotypic and neutral genetic markers (Magalhaes et al. 2010).

A METHOD FOR ASSESSING AND COMPARING EFFECTS OF FINE GRAIN SPATIAL STRUCTURE

In order to compare the relative importance to speciation of spatial ecological structure between the three common types of polymorphisms, we calculated for each replicate phenotype pair the correlation between phenotype and water depth. A correlation coefficient (R) of 1 is the equivalent of saying that phenotypes have little spatial overlap. A correlation coefficient of 0 is the equivalent of saying that phenotypes have complete spatial overlap. We then asked if variation in the progress of speciation (measured by the global µsat F_{ST}), was explained by the variation in spatial overlap, by the kind of phenotypic polymorphism, or the interaction between these. We first calculated Pearson's Correlation Coefficients between global F_{ST} and the phenotype*depth correlation. We then calculated a general linear model with global F_{ST} as dependent variable, type of polymorphism (3 types) as factor, and the correlation coefficient (R) between phenotype and water depth as a covariate.

To express spatial overlap in a general currency that is comparable across studies, we propose to express the correlation coefficient between phenotype and water depth in terms of an estimate of spatial opportunity for cross-mating between morphs. We can calculate potential m as

$$m = \frac{(1-R)}{2}$$

where m ranges from 0 (two spatially isolated populations) to 0.5 (a spatially panmictic population). Importantly, our estimate of m is not an estimate of actual gene flow, but one of the potential for gene flow between morphs if there was no behavioral assortative mating (note

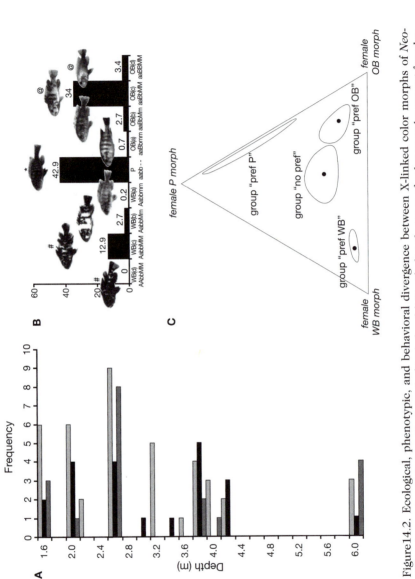

Figure 14.2. Ecological, phenotypic, and behavioral divergence between X-linked color morphs of *Neochromis omnicaeruleus* at the island of Makobe. (a) Depth distributions of color morphs among females. (b) Frequency distributions of color morphs among females. (c) Male mating preference classes, generated by a hierarchical cluster analysis of individual male preferences for female color morphs. Compositional geometric mean and confidence region for each preference class are plotted in a ternary diagram.

that seasonality does not contribute to reproductive isolation between these incipient species).

RESULTS

We will first review our results for each of the three polymorphisms, followed by a discussion on the quantitative comparative analysis of all phenotype pairs.

1. Male Nuptial Coloration

Evidence for incipient speciation among geographically sympatric male nuptial color morphs is seen in significant neutral marker differentiation (we used 11 microsatellites) at three of our five islands (plate 13a). Evidence for $m > 0$ is seen in replicate pairs studied at five islands: sympatric populations of different phenotype are often more similar to each other in their allele frequencies at neutral loci (11 unlinked microsatellites) than allopatric populations of same phenotype (fig. 14.3a). The extent of sympatric differentiation at neutral loci exhibits continuous variation between replicate phenotype pairs (plate 13a). The extent of sympatric differentiation in male nuptial coloration too exhibits continuous variation (plate 12b). In contrast, the extent of sympatric differentiation in a visual pigment gene exhibits rather discontinuous variation (plate 13b). Where phenotypes are differentiated incipient species, "Blue" lives in more shallow waters than "Red" and in more blue shifted ambient light (plate 12a). Haplochromine cichlids of Lake Victoria express four or five different opsin genes, each coding a visual pigment with peak sensitivities at different wavelength. The most variable of these genes among the cichlids of Lake Victoria is the long wavelength sensitive opsin gene (*LWS*). *LWS* alleles of individuals of the blue *Pundamilia* species are blue shifted by 15 nm relative to those of fish of the red sister species. This was determined through in-vitro mutagenesis of the red shifted allele, expression of both proteins and absorbance measurements (Seehausen et al. 2008; Terai et al. 2006).

Ambient subsurface light composition changes markedly with increasing water depth, not just in its intensity but in its wavelength composition. The slope of this light gradient varies with water transparency and physical shore slope (steep when water is turbid and when

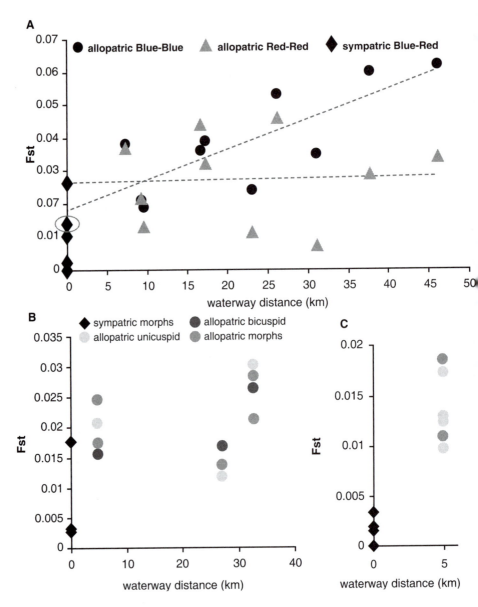

Figure 14.3. Evidence for divergence with gene flow. Microsatellite-based estimates of F_{ST} between sympatric phenotypes (black diamonds) are generally smaller than those between allopatric populations of the same phenotype, here plotted against the geographical distance between sampling sites. (a) Blue and red male nuptial color morphs of *Pundamilia*. (b) Tooth shape morphs of *Neochromis*. (c) X-linked color morphs of *Neochromis* (light gray = allopatric different color, dark gray = allopatric same color).

the shore slopes steeply), whereas its magnitude (i.e., the difference in spectral composition between the ends of the gradient) was very similar at all studied sites. Given constant gradient magnitude among the five islands studied, the length of the gradient correlates negatively with its slope. It follows that the physical distance in meters between the extreme environments becomes shorter as the slope gets steeper. This variation in the gradient slope explains 87 percent of the variance in the extent of phenotypic differentiation along water depth, 65 percent of that in genotypic differentiation at the *LWS* locus, and 90 percent of that in genotypic differentiation at eleven microsatellites. If the gradient is very short and steep, progress in speciation is not observed, despite the presence of considerable heritable variation in color and mating preferences. Speciation is hence not constrained by lack of suitable genetic variation (Seehausen et al. 2008).

Earlier we had studied similar variation in *LWS* genes and male nuptial coloration in *Neochromis greenwoodi*, but in this case, divergently adapted populations occur in geographical parapatry, namely on different islands with different light conditions. In this case we again saw strong adaptation in the *LWS* gene, but not much progress toward speciation. Only a very incomplete association of *LWS* alleles with color was observed (Terai et al. 2006). Taking the two investigations together, we observed adaptation whenever migration between contrasting environments was reduced ($0.5 > m = 0$), but we observed speciation only when $0.5 > m > 0$, whereas neither was observed when migration between contrasting environments was unreduced (i.e., $m \sim 0.5$), nor when it was close to zero.

2. Trophic Morphology

Evidence for incipient speciation among geographically sympatric trophic ecomorphs is seen in significant neutral marker differentiation at one of our three islands (Igombe Island, $F_{ST} = 0.018$, $p < 0.001$). Ecomorphs occupying the same island (sympatric morphs) are often more similar to each other in their allele frequencies at neutral loci (9 unlinked microsatellites) than allopatric populations belonging to the same ecomorph (fig. 14.3b), implying $m > 0$ between the ecomorphs. Islands vary in the extent of phenotypic differentiation between ecomorphs, from single traits to more than ten different traits. Just like among the five replicate pairs in male nuptial coloration, the extent of differentiation between sympatric ecomorphs exhibits continuous variation, both in the number of traits that are divergent, and in the magnitude of their divergence (fig. 14.1d).

Phenotypic differentiation (P_{ST}) between ecomorphs was generally stronger than the neutral genetic differentiation (F_{ST}) (fig. 14.1d). At Makobe Island, divergence in quantitative functional traits, as measured by P_{ST}, was high for several morphological traits. The P_{ST} values and their confidence intervals for PC1 ($P_{ST} = 0.433$), number of tooth rows ($P_{ST} = 0.471$), and percentage of unicuspid teeth ($P_{ST} = 0.3$) did not overlap with the confidence interval of the F_{ST} value, indicating that differentiation along these axes was higher than expected by drift alone, and that several traits might be under divergent selection. When compared to the results from Makobe, the divergence in phenotypic traits between ecomorphs at Igombe Island was low. Only for PC1 and the percentage of unicuspid teeth did the C.I.s not overlap with the C.I. of the F_{ST} estimate, indicating divergent selection acted on fewer morphological traits than at Makobe. At Bihiru Island, P_{ST} values were even lower. With the exception of the percentage of unicuspid teeth, all confidence intervals overlapped fully with the confidence interval of the F_{ST} value.

The three islands had different combinations of water clarity, shore slope steepness, and boulder size, generating environmental gradients that differed in steepness and in spatial linearity. Clear water and gentle lake floor slopes make the transition from algae growth dominated shallow water rocky substrate to poorly illuminated deeper water long and gradual. Small boulders generate a locally homogeneous light environment, allowing for a spatially linear gradient of light-dependent resources that only depends on water depth. Large boulders, to the contrary, create a spatial mosaic of light habitats and associated resource abundances. The extent of phenotypic differentiation between trophic ecomorphs was predicted by the slope and length of the environmental gradient (in this case simply water depth; fig. 14.1d), similar to the case of male nuptial color polymorphism and visual adaptation. However, neutral genetic differentiation was generally weaker, and variation in the extent of neutral genetic differentiation was not well predicted by the gradient slopes, nor did it correlate with the extent of phenotypic differentiation (Magalhaes et al. unpubl. data).

3. X-linked (Female) Color

We found evidence for non-random mating among the OB and P color morphs of *Neochromis omnicaeruleus* at Makobe Island, as indicated by significant assignment tests based on neutral markers (9 microsatellites), linkage disequilibrium at many loci, and significant F_{ST} at two of the 9 loci. However, it appears as if speciation had not progressed beyond the stage of incompletely assortative mating, since multilocus F_{ST} did not differ significantly between any of the color morph pairs (Magalhaes et al. 2010). This was so despite the presence of slight morphological differences, and despite the presence of matching and quite discrete female color polymorphisms (Seehausen et al. 1999) and male mating preference polymorphisms (Pierotti et al. 2009 and fig. 14.2), which most likely are maintained by some form of disruptive selection. Hence, here too, neither lack of suitable genetic variation nor the form of selection constrain the progress in speciation.

Different color morphs from the same island were generally more similar to each other in their allele frequencies than the same (or different) color morphs from different islands (fig. 14.3c), implying that incipient differentiation between sympatric morphs is maintained despite gene flow. We could not detect any significant differences in depth distribution between these morphs at Makobe Island (fig. 14.2a), nor any other microdistribution differences.

In another investigation of a WB/P color polymorphism in the distantly related *Paralabidochromis chilotes* (plate 11), we did observe population genetic and behavioral evidence for complete speciation into sister species fixed for different X-linked female color. However, in this case the incipient species occupy different islands, though fully nested geographically with evidence for gene flow, suggesting parapatric speciation with a large component of geographical isolation (M. Pierotti et al. unpublished data).

4. Effects of Spatially Fine Grained Environmental Structure

The correlation between prevalence of either phenotype and water depth was significant for the male nuptial color morphs in *Pundamilia* at three of five islands, and for the dentition morphs of *Neochromis* at one of three islands (Makobe Island), but was not significant in the

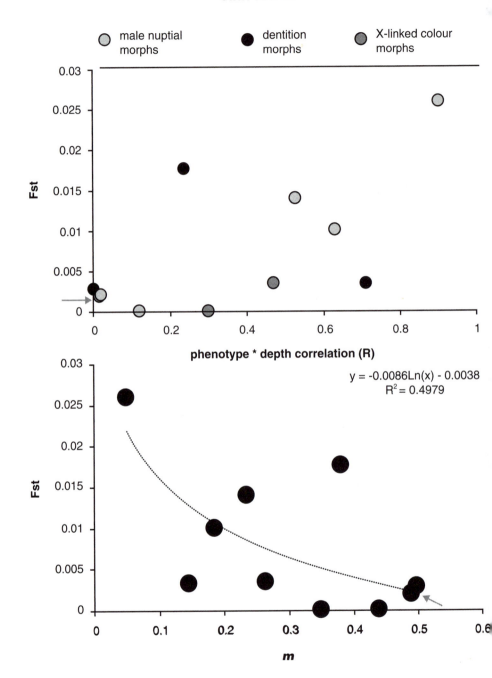

other cases, indicating no or only weak associations between phenotype and water depth. Over all eleven morph pairs, there was a significant positive correlation between the multilocus F_{ST} and the magnitude of the phenotype*water depth correlation ($R = 0.62$, p (1-tailed) = 0.02; fig. 14.4). In a general linear model with multilocus F_{ST} as dependent variable, the type of polymorphism did not explain variation in F_{ST} (2df, F = 0.63, $p = 0.56$), whereas there was a trend for the magnitude of the phenotype*water depth correlation to explain variation in F_{ST} (1df, F = 3.8, $p = 0.09$).

Calculating the spatial opportunity for gene flow between phenotypes, we find that significant progress in the speciation process was observed in the range $0 \leq m \leq 0.4$ and that three of four cases were in the range $0 \leq m \leq 0.25$. The spatial opportunity for gene flow (m) predicts F_{ST} as

$$F_{ST} = -0.0086 Ln(m) - 0.0038$$

DISCUSSION
Ecologically Parapatric Speciation along Depth-mediated Habitat Gradients

The population genetics and ecology of morph differentiation in eleven different pairs of geographically sympatric morphs of Lake Victoria cichlid fish suggest that the extent of spatial segregation of spawning sites along the sloping lake floor predicts the progress toward speciation. Such speciation, even though completely sympatric at coarse geographical scale (plates 10 and 11), is ecologically parapatric at fine grain. We estimate that speciation is possible with m, here defined as the spatial opportunity for cross-mating between morphs, between 0 and 0.4, and when m is below 0.25 perhaps even likely. If the spatial

---

Figure 14.4. Progress in speciation as a function of the phenotype-depth correlation and m. Progress in speciation as the multilocus F_{ST} value, measured using 9–11 microsatellites, in 11 pairs of sympatric morphs and incipient species of Lake Victoria cichlids. (a) Multilocus F_{ST} as a function of the phenotype-depth correlation. See legend for symbol explanation. Arrowhead points to one dark gray data point hidden behind a light gray data point. (b) Multilocus F_{ST} as a function of the potential for gene flow, m.

opportunity for cross-mating between morphs is similar now and in the past, speciation is close to sympatric even in its population genetic definition. We observed several morph pairs that were not spatially segregated at all, hence $m \sim 0.5$. None of these pairs revealed significantly differentiated allele frequencies at nine to eleven microsatellite loci. This applied to all three major classes of phenotypic polymorphisms that have been associated with incipient speciation in Lake Victoria cichlids; male nuptial coloration, trophic (dental) morphology, and X-linked (female) coloration.

Despite absence of differentiation at neutral DNA loci, several of these fully sympatric morph pairs were phenotypically nevertheless strongly differentiated. Failure to proceed in speciation can hence not be explained by lack of suitable genetic variation. Neither can it be explained by lack of selection. The frequency distribution of tooth shape morphs revealed evidence for response to disruptive selection in a case of $m \sim 0.5$. Similarly, the frequency distribution of the X-linked female coloration at Makobe Island was consistent with disruptive selection. Finally, female preference polymorphism is expected to exert disruptive selection on male nuptial coloration in *Pundamilia* even at the sites where $m \sim 0.5$ (Stelkens et al. 2008).

Where gene flow between the morphs is restricted by even small amounts of fine grain spatial segregation, these polymorphisms appear to often but not always achieve the stage of incipient species with isolation-by-adaptation (Nosil et al. 2008) permitting the onset of divergence at neutral unlinked DNA loci. The polymorphisms that we studied are widespread and common in Lake Victoria. Hence, geographically sympatric ecologically parapatric speciation may be widespread too in haplochromine cichlid fish of Lake Victoria, but ecologically fully sympatric speciation may be rare or absent. The very weak population differentiation that we observe between allopatric populations of several Lake Victoria cichlid species, occupying habitat patches that are separated by many kilometers of unsuitable habitat (fig. 14.3), suggests that the scope for completely allopatric speciation may be limited. Interestingly, we observed complete speciation between populations with divergent visual adaptations only in geographical sympatry / ecological parapatry (Seehausen et al. 2008). In geographical allopatry, in contrast, divergent adaptation was observed, but no speciation (measured as the completeness of association between opsin alleles and male nuptial coloration) (Terai et al. 2006). It is tempting to speculate that speciation at the rates seen in the Lake Victoria cichlid fish radiation may sometimes require migration and gene-flow to generate selection for assortative mating in a reinforcement-like process.

Support from Evolutionary Species Area Relationships

We are not suggesting that geographically sympatric speciation is the most frequent mode of speciation in haplochromine cichlids. The shape of the evolutionary species area relationship for African cichlid fish (Seehausen 2006) suggests a combination of more sympatric-like and more allopatric-like speciation in the evolution of cichlid species flocks. The number of endemic species has a flat relationship with lake size across several orders of magnitude of lake size variation, but suddenly increases steeply when lakes become larger than 1000 km^2 surface area. It seems that speciation did indeed occur in many lakes without opportunity for geographical isolation, but that many more species are produced in large lakes where there is greater opportunity for geographical isolation. The two arms of the curve are reminiscent of the relatively flat and steep slopes (z) respectively of intra- and interprovincial species area curves in island biogeography. Scope for allopatric speciation is significant between provinces (or islands), but not within (Rosenzweig 1997; Rosenzweig 2001). However, we note that there is an alternative explanation: geographical structure may allow parallel sympatric speciation in different sections of the same lake.

Is Divergent Adaptation along Aquatic Depth Gradients a "Magic trait"?

Sympatric speciation has been reported in lake-dwelling fishes probably more often than in other vertebrates. Lakes (and the sea) differ from many terrestrial vertebrate environments (and many river environments) by the additional spatial dimension of water depth and the associated strong environmental gradients that exist within even small habitat patches. Water depth mediates gradients in light intensity and composition, oxygen concentration, and temperature. These in turn affect resources, predators, parasites, sensory and signaling requirements, and hence call for divergent adaptation along the gradients within aquatic populations. If variable depth adaptation causes deviation from random mating, adaptation to depth may act like a "magic trait" in speciation. This scenario has a lot in common with host-plant-based mating of phytophagous insects, but with the difference that the discreteness of habitat patches may make speciation in phytophagous insects microallopatric, whereas the gradient-like nature of the habitat makes speciation in lacustrine fish ecologically parapatric. Our data suggest that such speciation may be typical in the rapid adaptive radiation of Lake Victoria cichlid fish.

A non-exhaustive literature survey on speciation in other fish suggests that divergent adaptation along water-depth-mediated habitat gradients may explain the apparent propensity for geographically sympatric speciation in lacustrine fish more generally. Divergent adaptation along depth gradients (to feeding and spawning requirements) appears to be critical to ecological speciation of whitefish (Vonlanthen et al. 2008), and ciscoes (Ohlberger et al. 2008; Ohlberger et al. 2009), and is involved in speciation of sticklebacks (Boughman 2001), and silversides (Herder et al. 2008). The distribution of spawning sites suggests some depth segregation of sister species also in classical cases of sympatric speciation in lakes (Barluenga et al. 2006; Schliewen et al. 2001). Finally, similar divergent adaptation along depth gradients is likely to drive ecologically parapatric speciation in the sea too, both in fish (Hyde et al. 2008) and in invertebrates (Grahame et al. 2006).

While we here conclude that geographically sympatric speciation may be more common than previously thought, a corollary of our observations is that spatial environmental structure does indeed have strong impact on the likelihood of ecological speciation with gene flow through permitting fine grain parapatric conditions nested in coarser grain sympatric conditions. Three predictions follow: (1) geographically sympatric speciation is expected to be less common in terrestrial systems that lack the strong third dimension at least for vertebrates. This prediction finds circumstantial support in existing reviews of speciation in terrestrial vertebrates (Coyne and Price 2000). (2) Classical cases of sympatric speciation in lacustrine fish may reveal themselves as ecologically parapatric at fine grain upon closer examination. The test of this prediction awaits quantitative phenotype distribution mapping in the Cameroonian and Nicaraguan crater lakes. (3) When disruptive selection on traits not associated with water depth nevertheless is associated with significant progress in speciation, other yet hidden phenotype-space associations may often be discovered on closer inspection.

REFERENCES

Abbott, R. J., M. G. Ritchie, and P. M. Hollingsworth. 2008. Speciation in plants and animals: pattern and ocess—Introduction. *Philos. Trans. R. Soc. B* 363: 2965–2969.

Barluenga, M., K. N. Stolting, W. Salzburger, M. Muschick, and A. Meyer. 2006. Sympatric speciation in N

Bolnick, D. I., and B. M. Fitzpatrick. 2007. Sympatric speciation: Models and empirical evidence. *Annu. Rev. Ecol. Evol. Syst.* 38: 459–487.

Boughman, J. W. 2001. Divergent sexual selection enhances reproductive isolation in sticklebacks. *Nature* 411: 944–948.

Bouton, N., O. Seehausen, and J.J.M. van Alphen. 1997. Resource partitioning among rock-dwelling haplochromines (Pisces : Cichlidae) from Lake Victoria. *Ecology of Freshwater Fish* 6: 225–240.

Bouton, N., F. Witte, J.J.M. van Alphen, A. Schenk, and O. Seehausen. 1999. Local adaptations in populations of rock-dwelling haplochromines (Pisces : Cichlidae) from southern Lake Victoria. *Proc. R. Soc. London B* 266: 355–360.

Bush, G. ed. 1998. *The Conceptual Radicalization of an Evolutionary Biologist. Endless Forms: Species and Speciation.* New York: Oxford University Press.

Bush, G. L. 1969. Sympatric host race formation and speciation in frugivorous flies of genus Rhagoletis (Diptera, Tephritidae). *Evolution* 23: 237–.

Butlin, R. K., J. Galindo, and J. W. Grahame. 2008. Sympatric, parapatric or allopatric: the most important way to classify speciation? *Philosophical Transactions of the Royal Society B-Biological Sciences* 363: 2997–3007.

Carleton, K. L., J.W.L. Parry, J. K. Bowmaker, D. M. Hunt, and O. Seehausen. 2005. Colour vision and speciation in Lake Victoria cichlids of the genus Pundamilia. *Mol. Ecol.* 14: 4341–4353.

Coyne, J., and H. A. Orr. 2004. *Speciation.* Sunderland, MA: Sinauer Associates.

Coyne, J. A., and T. Price. 2000. Little evidence for sympatric speciation in island birds. *Evolution* 54: 2166–2171.

Dijkstra, P. D., C. Hemelrijk, O. Seehausen, and T.G.G. Groothuis. 2008. Colour polymorphism and intrasexual competition in assemblages of cichlid fish. *Behav. Ecol.*: 138–144.

Dijkstra, P. D., O. Seehausen, B.L.A. Gricar, M. E. Maan, and T.G.G. Groothuis. 2006. Can male-male competition stabilize speciation? A test in Lake Victoria haplochromine cichlid fish. *Behav. Ecol. Sociobiol.* 59: 704–713.

Dijkstra, P. D., O. Seehausen, and T.G.G. Groothuis. 2005. Direct male-male competition can facilitate invasion of new colour types in Lake Victoria cichlids. *Behav. Ecol. Sociobiol.* 58: 136–143.

———. 2008. Intrasexual competition among females and the stabilization of a conspicuous colour polymorphism in a Lake Victoria cichlid fish. *Proceedings of the Royal Society B-Biological Sciences* 275: 519–526.

Dijkstra, P. D., O. Seehausen, M.E.R. Pierotti, and T.G.G. Groothuis. 2007. Male-male competition and speciation: aggression bias towards differently coloured rivals varies between stages of speciation in a Lake Victoria cichlid species complex. *J. Evol. Biol.* 20: 496–502.

Fitzpatrick, B. M., J. A. Fordyce, and S. Gavrilets. 2008. What, if anything, is sympatric speciation? *J. Evol. Biol.* 21: 1452–1459.

Gavrilets, S. 2004. *Fitness Landscapes and the Origin of Species.* Princeton, NJ: Princeton University Press.

Genner, M. J., P. Nichols, G. Carvalho, R. L. Robinson, P. W. Shaw, A. Smith, and G. F. Turner. 2007. Evolution of a cichlid fish in a Lake Malawi satellite lake. *Proc. R. Soc. London B* 274: 2249–2257.

Grahame, J. W., C. S. Wilding, and R. K. Butlin. 2006. Adaptation to a steep environmental gradient and an associated barrier to gene exchange in Littorina saxatilis. *Evolution* 60: 268–278.

Grant, B. R., and P. R. Grant. 2008. Fission and fusion of Darwin's finches populations. *Philos. Trans. R. Soc. B*, doi: 10.1098/rstb.2008.0051.

Grant, P. R., and B. R. Grant. 2006. Evolution of character displacement in Darwin's finches. *Science* 313: 224–226.

Greenwood, P. 1965. The cichlid fishes of Lake Nabugabo, Uganda. *Bull. Br. Mus. nat. Hist. (Zool.)* 12: 315–357.

Haesler, M. P., and O. Seehausen. 2005. Inheritance of female mating preference in a sympatric sibling species pair of Lake Victoria cichlids: implications for speciation. *Proc. R. Soc. London B* 272: 237–245.

Herder, F., J. Pfaender, and U. K. Schliewen. 2008. Adaptive sympatric speciation of polychromatic "roundfin" sailfin silverside fish in Lake Matano (Sulawesi). *Evolution* 62: 2178–2195.

Hyde, J. R., C. A. Kimbrell, J. E. Budrick, E. A. Lynn, and R. D. Vetter. 2008. Cryptic speciation in the vermilion rockfish (Sebastes miniatus) and the role of bathymetry in the speciation process. *Mol. Ecol.* 17: 1122–1136.

Kocher, T. D. 2004. Adaptive evolution and explosive speciation: The cichlid fish model. *Nature Reviews Genetics* 5: 288–298.

Lande, R., O. Seehausen, and J.J.M. van Alphen. 2001. Mechanisms of rapid sympatric speciation by sex reversal and sexual selection in cichlid fish. *Genetica* 112: 435–443.

Leinonen, T., J. M. Cano, H. Makinen, and J. Merilä. 2006. Contrasting patterns of body shape and neutral genetic divergence in marine and lake populations of threespine sticklebacks. *Journal of Evolutionary Biology* 19: 1803–1812.

Maan, M. E., B. Eshuis, M. P. Haesler, M. V. Schneider, J.J.M. van Alphen, and O. Seehausen. 2008. Color polymorphism and predation in a Lake Victoria cichlid fish. *Copeia*: 621–629.

Maan, M. E., K. D. Hofker, J.J.M. van Alphen, and O. Seehausen. 2006. Sensory drive in cichlid speciation. *Am. Nat.* 167: 947–954.

Maan, M. E., A.M.C. Van Rooijen, J.J.M. van Alphen, and O. Seehausen. 2008. Parasite-mediated sexual selection and species divergence in Lake Victoria cichlid fish. *Biol. J. Linn. Soc.* 94: 53–60.

Magalhaes, I., B. Lundsgaard-Hansen, S. Mwaiko, and O. Seehausen. Unpubl. data. Eco-morphological but not genetic differentiation within cichlid fish populations correlates with the slope of resource gradients.

Magalhaes, I., S. Mwaiko, and O. Seehausen. 2010. Sympatric colour polymorphisms associated with non-random gene flow in cichlid fish of Lake Victoria. *Mol. Ecol.* in press.

Mallet, J. 1995. A species definition for the modern synthesis. *Trends Ecol. Evol.* 10: 294–299.

Mayr, E. 1942. *Systematics and the Origin of Species*, New York: Columbia University Press.

———. 1963. *Animal Species and Evolution*. Cambridge, MA: Belknap Press.

———. 1982. *The Growth of Biological Thought: Diversity, Evolution and Inheritance*. Cambridge, MA: Belknap Press.

Merilä, J. 1997 Quantitative trait and allozyme divergence in the greenfinch (Carduelis chloris, Aves: Fringillidae). *Biological Journal of the Linnean Society* 61: 243–266.

Meyer, A., T. D. Kocher, P. Basasibwaki, and A. C. Wilson. 1990. Monophyletic origin of Lake Victoria cichlid fishes suggested by mitochondrial DNA sequences. *Nature* 347: 550–553.

Nosil, P., and B. J. Crespi. 2004. Does gene flow constrain adaptive divergence or vice versa? A test using ecomorphology and sexual isolation in Timema cristinae walking-sticks. *Evolution* 58: 102–112.

305

Nosil, P., S. R. Egan, and D. J. Funk. 2008. Heterogeneous genomic differentiation between walking-stick ecotypes: "Isolation by adaptation" and multiple roles for divergent selection. *Evolution* 62: 316–336.

Nosil, P., and C. P. Sandoval. 2008. Ecological niche dimensionality and the evolutionary diversification of stick insects. *PlosOne* 3: e1907.

Ohlberger, J., T. Mehner, G. Staaks, and F. Hölker, F. 2008. Temperature-related physiological adaptations promote ecological divergence in a sympatric species pair of temperate freshwater fish, *Coregonus* spp. *Funct. Ecol.* 22: 501–508.

Ohlberger, J., G. Staaks, T. Petzoldt, T. Mehner, and F. Hölker, F. 2008. Physiological specialization by thermal adaptation drives ecological divergence in a sympatric fish species pair. *Evol. Ecol. Res.* 10: 1173–1185.

Ostbye, K., T. F. Naesje, L. Bernatchez, O. T. Sandlund, and K. Hindar. 2005 Morphological divergence and origin of sympatric populations of European whitefish (*Coregonus lavaretus L.*) in Lake Femund, Norway. *Journal of Evolutionary Biology* 18: 683–702.

Pierotti, M.E.R., J. A. Martín-Fernández, and O. Seehausen. 2009. Mapping individual variation in male mating preference space: multiple choice in a colour polymorphic cichlid fish. *Evolution*. doi:10.1111/j.1558-5646.2009.00716.x

Rosenzweig, M. L. 1997. Tempo and mode of speciation. *Science* 277: 1622–1623.

———. 2001 Loss of speciation rate will impoverish future diversity. *Proc. Natl Acad. Sci. USA* 98: 5404–5410.

Rundle, H. D., L. Nagel, J. W. Boughman, and D. Schluter. 2000. Natural selection and parallel speciation in sympatric sticklebacks. *Science* 287: 306–308.

Saint-Laurent, R., M. Legault, and L. Bernatchez. 2003. Divergent selection maintains adaptive differentiation despite high gene flow between sympatric rainbow smelt ecotypes (Osmerus mordax Mitchill). *Mol. Ecol.* 12: 315–330.

Schliewen, U., and B. Klee. 2004. Reticulate sympatric speciation in Cameroonian crater lake cichlids. *Frontiers in Zoology* 1, doi: 10.1186/1742-9994-1-5.

Schliewen, U., K. Rassmann, M. Markmann, J. Markert, T. D. Kocher, and D. Tautz. 2001. Genetic and ecological divergence of a monophyletic cichlid species pair under fully sympatric conditions in Lake Ejagham, Cameroon. *Mol. Ecol.* 10: 1471–1488.

Seehausen, O. 1996. *Lake Victoria Rock Cichlids. Taxonomy, Ecology and Distribution.*: Zevenhuizen, NL: Cichlid Press.

———. 1997. Distribution of and reproductive isolation among color morphs of a rock-dwelling Lake Victoria cichlid (Haplochromis nyererei). *Ecology of Freshwater Fish* 6: 57–.

———. 2006. African cichlid fish: a model system in adaptive radiation research. *Proc. R. Soc. London B* 273: 1987–1998.

———. 2009. *Progressive levels of trait divergence along a "speciation transect" in the Lake Victoria cichlid fish Pundamilia.* Ecology and Speciation. Cambridge: Cambridge University Press.

Seehausen, O., and N. Bouton. 1997. Microdistribution and fluctuations in niche overlap in a rocky shore cichlid community in Lake Victoria. *Ecology of Freshwater Fish* 6: 161–173.

Seehausen, O., Y. Terai, I. S. Magalhaes, K. L. Carleton, H.D.J. Mrosso, R. Miyagi, I. van der Sluijs, et al. 2008. Speciation through sensory drive in cichlid fish. *Nature* 455: 620–623.

Seehausen, O., and J.J.M. van Alphen. 1998. The effect of male coloration on female mate choice in closely related Lake Victoria cichlids (Haplochromis nyererei complex). *Behav. Ecol. Sociobiol.* 42: 1–8.

———. 1999. Can sympatric speciation by disruptive sexual selection explain rapid evolution of cichlid diversity in Lake Victoria? *Ecology Letters* 2: 262–271.

Seehausen, O., J.J.M. van Alphen, and R. Lande. 1999. Color polymorphism and sex ratio distortion in a cichlid fish as an incipient stage in sympatric speciation by sexual selection. *Ecology Letters* 2: 367–378.

Spitze, K. 1993. Population-structure in Daphnia-Obtusa—Quantitative genetic and allozymic variation. *Genetics* 135: 367–374.

Stager, J. C., and T. C. Johnson. 2008. The late Pleistocene desiccation of Lake Victoria and the origin of its endemic biota. *Hydrobiologia* 596: 5–16.

Stelkens, R. B., M.E.R. Pierotti, D. A. Joyce, A. M. Smith, I. Van der Sluijs, and O. Seehausen. 2008. Female mating preferences facilitate disruptive sexual selection on male nuptial colouration in hybrid cichlid fish. *Phil. Trans. Roy. Soc. Series B* 363: 2861–2870.

Sturmbauer, C. 1998. Explosive speciation in cichlid fishes of the African Great Lakes: a dynamic model of adaptive radiation. *J. Fish Biol.* 53: 18–36.

Sturmbauer, C., U. Hainz, S. Baric, E. Verheyen, and W. Salzburger. 2003. Evolution of the tribe Tropheini from Lake Tanganyika: synchronized explosive speciation producing multiple evolutionary parallelism. *Hydrobiologia* 500: 51–64.

Terai, Y., O. Seehausen, T. Sasaki, K. Takahashi, S. Mizoiri, T. Sugawara, T. Sato, et al. 2006. Divergent selection on opsins drives incipient speciation in Lake Victoria cichlids. *Plos Biology* 4: 2244–2251.

Thibert-Plante, X., and A. P. Hendry. 2009. Five questions on ecological speciation addressed with individual-based simulations. *Journal of Evolutionary Biology* 22: 109–123.

van der Sluijs, I., T.J.M. Van Dooren, K. D. Hofker, J.J.M. van Alphen, R. B. Stelkens, and O. Seehausen, O. 2008. Female mating preference functions predict sexual selection against hybrids between sibling species of cichlid fish. *Philos. Trans. R. Soc. B* 363: 2871–2877.

Verzijden, M. N., J. Zwinkels, and C. ten Cate. 2009. Cross-fostering does not influence the mate preferences and territorial behaviour of males in Lake Victoria cichlids. *Ethology* 115: 39–48.

Vonlanthen, P., D. Roy, A. G. Hudson, C. R. Largiader, D. Bittner, and O. Seehausen. 2008. Divergence along a steep ecological gradient in Lake whitefish (*Coregonus sp.*). *J. Evol. Biol.*

Chapter Fifteen

A Tale of Two Radiations: Similarities and Differences in the Evolutionary Diversification of Darwin's Finches and Greater Antillean *Anolis* Lizards

Jonathan B. Losos

Recent years have seen renewed interest in adaptive radiation, the phenomenon in which a single ancestral species gives rise to descendent species adapted to use a wide variety of different ecological niches (e.g., Givnish and Sytsma 1997; Schluter 2000; Gavrilets and Losos 2009). Researchers are taking a wide variety of approaches that span fields as disparate as phylogenetics, behavior, functional morphology, ecology, and developmental biology and use methods as distinct as ecological field experiments and molecular laboratory studies (Schluter 2000).

Being a historical science, evolutionary biology is inductive; generalizations about how evolution occurred are typically not deduced from axioms or general principles, but are synthesized from the examination of many, usually non-experimental, case studies (Mayr 2004). In this respect, the study of adaptive radiation benefits especially greatly from case studies that are broadly synthetic, that integrate a wide range of different approaches, and that synthesize across timescales and methods of inquiry to develop the fullest understanding of the patterns and processes of evolutionary diversification in a particular clade. With this in mind, I have chosen to examine two of the best-known and most broadly integrative case studies of adaptive radiation, Darwin's finches and Greater Antillean *Anolis* lizards, with the goal of identifying which aspects of evolutionary diversification are general to these radiations and which aspects differ between them.

Darwin's finches and Greater Antillean *Anolis* lizards are two of, and perhaps *the two*, best known examples of adaptive radiation in terms of the breadth and depth of study they have received.[1] Since the seminal work of Lack and Bowman on Darwin's finches, and Williams, Schoener, Roughgarden, and others on *Anolis*, these two groups have played an important role in the development of ecological theory. In addition, both groups have been studied extensively from many other perspectives, including behavior, functional morphology, phylogenetics, population biology and, even recently, developmental biology (reviewed in P. R. Grant and B. R. Grant 2008; Losos 2009). As a result, I suggest that we have a broader and more complete understanding of the natural history and evolutionary framework of anoles and Darwin's finches than of any comparably diverse clade.

In this light, the goal of this chapter is to compare the adaptive radiations of Greater Antillean anoles and Darwin's finches to identify both similarities and differences in how these radiations have proceeded and what have been the ecological and evolutionary processes driving them. My hope is that this simple, two-case comparison will serve as a template to stimulate further comparative work on other taxa, with the ultimate goal of developing a broad-based, synthetic understanding of adaptive radiation.[2]

EVOLUTIONARY BACKGROUND

At first pass, one might see the Greater Antillean anole and Darwin's finch radiations as two similar evolutionary stories. However, aside from the fact that both have radiated in island settings, descendents in each case of a single ancestral species, the historical and geographical

[1] This begs the question of what it takes for a clade to be considered an adaptive radiation. Without getting into that topic here (see, e.g., Givnish 1997; Schluter 2000; Losos and Miles 2002; Losos and Mahler 2010), I would point out that some well-studied groups, such as sticklebacks, though providing great insight into the evolutionary process, do not contain enough ecological and phenotypic diversity to be considered adaptive radiations.
[2] Note that these two radiations have recently been the subject of comprehensive reviews, by B. R. Grant and P. R. Grant (2008) for the finches and by me for anoles (Losos 2009). These two treatments provide a reasonably comprehensive entrée to the literature on these taxa. Consequently, except where otherwise attributed, support for general statements about these clades comes from these two works.

backgrounds of the two groups are quite different. In particular, Darwin's finches occur in an archipelago of small and for the most part closely situated islands. By contrast, the Greater Antilles (Cuba, Hispaniola, Jamaica, and Puerto Rico) are much larger islands (the smallest, Jamaica, being nearly half again as large as the landmass of all the Galápagos combined) with greater distances separating them—the biological significance of this greater difference is magnified, of course, by the substantially greater ability of birds to disperse across water compared to that of lizards.

In addition, anoles have been radiating in the Greater Antilles for a considerably longer period of time. The Galápagos are relatively young, volcanic islands; few existing islands are greater than 3 million years old, although including now submerged islands, the archipelago had its start more than 10 million years ago (Ma). Molecular dating of Darwin's finches is concordant with this chronology, placing colonization of the islands at about 2–3 Ma. In contrast, the Greater Antilles are much older landmasses, which originated in what is now the Pacific Ocean during the Cretaceous before drifting, colliding, fragmenting, and otherwise making their way to their current location. When anoles colonized the islands from the mainland is unclear: amber fossils from the Dominican Republic have a minimum age of 15 Myr and several molecular clocks with highly uncertain calibrations place this event in the range of 40 Ma or even older.

These differences in age and geography may explain some of the differences in the evolutionary patterns and outcomes exhibited by these two radiations.

Adaptive Radiation

Darwin's finches are the poster child of adaptive radiation, having diversified to produce species filling a wide range of ecological niches—usually occupied by different bird families on the mainland—including granivores, frugivores, insectivores, and woodpeckers (fig. 15.1a). Bowman (1963) initially argued for the adaptive basis of the morphological diversity seen among these birds on engineering grounds, suggesting that the different sizes and shapes of beaks are well-suited for the diverse feeding modes of different species. Subsequent work on

A

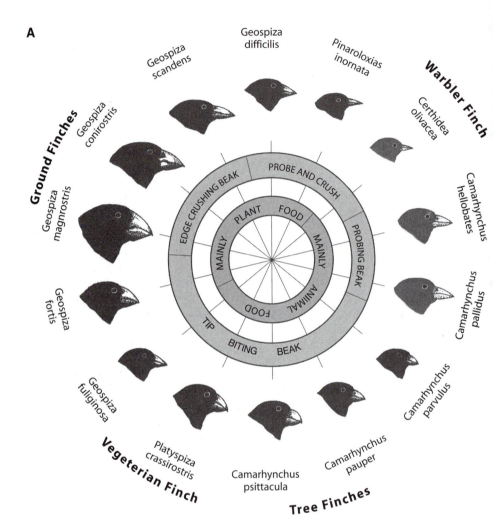

Figure 15.1. Adaptive radiation. (a) Darwin's finches (from P. R. Grant and B. R. Grant 2008). *Pinarolaxias inorta* is the Cocos Island finch; (b) Greater Antillean Anolis (from Losos 2009).

seed-eating finches confirmed this hypothesis by demonstrating both that species vary in the amount of force their beaks can produce (Herrel et al. 2005, 2009) and that a match exists between the cracking power of the beaks of different species and the toughness of the seeds they eat (Abbott et al. 1977; Schluter and Grant 1984).

B

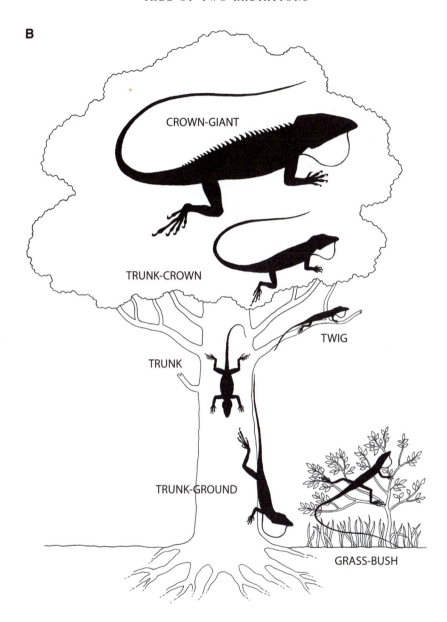

Figure 15.1. (*Continued*)

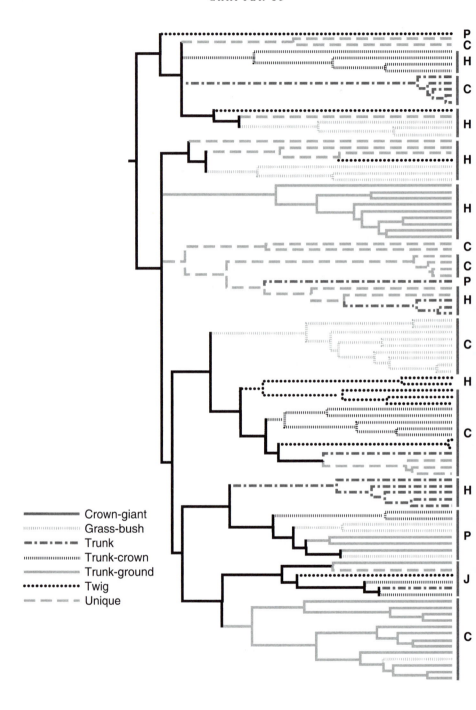

Anoles, too, are ecologically diverse, with species adapted to use different parts of the structural environment, such as grass, twigs, tree trunks, and the canopy (fig. 15.1b). Here, too, the morphological differences exhibited by different species correspond with their habitat use and behavior. For example, species that occur near the ground and run and jump frequently have long legs that maximize those capabilities, whereas twig species have shorter legs that allow these lizards to move nimbly over narrow and irregular surfaces. Similarly, more arboreal species have large toepads, allowing them to hang on to smooth surfaces high in the canopy, whereas the toepads of more terrestrial species are not as well developed and confer less clinging ability.

Intra- versus Inter-Island Diversification

Phylogenetic analysis reveals both differences and similarities in the historical records of these two groups. The primary difference is the extent of evolutionary independence between islands. In the Galápagos, the theater of evolution has been the entire archipelago, and adaptive radiation has not proceeded independently on different islands. In other words, similar niches on different islands are occupied by closely related taxa. By contrast, in anoles, adaptive radiation has occurred independently on each island in the Greater Antilles (fig. 15.2). With only one or two exceptions, habitat specialists on one island are not closely related to the corresponding habitat specialist on another island; although only the Jamaican radiation is the result of a single ancestral colonist, each of the other island radiations has diversified from few ancestral species.

The explanation for this discrepancy is obvious. The combination of the much closer proximity of islands in the Galápagos than those in the Greater Antilles, combined with the much greater overwater dispersal ability of birds than lizards, can explain why structural habitat

Figure 15.2. Anolis Ecomorph evolution in the Greater Antilles (from Losos 2009). Letters represent islands (Cuba, Hispaniola, Jamaica, Puerto Rico). Species in the same ecomorph class on different islands are not closely related and evolved convergently. For the most part, ecomorphs originate deep in anole phylogeny; subsequent intra-ecomorph clade diversification produces species adapted to using parts of the ecomorph niche (e.g., adapted to use different thermal habitats or different prey). "Unique" refers to species which do not conform to any of the ecomorph classes.

niches have been filled by colonization in Darwin's finches and by repeated evolution in anoles (indeed, a number of Darwin's finch species are found on many Galápagos islands, whereas only one species of *Anolis* is found on more than one Greater Antillean island).[3]

Species Richness and Diversification within a Habitat Specialist Clade

The difference in geographic context is also related to differences in species richness. There are only fifteen species of Darwin's finches compared to approximately 120 Greater Antillean *Anolis*. One reason for the difference is that species of anoles are not shared between islands. A second reason is the extensive intra-island radiation of anoles, virtually unknown in Darwin's finches, which has produced forty-one species on Hispaniola and sixty-three on Cuba. No doubt, difference in island size has played a role (Cuba: 111,000 km²; Hispaniola: 76,000 km²; Isabela, larger than all other Galápagos islands combined: 5,000 km²), as has the much greater age of the anole radiation (see above).

If we treat the anoles on each of the Greater Antillean islands as separate evolutionary radiations, then the pattern of adaptive diversification has been very similar in both anoles and finches. In both groups, the phylogenetically deepest divergence events are characterized by adaptation to using different parts of the structural habitat (figs. 15.2 and 15.3). In the case of the finches, these early events represent differentiation of the warbler finch,[4] the parrot-like vegetarian finch, and the ground and tree finch clades. Similarly, in the anole radiation, the habitat specialists, termed "ecomorphs" (Williams 1972, 1983), diverge deep in the phylogeny (Losos 2009). In both groups, once a habitat specialist evolves on an island, it never (finches) or almost never (anoles) independently evolves a second time on that island.

Also in both groups, subsequent evolution within a clade of a particular type of habitat specialist produces a set of species that diversify

---

[3] And that species, *A. sagrei*, may have been introduced from Cuba to Jamaica by humans (Kolbe et al. 2004).

[4] Or, possibly, warbler finches. Surprisingly, populations of the warbler finch are not monophyletic, but form a paraphyletic group at the base of the Geospizidae. Whether the genetically distinct lineages of the warbler finch are reproductively isolated is unknown, although playback experiments suggest that they may not be (Grant and Grant 2002a).

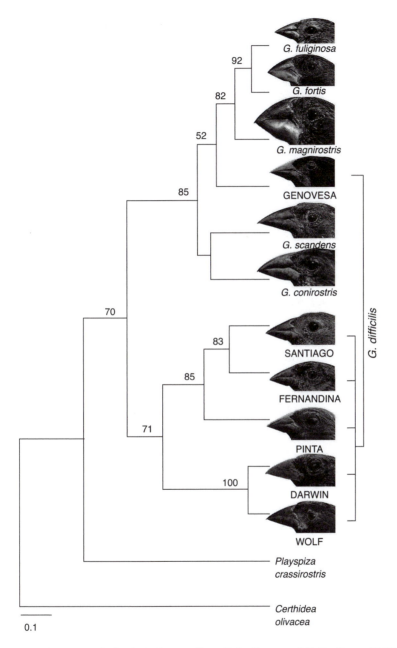

Figure 15.3. Darwin's finch evolution (from P. R. Grant and B. R. Grant 2008). The major habitat types—warbler finches (*Certhidea*), the vegetarian finch (*Platyspiza*), ground finches (*Geospiza*), tree finches (*Camarhynchus*)—evolved early, followed by subsequent niche subdivision in ground and tree finches.

317

along some other resource axis. In finches, this diversification occurs in terms of seed size in ground finches and in invertebrate food type and size in tree finches (Bowman 1961). In anoles, the situation is a little more complicated; in arboreal habitat specialists, diversification parallels that in ground finches and leads to species that differ in body size, and consequently in prey size. By contrast, in more terrestrial habitat specialists, intra-clade diversification produced same-sized species that differ in their thermal microhabitat—that is, species that use the same structural habitat (e.g., grass and low-lying narrow vegetation), but vary in whether they are in relatively cool (shaded) or warm (sunny) areas.

Timing and Pace of Diversification

Darwin's finches are known for the rapidity of their evolutionary radiation, producing such a great variety of species over a relatively short period of time. As I've just discussed, patterns of diversification in anoles are very similar, yet the anole radiation has been ongoing for a substantially longer period of time—indeed, geographically differentiated populations of some anole species (discussed below) may be as old as species of Darwin's finches.

Given this vast difference in age of radiations, two conclusions are possible: first, the anole radiations may have occurred much more slowly than that of Darwin's finches. Alternatively, the diversification of anoles into different niches may have occurred just as rapidly early on, and since then little subsequent ecological diversification has occurred. The truth appears to lie somewhere in the middle. The anole ecomorphs evolved early in the radiations on each of the Greater Antillean islands—although dating is not very precise, this part of the anole radiations probably occurred very quickly. On the other hand, the ecomorphs, once they originated, proliferated greatly, producing clades of as many as fourteen ecomorphologically similar species on a single island. Some of these species are simply allospecies, ecologically similar species found in different places (Mayr 1963). However, other species have differentiated in other niche dimensions, such as food size or thermal microclimate, as discussed above. Although the phylogenetic history of within-ecomorph diversification is just now being studied (e.g., Glor et al. 2003), it appears that such differentiation has gone on throughout the anole radiations, rather than just in its early phases. Thus, the process of ecological differentiation of anoles has occurred over a longer time period in anoles than in Darwin's finches—and has produced a correspondingly greater degree of ecological diversity and species richness (see also chapter 13).

Ecological Processes Driving Adaptive Radiation

David Lack (1947) outlined what has become the standard model of adaptive radiation (see also Schluter 2000 and Losos, J. B. 2010): an ancestral species occupies an environment rich in resources; speciation occurs for some reason, leading to the coexistence of multiple species; as the species increase in abundance, resources become limiting; in response, the species shift their use of the environment, thus minimizing interspecific overlap in resource use; subsequently, the species evolve adaptations to their new ecological condition, and voilà, the result is an adaptive radiation.

This model has three components which we can evaluate for anoles and Darwin's finches:

1. Interspecific interactions occur between sympatric species.
2. Species respond to the presence of other species by shifting resource use.
3. These shifts lead to natural selection producing evolutionary change.

With regard to the first two components, a large body of literature—experimental and observational—documents the existence of interspecific interactions, primarily competition, among sympatric anole species (reviewed in Roughgarden 1995; Losos 2009). In anole communities, coexisting species always differ in either habitat or prey use (or both); sympatric species that are similar in position along one resource axis are greatly dissimilar along other axes (Schoener 1974; Roughgarden 1995); the presence of other species causes lower population densities, growth and feeding rates, and body condition (e.g., Pacala and Roughgarden 1985; Leal et al. 1998; Losos and Spiller 1999); and species alter their habitat use in the presence of congeners (e.g., Jenssen 1973; Schoener 1975; Lister 1976). The third component, that populations adapt to new conditions subsequent to habitat shifts, is supported by the morphological differences seen among geographical populations of species that differ in the number and identity of anole species with which they coexist (e.g., Lister 1976; Losos et al. 1994).[5] Studies directly measuring selection in anoles have just begun, but already one experimental study has demonstrated that the presence of a predatory lizard leads to both habitat shifts and different patterns of natural selection on limb length in *Anolis sagrei* (Losos et al. 2004, 2006).

Experimental ecological studies are not possible for Darwin's finches. Nonetheless, a wealth of observational data strongly indicates the existence of interspecific competition. For example, populations of some species vary in habitat use or diet depending on the other species with which they are sympatric. Moreover, populations also vary in bill dimensions and body size in relation to local community composition, presumably because the populations adapt to using different food resources as a result of the presence of other species. In fact, sympatric species are always more dissimilar in bill dimensions than would be expected by chance, a non-random pattern likely the outcome of interspecific interactions.

Darwin's finches are renowned for the long-term studies of natural selection conducted over a period of thirty years by the Grants. These studies have clearly indicated how environmental changes drive adaptation. Several studies have documented that as the distribution of available resources shifts, strong directional selection leads to evolutionary change. For example, when small seeds became scarce during a drought in 1977, large-beaked members of the *G. fortis* population on Daphne Major were favored, and the mean value for beak size increased in the population in the next generation (fig. 15.4; Boag and Grant, 1981; P. R. Grant and B. R. Grant 2008). Moreover, this long-running study also has demonstrated the role of interspecific interactions. A similar drought in 2004 also led to a scarcity of seeds. In this case, however, *G. magnirostris*, which is larger than *G. fortis* and has

---

[5]More generally, correlations between habitat use and morphology among populations suggest that anoles readily adapt to differences in habitat use (Losos et al. 1994; Thorpe et al. 2004; Calsbeek et al. 2007).

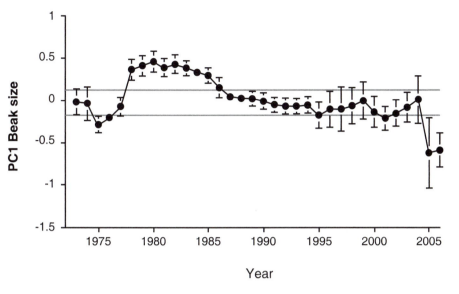

Figure 15.4. Beak size evolution in the medium ground finch, *Geospiza fortis*, on Daphne Major. After a drought in 1977, beak size increased; however, after another drought in 2004, when the larger-beaked G. *magnirostris* was common, beak size decreased (from Grant and Grant 2006a).

a substantially more massive beak, had become common on the island, and these birds monopolized the larger seeds. As a result, large-beaked G. *fortis* were at a disadvantage, and natural selection strongly favored small-beaked birds despite the paucity of small seeds, leading to an evolutionary decrease in beak size in the next generation (Grant and Grant 2006a), exactly the opposite of what occurred in the previous drought (fig. 15.4). This is perhaps the best documented example in nature of natural selection driving character displacement.

Overall, these similarities suggest that Darwin's finches and anoles have radiated adaptively in much the same way. Interspecific competition is a strong force that operates even in modern times among species that already are differentiated ecologically and morphologically. It seems likely that competition played an important role leading to the divergence of initially similar species, producing the diverse communities of finches and anoles observed today. Moreover, the phylogenetic congruence in pattern of radiation suggests that early in diversification, species partition the environment by using different habitats or

321

food types; subsequently, diversification proceeds not by repeated adaptation to these niches, but rather by subdividing them.

Predation and Adaptive Radiation

Interspecific competition has long been considered the driving force in adaptive radiation, but other forms of ecological interaction potentially could play a role as well. Schluter (2000) noted that species in some adaptive radiations occupy multiple trophic levels, and thus predation could be an important driver of adaptive diversification as well. In addition, competitors can also prey on each other, the phenomenon of intra-guild predation. The role of these processes in adaptive radiation has been little studied (Vamosi 2005).

In Darwin's finches, this consideration is not relevant. Vampire finches notwithstanding, finches are not carnivorous and thus intra-clade predation has not been a factor in finch diversification. By contrast, anoles differ greatly in body size, and larger anoles regularly eat smaller ones (Gerber 1999). The role that such predation may play both in structuring ecological communities (cf. Schoener et al. 2002) and in driving evolutionary divergence (Losos et al. 2004) would be a fertile area for future work.

Differences in Patterns and Processes of Adaptive Radiation
Species Recognition and Speciation

In Darwin's finches, females recognize members of their species by the size of the beak. Moreover, species-specific songs are also used in species-recognition, and these songs may also be affected by beak size and shape (Podos 2001; see also Grant and Grant 2002b). The importance of the beak both for resource utilization and species recognition sets up the possibility of "ecological speciation" (Schluter 2000; see also chapters 13 and 14); populations that diverge to adapt to different seed availability may incidentally end up being reproductively isolated as they diverge in beak shape.

An analogous situation could occur in anoles. Anoles use the color of the dewlap (plate 14) and species-specific bobbing patterns to distinguish conspecifics from heterospecifics. The communicatory effectiveness of particular dewlap configurations and signaling patterns is a function of the light environment and background: in dark areas, for example, dewlaps that are light colored reflect more light and stand out more; by contrast, in open environments, darker colors are more effective (Fleishman 1992; Leal and Fleishman 2004). Similarly, the vegetative background and the extent to which the background moves probably affects the form of the head-bobbing display that can be detected best (Fleishman 1992; Ord et al. 2007). For these reasons, anoles occupying new habitats may adapt their signaling behavior and morphology to maximize communication effectiveness, with the incidental effect of increasing reproductive isolation between populations in different environments.

Thus, the possibility exists for both Darwin's finches and anoles that adaptive divergence may have the incidental effect of promoting speciation. In theory, such adaptive divergence could occur either in sympatry, parapatry, or allopatry (see chapter 14). Sympatric speciation has been suggested for both finches and anoles, though little compelling data support this hypothesis in either case (P. R. Grant and B. R. Grant 2008; Losos 2009; but see Hendry et al. 2006, 2009). Moreover, the observation that speciation has not occurred on small and isolated islands that seem to have the resource and habitat heterogeneity to support multiple species suggests that allopatry may be necessary for speciation to occur (e.g., Cocos Island, occupied by one species of Darwin's finches and one species of anole, and many Lesser Antillean islands—e.g., Dominica, Guadeloupe—with only one anole species).

Most of Darwin's finch species have populations on multiple islands, and often these populations have differentiated morphologically as they have adapted to local conditions. Consequently, the scenario that speciation is promoted by adaptive divergence in allopatry is easy to envision for these birds. By contrast, envisioning how allopatric speciation occurs in Greater Antillean anoles is not so clear. Many Greater Antillean anoles occur island-wide, which raises the question of how allopatry might be produced. Vicariant fragmentation and offshore islands offer explanation in some cases (e.g., Glor et al. 2004; Lazell 1996), but in many cases such scenarios are entirely speculative.

Species Boundaries and Hybridization

An important difference between the two clades concerns species boundaries and the extent of hybridization. Research in recent years has shown that hybridization occurs at relatively high rates among a few sympatric Darwin's finch species and at lower levels among many others. The result is that some Darwin's finch species are not well-defined genetically; rather, populations of these species often are more genetically similar to sympatric populations of other species than to populations of their own species on other islands (Grant et al. 2005; Petren et al. 2005; also see discussion in Zink 2002 and Grant and Grant 2006b). The situation for Greater Antillean anoles is exactly the opposite. Very few cases of hybridization have been reported, and most of these did not lead to the production of fertile offspring (reviewed in Losos 2009).

One factor that might be responsible for the difference in hybridization and species boundaries is the difference in age of the two radiations—perhaps sympatric finches are too recently diverged for reproductive isolation to be complete, whereas most sympatric anoles have been evolving separately for much longer periods of time (B. R. Grant and P. R. Grant 2008). Another possible reason why hybridization is so much more prevalent in the finches is that bill size is linked to a resource that fluctuates greatly over short time periods, leading to intense directional selection and corresponding shifts in morphology, which raises the possibility that the degree of reproductive isolation between species may also fluctuate. Moreover, finches do not have the option of not responding to the distribution of seed size availability. By contrast, species recognition in anoles is linked to light environment. Light environment, in turn, is related to thermal environment (e.g., darker areas tend to be cooler), and anoles have evolved physiological adaptations to differences in micro-climate. The result is that anoles may be more likely to track environmental shifts behaviorally (Huey et al. 2003), altering their habitat use and maintaining interspecific differences, an option not open to finches when seed size distributions shift.

In addition to the lack of hybridization, anole species exhibit another difference from Darwin's finches: recent research has revealed that species tend to be highly subdivided genetically, such that a species is composed of parapatrically distributed clades that often differ by as much as 10 percent in uncorrected distance in mitochondrial DNA (e.g., Glor et al. 2003; Kolbe et al. 2004, 2007). These differences are so great that we must entertain the possibility that what are recognized as island-wide species are actually species complexes whose members replace each other across the geographic landscape; data from other genetic loci, particularly from the contact zone between different mitochondrial clades, are needed to test this hypothesis (e.g., Glor et al. 2004).

Recent molecular work also has shown that anole species are sometimes not monophyletic, but for a different reason than Darwin's finches. In a number of cases, species with small geographic ranges appear to have arisen from within widespread ranges of other species (e.g., Glor et al. 2003).

CONCLUSIONS

Overall, adaptive radiation in Darwin's finches and Greater Antillean anoles has occurred in very much the same way. Interspecific competition appears to have been the driving force leading to resource partitioning and subsequently adaptation to different niches, and speciation is probably primarily allopatric and may be promoted as an incidental consequence of adaptation to different environments. Differences exist as well, such as the extent of hybridization and of independent evolution on different islands; many of these differences probably result because the radiations differ in age and aspects of natural history.

Of course, a sample size of two is too small to draw general conclusions. What we need now are similar comparisons to other radiations, ideally a diversity of different types of organisms in different settings. Many putative adaptive radiations have been identified (Givnish 1997; Gillespie et al. 2001; see chapter 14), some more thoroughly studied than others. Most likely, the more radiations we study, the greater the diversity in patterns and processes we will discover, particularly if the radiations are diverse in taxa and setting. Nonetheless, my guess is that overarching generalities will emerge as well, such as the roles of ecological opportunity and interspecific competition and the pattern that most niche divergence occurs early in a radiation (Gavrilets and Losos 2009; Losos and Mahler 2010). These are just predictions, however, waiting to be tested.

One exciting further aspect is the ability to study adaptive radiation, and to compare adaptive radiations, at the molecular developmental and genomic levels. Already, comparative developmental work is shedding light on the precise genetic changes leading to differences in beak shape among Darwin's finches (Abzhanov et al. 2004, 2006; Grant et al. 2006c). The genome of *Anolis carolinensis* has just been sequenced, potentially opening many doors for investigation of genomic pathways to evolutionary diversification. Particularly interesting will be investigation of whether convergent evolution has been accomplished through the same or different genetic means (Schneider 2008; see also chapter 7). More generally, the dawn of the genomic era will allow comparative studies of adaptive radiation to be taken to an entirely new level, adding considerations such as genetic constraints, genetic architecture, and genome organization to the many other factors that are important in shaping evolutionary diversification.

CODA

We know so much about these two evolutionary clades as a result of intense, long-term, and broad-based investigations, each developed from a single research program. This volume is dedicated to honoring the remarkable careers of Peter and Rosemary Grant. Despite the historical pedigree of research tracing through Robert Bowman, David Lack, Erwin Stresemann, Harry Swarth, John Gould, all the way back to Charles Darwin, there can be no doubt that the status of Darwin's finches as one of the premier examples of evolutionary research is the result of the broad-based, synthetic, and novel research of the Grants. Collaborating with an exemplary group of graduate students (many of whom have gone on to become leaders in the field in their own right) and with others, *El Grupo Grant* has single-handedly established Darwin's finches as one of the best known case studies in evolutionary biology today.

Our knowledge of *Anolis*, too, is to a great extent the result of a single research program, in this case that of Ernest Williams at the Museum of Comparative Zoology at Harvard. Primarily a systematist by training, Williams attracted the best and brightest Harvard graduate students to his lab, developing a synthetic approach to the study of evolutionary diversification that encompassed ecology, behavior, and functional morphology and presaged the phylogenetic approach to the study of evolutionary diversification that would not develop for another twenty years. Along the way, his lab produced a series of leaders in the fields of ecology, behavior, and functional morphology, including Carl Gans, Robert Holt, Ray Huey, Jonathan Roughgarden, Thomas Schoener, and Robert Trivers.

We may wonder whether the exceptional extent of our knowledge of Darwin's finches and anoles has been gained because these two groups are in some ways extraordinary, both in terms of their evolutionary diversity and in their suitability for study in a variety of different ways, or because the researchers who have chosen to work on these two groups were exceptionally creative, synthetic, and forward-thinking. Of course, these two possibilities are not mutually exclusive: among the attributes of exceptional researchers is the ability to identify exceptional groups and to understand their potential for addressing questions of broad and general significance.

REFERENCES

Abbott, I., L. K. Abbott, and P. R. Grant. 1977. Comparative ecology of Galápagos ground finches (*Geospiza* Gould): evaluation of the importance of floristic diversity and interspecific competition. *Ecol. Monogr.* 47: 151–184.

Abzhanov, A., W. P. Kuo, C. Hartmann, B. R. Grant, P. R. Grant, and C. J. Tabin. 2006. The calmodulin pathway and evolution of elongated beak morphology in Darwin's finches. *Nature* 442: 563–567.

Abzhanov, A., M. Protas, B. R. Grant, P. R. Grant, and C. J. Tabin. 2004. *BMP4* and morphological variation of beaks in Darwin's finches. *Science* 305: 1462–1465.

Boag, P. T., and P. R. Grant. 1981. Intense natural selection in a population of Darwin's finches (Geospizinae) in the Galápagos. *Science* 214: 82–85.

Bowman, R. I. 1961. Adaptation and differentiation of the Galápagos finches. *Univ. Calif. Publs. Zool.* 58: 1–302.

———. 1963. Evolutionary patterns in Darwin's finches. *Occas. Papers Calif. Acad. Sci.* 44: 107–140.

Calsbeek, R., T. B. Smith, and C. Bardeleben. 2007. Intraspecific variation in *Anolis sagrei* mirrors the adaptive radiation of Greater Antillean anoles. *Biol. J. Linn. Soc.* 90: 189–199.

Fleishman, L. J. 1992. The influence of sensory system and the environment on motion patterns in the visual displays of anoline lizards and other vertebrates. *Am. Nat.* 139: S36–61.

Gavrilets, S., and J. B. Losos. 2009. Adaptive radiation: contrasting theory with data. *Science* 323: 732–737.

Gerber, G. P. 1999. A review of intraguild predation and cannibalism in *Anolis*. In J. B. Losos and M. Leal, eds., *Anolis Newsletter V.* 28–39. Saint Louis, MO: Washington University.

Gillespie R. G., F. G. Howarth, and G. K. Roderick. 2001. Adaptive Radiation. In S. Levin, ed., *Encyclopedia of Biodiversity, Vol. 1*, 25–44. New York: Academic Press.

Givnish, T. J. 1997. Adaptive radiation and molecular systematics: issues and approaches. In T. J. Givnish and K. J. Sytsma, eds., *Molecular Evolution and Adaptive Radiation*, 1–54. Cambridge: Cambridge University Press.

Givnish, T. J., and K. J. Sytsma, eds. 1997. *Molecular Evolution and Adaptive Radiation*. Cambridge: Cambridge University Press.

Glor, R. E., M. E. Gifford, A. Larson, J. B. Losos, L. Rodríguez Schettino, A. R. Chamizo Lara, and T. R. Jackman. 2004. Partial island submergence and speciation in an adaptive radiation: a multilocus analysis of the Cuban green anoles. *Proc. R. Soc. London B* 271: 2257–2265.

Glor, R. E., J. J. Kolbe, R. Powell, A. Larson, and J. B. Losos. 2003. Phylogenetic analysis of ecological and morphological diversification in Hispaniolan trunk-ground anoles (*Anolis cybotes* group). *Evolution* 57: 2383–2397.

Grant, B. R., and P. R. Grant. 2002a. Lack of premating isolation at the base of a phylogenetic tree. *Am. Nat.* 160: 1–19.

———. 2002b. Simulating secondary contact in allopatric speciation: an empirical test of premating isolation. *Biol. J. Linn. Soc.* 76: 542–556.

———. 2008. Fission and fusion of Darwin's finches populations. *Philos. Trans. R. Soc. B* 363: 2821–2829.

Grant, P. R., and B. R. Grant. 2006a. Evolution of character displacement in Darwin's finches. *Science* 313: 224–226.

———. 2006b. Species before speciation is complete. *Annals Missouri Botan. Gard.* 93: 94–102.

———. 2008. *How and Why Species Multiply: The Radiation of Darwin's Finches*. Princeton, NJ: Princeton University Press.

Grant, P. R., B. R. Grant, and A. Abzhanov. 2006. A developing paradigm for the development of bird beaks. *Biol. J. Linn. Soc.* 88: 17–22.

Grant, P. R., B. R. Grant, and K. Petren. 2005. Hybridization in the recent past. *Am Nat.* 166: 56–67.

Hendry, A. P., P. R. Grant, B. R. Grant, H. A. Ford, M. J. Brewer, and J. Podos. 2006. Possible human impacts on adaptive radiation: beak size bimodality in Darwin's finches. *Proc. R. Soc. B* 273: 1887–1894.

Hendry, A. P., S. K. Huber, L. F. De León, A. Herrel, and J. Podos. 2009. Disruptive selection in a bimodal populations of Darwin's finches. *Proc. R. Soc. B* 276: 753–759.

Herrel, A., J. Podos, S. K. Huber, and A. P. Hendry. 2005. Evolution of bite force in Darwin's finches: a key role for head width. *J. Evol. Biol.* 18: 669–675.

Herrel, A., J. Podos, B. Vanhooydonck, and A. P. Hendry. 2009. Force-veolcity trade-off in Darwin's finch jaw function: a biomechanical basis for ecological speciation? *Funct. Ecol.* 23: 119–125.

Huey, R. B., P. E. Hertz, and B. Sinervo. 2003. Behavioral drive versus behavioral inertia in evolution: a null model approach. *Am Nat.* 161: 357–366.

Jenssen, T. A. 1973. Shift in the structural habitat of *Anolis opalinus* due to congeneric competition. *Ecology* 54: 863–869.

Kolbe, J. J., R. E. Glor, L. Rodríguez Schettino, A. Chamizo Lara, A. Larson, and J. B. Losos. 2004. Genetic variation increases during biological invasion by a Cuban lizard. *Nature* 431: 177–181.

———. 2007. Multiple sources, admixture, and genetic variation in introduced *Anolis* lizard populations. *Conserv. Biol.* 21: 1612–1625.

Lack, D. 1947. *Darwin's Finches*. Cambridge: Cambridge University Press.

Lazell, J. 1996. Careening Island and the Goat Islands: evidence for the arid-insular invasion wave theory of dichopatric speciation in Jamaica. In R. Powell and R. W. Henderson, eds., *Contributions to West Indian Herpetology: A Tribute to Albert Schwartz*, 195–205. Ithaca, NY: Society for the Study of Amphibians and Reptiles.

Leal, M., and L. J. Fleishman. 2004. Differences in visual signal design and detectability between allopatric populations of *Anolis* lizards. *Am. Nat.* 163: 26–39.

Leal, M., J. A. Rodríguez-Robles, and J. B. Losos. 1998. An experimental study of interspecific interactions between two Puerto Rican *Anolis* lizards. *Oecologia* 117: 273–278.

Lister, B. C. 1976. The nature of niche expansion in West Indian *Anolis* lizards II. Evolutionary components. *Evolution* 30: 677–692.

Losos, J. B. 2009. *Lizards in an Evolutionary Tree: Ecology and Adaptive Radiation of Anoles*. Berkeley: University of California Press.

Losos, J. B. 2010. Adaptive radiation, ecological opportunity, and evolutionary determinism. *American Naturalist* 175: 623–639.

Losos, J. B., D. J. Irschick, and T. W. Schoener. 1994. Adaptation and constraint in the evolution of specialization of Bahamian *Anolis* lizards. *Evolution* 48: 1786–1798.

Losos, J. B., and D. L. Mahler. 2010. Adaptive Radiation: The Interaction of Ecological Opportunity, Adaptation, and Speciation. In M. A. Bell, D. J. Futuyma, W. F. Eanes, and J. S. Levinton, Eds., *Evolution Since Darwin: The First 150 Years*. Sinauer Assoc.: Sunderland, MA.

Losos, J. B., and D. B. Miles. 2002. Testing the hypothesis that a clade has adaptively radiated: iguanid lizard clades as a case study. *Am. Nat.* 160: 147–157.

Losos, J. B., T. W. Schoener, R. B. Langerhans, and D. A. Spiller. 2006. Rapid temporal reversal in predator-driven natural selection. *Science* 314: 1111.

Losos, J. B., T. W. Schoener, and D. A. Spiller. 2004. Predator-induced behaviour shifts and natural selection in field-experimental lizard populations. *Nature* 432: 505–508.

Losos, J. B., and D. A. Spiller. 1999. Differential colonization success and asymmetrical interactions between two lizard species. *Ecology* 80: 252–258.

Mayr, E. 1963. *Animal Species and Evolution*. Cambridge, MA: Belknap Press.

———. 2004. *What Makes Biology Unique?* Cambridge: Cambridge University Press.

Ord, T. J., R. A. Peters, B. Clucas, and J. A. Stamps. 2007. Lizards speed up visual displays in noisy motion habitats. *Proc. R. Soc. B* 274: 1057–1062.

Pacala, S. W., and J. Roughgarden. 1985. Population experiments with the *Anolis* lizards of St. Maarten and St. Eustatius. *Ecology* 66: 129–141.

Petren, K., P. R. Grant, B. R. Grant, and L .F. Keller. 2005. Comparative landscape genetics and the adaptive radiation of Darwin's finches: the role of peripheral isolation. *Mol. Ecol.* 14: 2943–2957.

Podos, J. 2001. Correlated evolution of morphology and vocal signal structure in Darwin's finches. *Nature* 409: 185–188.

Roughgarden, J. 1995. Anolis *Lizards of the Caribbean: Ecology, Evolution, and Plate Tectonics*. Oxford: Oxford University Press.

Schluter, D. 2000. *The Ecology of Adaptive Radiation*. Oxford: Oxford University Press.

Schluter, D., and P. R. Grant. 1984. Determinants of morphological patterns in communities of Darwin's finches. *Am Nat.* 123: 175–196.

Schneider, C. J. 2008. Exploiting genomic resources in studies of speciation and adaptive radiation of lizards in the genus *Anolis*. *Int. Comp. Biol.* 98: 520–526.

Schoener, T. W. 1974. Resource partitioning in ecological communities. *Science* 185: 27–39.

————. 1975. Presence and absence of habitat shift in some widespread lizard species. *Ecol. Monogr.* 45: 233–258.

Schoener, T. W., D. A. Spiller, and J. B. Losos. 2002. Predation on a common *Anolis* lizard: can the food-web effects of a devastating predator be reversed? *Ecol. Monogr.* 72: 383–408.

Thorpe, R. S., A. Malhotra, A. G. Stenson, and J. T. Reardon. 2004. Adaptation and speciation in Lesser Antillean anoles. In U. Dieckmann, M. Doebeli, J.A.J. Metz, and D. Tautz. eds., *Adaptive Speciation*, 322–344. Cambridge: Cambridge University Press.

Vamosi, S. M. 2005. On the role of enemies in divergence and diversification of prey: a review and synthesis. *Canad. J. Zool.* 83: 894–910.

Williams, E. E. 1972. The origin of faunas. Evolution of lizard congeners in a complex island fauna: a trial analysis. *Evol. Biol.* 6: 47–89.

————. 1983. Ecomorphs, faunas, island size, and diverse end points in island radiations of *Anolis*. In R. B. Huey, E. R. Pianka, and T. W. Schoener, eds., *Lizard Ecology: Studies of a Model Organism*, 326–370. Cambridge, MA: Harvard University Press.

Zink, R. M. 2002. A new perspective on the evolutionary history of Darwin's finches. *Auk* 119: 864–871.

Chapter Sixteen

Clarifying the Mechanisms of Evolution in Sticklebacks Using Field Studies of Natural Selection on Genes

Rowan D. H. Barrett and Dolph Schluter

THE SYNTHESIS OF POPULATION BIOLOGY AND GENOMICS

Natural selection is recognized as the principal evolutionary force shaping adaptation, yet directly testing the mechanisms of selection remains a difficult task. A comprehensive understanding of adaptation requires that we determine how the selection pressures on phenotypes lead to changes in gene frequency, and how allelic variation at genes interacts with the environment to cause differences in phenotypes and fitness. Pioneering studies of natural selection in wild populations made the link between variation in phenotypic traits, fluctuations in the ecological environment, selection, and evolution (Boag and Grant 1981; Price et al. 1984; Gibbs and Grant 1987). This work continues to be a catalyst for a burgeoning research area concerned with measuring natural selection and its consequences. The accumulated efforts to date have provided crucial information about the form and strength of selection commonly found in the wild (reviewed in Endler 1986; Hoekstra et al. 2001; Kingsolver et al. 2001; Rieseberg et al. 2002).

The recent emergence of genomics now promises to allow measurements of selection on phenotypes to be combined with knowledge of the underlying genes that produce them (Feder and Mitchell-Olds 2003; Stinchcombe and Hoekstra 2007). A powerful addition to the approach, once the genes responsible for putatively adaptive traits have been identified (Bradshaw et al. 1998; Abzhanov et al. 2004; Shapiro et al. 2004; Albertson et al. 2005; Colosimo et al. 2005; Hoekstra et al. 2006; Rogers and Bernatchez 2007), is to conduct field experiments with selected genotypes to evaluate directly the fitness consequences arising from the phenotypic effects of specific alleles (Bradshaw and Schemske 2003; Lexer et al. 2003; Baack et al. 2008; Barrett et al. 2008). This approach would represent a step toward the synthesis of genomics and population biology that will generate insights not possible from the study of genetics and selection on phenotypes in isolation. Progress toward such a synthesis has been hampered by a shortage of model species for which we have both excellent genomic resources and a rich body of ecological information available. However, full genome sequences for species of ecological interest are now rapidly accumulating (reviewed in Abzhanov et al. 2008), and the coming years hold much promise for connecting field observations with the genetic mechanisms underlying natural variation. In this chapter we describe the start of such a program of research using experimental studies of natural selection on genotypes at a major locus in threespine stickleback.

PARALLEL EVOLUTION OF ARMOR REDUCTION IN THREESPINE STICKLEBACKS

The dramatic radiation of threespine stickleback fish in post-glacial freshwater environments provides a unique opportunity to study the molecular mechanisms that underlie adaptive evolution. Development of a number of genetic and genomic tools has made it possible to identify the genes underlying several putatively adaptive traits (Colosimo et al. 2004; Shapiro et al. 2004; Colosimo et al. 2005; Miller et al. 2007). One of the most obvious phenotypic differences between marine and freshwater sticklebacks is the significantly reduced number of bony lateral plates in freshwater populations. Using a combination of mapping and transgenic techniques, Colosimo et al. demonstrated that the gene *Ectodysplasin* (*Eda*) is responsible for the majority of the marine-freshwater difference in armor plates (2005). Fish homozygous for the "complete" allele, which predominates in the sea, typically possess a row of 30 to 36 plates down each side of the body (complete morph), whereas homozygotes for the "low" allele typically possess 0 to 9 plates (low morph). Heterozygotes typically possess an intermediate number of plates on each side (partial morph) (Hagen and Gilbertson 1972; Bell 1977; Bell and Foster 1994). Phylogenetic analysis of sequence variation at this gene in numerous freshwater and marine populations determined that virtually all of the "low" alleles commonly found in armor-reduced freshwater populations are all descended from a low allele that originated more than 2 million years ago. The finding is paradoxical because most of the lakes and streams inhabited by low-armor sticklebacks have existed for only 10,000 years since the last ice age. However, it is now clear that evolution of low armor in freshwater has occurred by the fixation of low armor alleles at the *Eda* locus brought repeatedly into freshwater environments by the marine colonizers. Indeed, sticklebacks with reduced armor can be found at low frequency in the ocean today (plate 4). Recurrent fixation of the low allele in so many lakes implies that it undergoes positive selection in freshwater, since it is unlikely that stochastic processes would repeatedly increase the frequency of the allele in different populations (Simpson 1953; Rundle et al. 2000; Schluter et al. 2004). This represents our "signature of selection," and like most signatures of selection it does not by itself identify the mechanism.

Testing a Mechanism for Armor Evolution in Threespine Stickleback

We carried out an experiment to quantify selection on the major gene underlying variation in lateral plates with the aim of clarifying the mechanisms driving evolution of reduced armor in freshwater stickleback. Lateral plates play a defensive role in stickleback, not only increasing the difficulty of ingestion by predatory vertebrates (Reimchen 1983), but also improving the probability of escape and survival after capture (Reimchen 1992, 2000). It is thought that the complete allele will be favored in oceanic habitats, where sticklebacks are often far from cover and experience intense vertebrate predation pressure with little chance of evasion (Reimchen 2000; Bell 2001; Colosimo et al. 2004; Marchinko 2009). Several hypotheses have been proposed for lateral plate reduction in freshwater, including increased cost of bone because of low calcium levels (Giles 1983), an indirect effect of tolerance of low salinity (Heuts 1947), changes in swimming performance (Bergstrom 2002), and reduced predation from piscivorous fish and diving birds (Hagen and Gilbertson 1973; Reimchen 1992, 2000).

While testing some of these ideas in the laboratory, Marchinko and Schluter (2007) found evidence that freshwater juvenile fish with reduced lateral plate armor have a growth advantage relative to juveniles with complete armor. This finding suggested that *Eda* may have antagonistic pleiotropic effects on armor and growth in freshwater (Bell et al. 1993; Arendt 1997; Arendt and Wilson 2000; Arendt et al. 2001). Under this scenario, the evolution of low plated populations in freshwater environments may be the result of a correlated response to positive selection for increased growth rate rather than negative selection on armor (Marchinko and Schluter 2007). The expectation is that in the wild, increased growth rate should, in turn, reduce predation by insects, which prey on the smallest size classes of stickleback (Foster et al. 1988; Marchinko 2009), and increase lipid stores, resulting in higher over-winter survival (Post and Parkinson 2001; Biro et al. 2004; Curry et al. 2005).

To test this hypothesis, we tracked adaptive evolution over a complete generation at the *Eda* locus in replicated transplants of marine stickleback to field and laboratory freshwater environments. We predicted that the low *Eda* allele would confer a growth advantage in both lab and field. If increased growth rate leads to improved survival as a result of ecological factors, then the higher growth rate associated with the low allele should result in stronger positive selection in the field than in the lab. Deviations from these expectations would suggest that *Eda* or linked genes have unexpected fitness effects.

Experimental Methods

We collected adult marine sticklebacks for our experiment that were heterozygous at the *Eda* locus, having one copy of the low allele and one copy of the complete allele. We chose fully marine stickleback rather than artificially crossed hybrids between marine and freshwater populations to minimize variation in the genetic background, to minimize effects of other genetic differences possibly linked to *Eda*, and in the field study to replicate on a smaller scale the events of 10,000 years ago when marine fish first colonized coastal freshwater in the region after the ice age.

The fish were obtained from a saltwater inlet north of Vancouver having salinity between 28 and 32 ppm (Barrett et al. 2008). Most marine populations breed in freshwater (i.e., they are anadromous) but we chose a saltwater breeding population to eliminate the possibility of including freshwater-resident fish in the sample. To find heterozygotes we live-sampled approximately 45,000 fish using unbaited minnow traps and kept individuals that did not possess a full number of lateral plates; these occurred at a frequency of about 0.01 in the Oyster Lagoon population. We returned all other captured fish to the lagoon. About 400 partially plated fish were brought to the lab and genotyped using diagnostic markers within the *Eda* gene to distinguish low and complete alleles, following the procedures described in (Colosimo et al. 2005). About 230 proved to be heterozygous at the *Eda* locus.

To initiate the field experiment, we released approximately equal numbers ($N = 45$ or 46) of *Eda* heterozygotes into each of four ponds located at the University of British Columbia, Vancouver, British Columbia (plate 15). These ponds are standing bodies of water constructed by excavating a site on the campus of the University of British Columbia about twenty years ago. Each measures 23 m $\times$ 23 m and has a maximum depth of 3 m in the center (Schluter 1994). The ponds are lined with sand and bordered with limestone. All ponds had been previously drained, cleaned, and refilled in 2001, allowing enough time for plant and invertebrate communities to re-establish, but remaining free of fish until this experiment. The plants and invertebrates used to seed the ponds were collected from Paxton Lake, Texada Island, British Columbia, an 11-ha lake that contains wild sticklebacks. Apart from their construction, initialization, and use in prior experiments, the ponds are unmanipulated environments. Within sixty days we observed larval fish in each pond, indicating that the marine colonizers had produced F_1 progeny. We then took periodic destructive samples of fifty fish from this cohort from each pond to track growth rates and changes in *Eda* allele frequency until the fish reached sexual maturity the following spring.

In the lab experiment, we used artificial fertilization to generate approximately 300 F_1 progeny in crosses between 36 *Eda* heterozygotes. We raised all fish in four blocks of four freshwater aquariums connected by a water and filtration recirculation system. We individually marked and genotyped every fish, and quantified growth rate and survival up to sexual maturity.

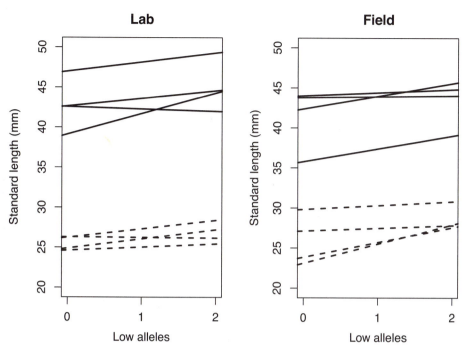

Figure 16.1. The association between standard length and *Eda* genotype in the sampled cohort in lab and field. Dashed lines represent juveniles at a mean length of 27 mm (October) and solid lines represent adults at a mean length of 43 mm (May). Each line represents a separate experimental replicate (blocks in the lab experiment and ponds in the field experiment). Fitted lines were obtained from linear regression of individual standard length on genotype scored by the number of low alleles (0, 1 or 2). A heterozygosity term (0 for homozygotes, 1 for heterozygotes) was also included in the linear model.

The Eda Low Allele Confers a Growth Advantage

We found convincing evidence that *Eda* has effects on growth rate in the field as well as the lab. In agreement with our predictions, fish carrying the low allele were larger than fish homozygous for the complete allele. The slope of the relationship between standard length and genotype (measured by the number of low alleles (0, 1 or 2)) was positive in all pond replicates and in three of four lab replicates (fig. 16.1). Final number of lateral plates is established fairly late in development, when the fish have attained a standard length of about 30 mm (Bell 1981; Bell et al. 1993; Bell 2001). In both lab and field the growth differential

TABLE 16.1

Relative Fitness (w) of Each Genotype during Juvenile and Adult Life History Stages

	Juveniles			Adults		
Genotype	CC	CL	LL	CC	CL	LL
Field w	1.00 ± 0.00	0.22 ± 0.05*	0.39 ± 0.09*	0.36 ± 0.11*	1.83 ± 0.67	1.00 ± 0.00
Lab w	0.88 ± 0.11	0.55 ± 0.06*	0.84 ± 0.09	0.998 ± 0.002	0.91 ± 0.10	0.77 ± 0.13

Relative fitness (w) indicates survival relative to that of the most fit homozygote. Juvenile stage is from birth to attainment of full lateral plate number (average length = 27 mm in both lab and field experiments). Adult stage is from average length 27 mm to 43 mm in both lab and field experiments. C refers to the complete allele and L to the low allele. Values shown are means fitness measurements from $n = 4$ ponds ± one standard error. *Indicates significantly different from one ($p < 0.05$) using a two-tailed t-test.

among genotypes was already evident at this time. The growth advantage of having low alleles persisted to the adult stage (fig. 16.1).

Selection on the Eda Locus

In the field experiment, we observed the predicted higher survival rates of fish carrying the low *Eda* allele. During adult growth, from the size of about 30 mm standard length, by which time the fish have attained their final number of lateral plates, to the start of the breeding season in May of the following year, by which time they reach about 50 mm standard length, the frequency of the complete allele rose from 33 percent to 51 percent, averaged over the four ponds. This change reflected the low relative survival of adult fish homozygous for the complete allele (table 16.1; fig. 16.2). Unexpectedly, we also detected strong selection favoring the complete allele during juvenile growth, before the fish had finalized their number of lateral plates during development (table 16.1; fig. 16.2). This selection early in life almost completely offset the gains by the low allele that occur later in life, resulting in much weaker net survival selection across the lifespan.

These sharp oscillations in allele frequencies seen in the field were largely absent in the laboratory (fig. 16.3). Whereas in the ponds a size advantage of the low allele translated to higher overwinter survival, we detected little or no survival advantage associated with large size in

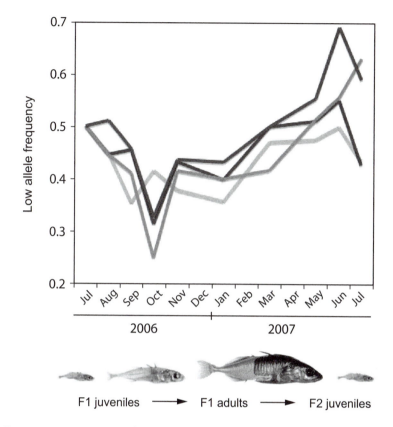

F1 juveniles ⟶ F1 adults ⟶ F2 juveniles

Figure 16.2. Frequency of the low allele in the field experiment. Each line represents a pond. All samples are from the first cohort of offspring, except the June and July 2007 samples, which are from the subsequent generation. From Barrett et al. (2008).

the lab. This makes sense if faster growth improved lipid stores and gave other advantages. Similarly, the drop in frequency of the low allele early in life did not happen in the lab. The contrast between survival effects in the lab and field suggests that selection in the field is driven by ecological mechanisms causing mortality. However, we are unable to propose a mechanism for the disadvantage of the low allele in juvenile stickleback.

Changes in the frequencies of heterozygotes in both lab and field are also difficult to explain in terms of the effects of armor on growth rate. The decline in low allele frequencies among juveniles in the ponds was

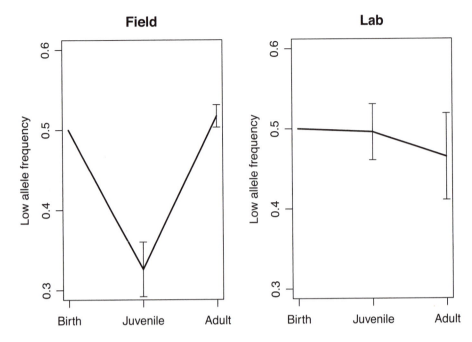

Figure 16.3. Frequency of the low allele in lab and field experiments. Lines represent the mean of four experimental replicates (blocks in the lab experiment and ponds in the field experiment). Error bars show ±1 standard error of the mean.

associated with a drop in the frequency of heterozygous fish and a rise in the frequency of the homozygous complete genotype (table 16.1). This heterozygote disadvantage for juvenile fitness was followed by strong heterozygote advantage in adult fitness. Although selection favored the low allele during adult growth, the frequency of heterozygotes increased more than the homozygous low genotype (table 16.1). We also observed heterozygote disadvantage for juvenile fitness in the lab (table 16.1), but in contrast to the field experiment, this period of juvenile heterozygote disadvantage was not followed by heterozygote advantage for adult fitness (table 16.1).

DISCUSSION

We conducted an experimental measurement of selection on the major locus underlying lateral plate armor in a replicated field setting. We used the timing of selection events to test a hypothesis for the mechanism of selection favoring one allele over the other at the *Eda* locus in freshwater. Our results partly confirm the hypothesized mechanism, but they also generated a surprising pattern of selection when we least expected it: before plates have completed development. This implies not only that *Eda* has effects on other traits under selection, but also that the evolution of plates is influenced by these effects of *Eda* on other traits. It would not have been possible to discover these selection events without knowing the major gene for the phenotypic trait of interest, since selection occurred before it was possible to distinguish between morphs. The work underscores the utility of field experiments to measure selection on genes, which can provide powerful tests of hypotheses for the mechanisms responsible and also lead to discovery of unanticipated fitness effects.

The difference in selection between adults in the lab and field is consistent with the hypothesis that the low allele is favored in the field because increased growth rate translates into improved overwinter survival. With a reduced burden of armor plates, fish with the low allele appear to devote more resources to growth. Increased growth typically translates into higher lipid stores and lower metabolic rate, which provides an advantage in surviving overwinter (Post and Parkinson 2001; Biro et al. 2004; Curry et al. 2005). In the lab, where survival was high regardless of size, we observed no rise in frequency of the low allele following plate development. This contrast between lab and field clearly demonstrates that the mechanism underlying the fitness advantage of the low allele is dependent on ecological context.

While the patterns of selection following development of lateral plates conformed to our predictions about the survival advantage conferred by increased growth rate, we also observed anomalous patterns that do not fit with the hypothesis that the low allele leads to improved fitness in freshwater. Strong selection against the low allele in juvenile fish in the pond experiment resulted in much weaker net positive selection over the full life cycle than would have occurred otherwise. This selection favoring the complete allele during juvenile growth is unexpected, because juveniles homozygous for the low allele are the fastest-growing genotype, and all else being equal, invertebrate predation is highest against the smallest fish (Foster et al. 1988; Marchinko

341

2009). This selection was evident in the field but not in the lab, again suggesting an extrinsic ecological mechanism. At this point we have not identified the mechanism of selection. Furthermore, we observed strong selection against heterozygotes during the juvenile life history stage in the ponds. Heterozygotes declined in frequency more rapidly than the homozygous low genotype. This pattern is also difficult to explain given our growth results, because heterozygotes are intermediate in size and plate number between the homozygotes. We observed a similar early decline in heterozygotes in the lab experiment, suggesting that this pattern is not caused by ecological factors.

These anomalous patterns imply that either variation at the *Eda* gene has direct or epistatic effects on other phenotypic traits contributing to fitness, or it is linked to another, unidentified locus affecting fitness. The evolution of plates will therefore be determined not solely by the selective consequences of armor, but also by the other effects of the major gene underlying this trait, and the effects of genes that are in linkage disequilibrium with it. Our results highlight how difficult it is to predict the evolutionary dynamics of a trait without being able to connect genotype and phenotype.

The utility of direct measurements of natural selection on genes is that they allow us to test hypotheses about the underlying mechanisms that drive phenotypic evolution. Through a combination of lab and field experiments, we have shown that the *Eda* low allele is favored in freshwater because it confers a growth advantage allowing improved overwinter survival. Because this effect was present in the field but not in the lab, we can be confident that ecological factors are probably responsible for the connection between increased growth rate and reduced mortality during the winter months. At the same time, anomalous patterns favoring the complete allele during juvenile growth, and oscillating fitness of heterozygotes point to unknown pleiotropic effects of *Eda* or genes linked to it.

In summary, our work shows that field experiments will be essential for understanding the genetics of adaptation. The ecological consequences of genetic variation at candidate loci may only be detected if they are studied under natural conditions. It thus appears that an early lesson from the nascent synthesis of population biology and genomics is that while increased molecular knowledge is invaluable, we also need to spend more time in the field to better understand the mechanisms of selection. We can think of no finer tribute to the Grants than to end on this recommendation.

REFERENCES

Abzhanov, A., C. G. Extavour, A. Groover, S. A. Hodges, H. E. Hoekstra, E. M. Kramer, and A. Moneiro. 2008. Are we there yet? Tracking the development of new model systems. *Trends in Genetics* 24: 353–360.

Abzhanov, A., M. Protas, B. R. Grant, P. R. Grant, and C. J. Tabin. 2004. Bmp4 and morphological variation of beaks in Darwin's finches. *Science* 305: 1462–1465.

Albertson, R. C., J. T. Streelman, T. D. Kocher, and P. C. Yelick. 2005. Integration and evolution of the cichlid mandible: The molecular basis of alternate feeding strategies. *Proc. Natl Acad. Sci. USA* 102: 16287–16292.

Arendt, J., D. S. Wilson, and E. Stark. 2001. Scale strength as a cost of rapid growth in sunfish. *Oikos* 93: 95–100.

Arendt, J. D. 1997. Adaptive intrinsic growth rates: an integration across taxa. *Quart. Rev. Biol.* 72: 149–177.

Arendt, J. D., and D. S. Wilson. 2000. Population differences in the onset of cranial ossification in pumpkinseed (*Lepomis gibbosus*), a potential cost of rapid growth. *Canad. J. Fish. Aquat. Sci.* 57: 351–356.

Baack, E. J., Y. Sapir, M. A. Chapman, J. M. Burke, and L. H. Rieseberg. 2008. Selection on domestication traits and quantitative trait loci in crop-wild sunflower hybrids. *Mol. Ecol.* 17: 666–677.

Barrett, R.D.H., S. M. Rogers, and D. Schluter. 2008. Natural selection on a major armor gene in threespine stickleback. *Science* 322: 255–257.

Bell, M. A. 1977. Late Miocene marine threespine stickleback, *Gasterosteus-Aculeatus-Aculeatus*, and its zoogeographic and evolutionary significance. *Copeia* 277–282.

———. 1981. Lateral plate polymorphism and ontogeny of the complete plate morph of threespine stickleback (*Gasterosteus aculeatus*). *Evolution* 38: 665–674.

———. 2001. Lateral plate evolution in the threespine stickleback: getting nowhere fast. *Genetica* 112: 445–461.

Bell, M. A., and S. A. Foster. 1994. *The Evolutionary Biology of the Threespine Stickleback*. Oxford: Oxford University Press.

Bell, M. A., G. Orti, J. A. Walker, and J. P. Koenings. 1993. Evolution of pelvic reduction in threespine stickleback fish—A test of competing hypotheses. *Evolution* 47: 906–914.

Bergstrom, C. A. 2002. Fast-start swimming performance and reduction in lateral plate number in threespine stickleback. *Can. J. Zool.-Rev. Can. Zool.* 80: 207–213.

Biro, P. A., A. E. Morton, J. R. Post, and E. A. Parkinson. 2004. Overwinter lipid depletion and mortality of age-0 rainbow trout (*Oncorhynchus mykiss*). *Canad. J. Fish. Aquat. Sci.* 61: 1512–1519.

Boag, T. B., and P. R. Grant. 1981. Intense natural selection in a population of Darwin's finches (Geospizinae) in the Galápagos. *Science* 214: 82–85.

Bradshaw, H. D., K. G. Otto, B. E. Frewen, J. K. McKay, and D. W. Schemske. 1998. Quantitative trait loci affecting differences in floral morphology between two species of monkeyflower (Mimulus). *Genetics* 149: 367–382.

Bradshaw, H. D., and D. W. Schemske. 2003. Allele substitution at a flower colour locus produces a pollinator shift in monkeyflowers. *Nature* 426: 176–178.

Colosimo, P. F., K. E. Hosemann, S. Balabhadra, G. Villarreal, M. Dickson, J. Grimwood, J. Schmutz, R. M. Myers, D. Schluter, and D. M. Kingsley. 2005. Widespread parallel evolution in sticklebacks by repeated fixation of ectodysplasin alleles. *Science* 307: 1928–1933.

Colosimo, P. F., C. L. Peichel, K. Nereng, B. K. Blackman, M. D. Shapiro, D. Schluter, and D. M. Kingsley. 2004. The genetic architecture of parallel armor plate reduction in threespine sticklebacks. *PLoS Biology* 2: e109.

Curry, R. A., S. L. Currie, S. K. Arndt, and A. T. Bielak. 2005. Winter survival of age-0 smallmouth bass, *Micropterus dolomieu*, in north eastern lakes. *Environ. Biol. Fishes* 72: 111–122.

Endler, J. A. 1986. *Natural Selection in the Wild*. Princeton, NJ: Princeton University Press.

Feder, M. E., and T. Mitchell-Olds. 2003. Evolutionary and ecological functional genomics. *Nat. Rev. Genet.* 4: 649–655.

Foster, S. A., V. B. Garcia, and M. Y. Town. 1988. Cannibalism as the cause of an ontogenetic niche shift in habitat use by fry of the threespine stickleback. *Oecologia* 74: 577–585.

Gibbs, H. L., and P. R. Grant. 1987. Oscillating selection on Darwin's finches. *Nature* 327: 511–513.

Giles, N. 1983. The possible role of environmental calcium levels during the evolution of phenotypic diversity in the outer Hebridean populations of the three-spined stickleback, *Gasterosteus aculeatus*. *J. Zool. London* 199: 535–544.

Hagen, D. W., and L. G. Gilbertson. 1972. Geographic variation and environmental selection in *Gasterosteus-aculeatus* in Pacific Northwest, America. *Evolution* 26: 32–43.

———. 1973. Selective predation and intensity of selection acting upon lateral plates of threespine sticklebacks. *Heredity* 30: 273–287.

Heuts, M. J. 1947. Experimental studies on adaptive evolution in *Gasterosteus aculeatus L. Evolution* 1: 89–102.

Hoekstra, H. E., R. J. Hirschmann, R. A. Bundey, P. A. Insel, and J. P. Crossland. 2006. A single amino acid mutation contributes to adaptive beach mouse color pattern. *Science* 313: 101–104.

Hoekstra, H. E., J. M. Hoekstra, D. Berrigan, S. N. Vignieri, A. Hoang, C. E. Hill, P. Beerli, and J. G. Kingsolver. 2001. Strength and tempo of directional selection in the wild. *Proc. Natl. Acad. Sci. USA* 98: 9157–9160.

Kingsolver, J. G., H. E. Hoekstra, J. M. Hoekstra, D. Berrigan, S. N. Vignieri, C. E. Hill, A. Hoang, P. Gibert, and P. Beerli. 2001. The strength of phenotypic selection in natural populations. *Am. Nat.* 157: 245–261.

Lexer, C., M. E. Welch, J. L. Durphy, and L. H. Rieseberg. 2003. Natural selection for salt tolerance quantitative trait loci (QTLs) in wild sunflower hybrids: Implications for the origin of *Helianthus paradoxus*, a diploid hybrid species. *Mol. Ecol.* 12: 1225–1235.

Marchinko, K. 2009. Predation's role in repeated phenotypic and genetic divergence of armor in threespine stickleback. *Evolution* 63: 127–138.

Marchinko, K. B., and D. Schluter. 2007. Parallel evolution by correlated response: lateral plate reduction in threespine stickleback. *Evolution* 61: 1084–1090.

Miller, C. T., S. Beleza, A. A. Pollen, D. Schluter, R. A. Kittles, M. D. Shriver, and D. M. Kingsley. 2007. Cis-regulatory changes in kit ligand expression and parallel evolution of pigmentation in sticklebacks and humans. *Cell* 131: 1179–1189.

Post, J. R., and E. A. Parkinson. 2001. Energy allocation strategy in young fish: Allometry and survival. *Ecology* 82: 1040–1051.

Price, T. D., P. R. Grant, and H. L. Gibbs. 1984. Recurrent patterns of natural selection in a population of Darwin's finches. *Nature* 309: 787–789.

Reimchen, T. E. 1983. Structural relationships between spines and lateral plates in threespine stickleback (*Gasterosteus-Aculeatus*). *Evolution* 37: 931–946.

———. 1992. Injuries on stickleback from attacks by a toothed predator (Oncorhynchus) and implications for the evolution of lateral plates. *Evolution* 46: 1224–1230.

————. 2000. Predator handling failures of lateral plate morphs in *Gasterosteus aculeatus*: Functional implications for the ancestral plate condition. *Behaviour* 137: 1081–1096.

Rieseberg, L. H., A. Widmer, A. M. Arntz, and J. M. Burke. 2002. Directional selection is the primary cause of phenotypic diversification. *Proc. Natl. Acad. Sci. USA* 99: 12242–12245.

Rogers, S. M., and L. Bernatchez. 2007. The genetic architecture of ecological speciation and the association with signatures of selection in natural lake whitefish (*Coregonus* sp. Salmonidae) species pairs. *Mol. Biol. Evol.* 24: 1423–1438.

Rundle, H. D., L. Nagel, J. W. Boughman, and D. Schluter. 2000. Natural selection and parallel speciation in sympatric sticklebacks. *Science* 287: 306–308.

Schluter, D. 1994. Experimental-evidence that competition promotes divergence in adaptive radiation. *Science* 266: 798–801.

Schluter, D., E. A. Clifford, M. Nemethy, and J. S. McKinnon. 2004. Parallel evolution and inheritance of quantitative traits. *Am. Nat.* 163: 809–822.

Shapiro, M. D., M. E. Marks, C. L. Peichel, B. K. Blackman, K. S. Nereng, B. Jonsson, D. Schluter, and D. M. Kingsley. 2004. Genetic and developmental basis of evolutionary pelvic reduction in threespine stickleback. *Nature* 428: 717–723.

Simpson, G. G. 1953. *The Major Features of Evolution*. New York: Columbia University Press.

Stinchcombe, J. R., and H. E. Hoekstra. 2007. Combining population genomics and quantitative genetics: finding the genes underlying ecologically important traits. *Heredity* 100: 158–170.

Chapter Seventeen

The Book and the Future: Perspective and Prospective

Peter R. Grant and B. Rosemary Grant

THE BIODIVERSITY PROBLEM

The study of evolution emerged from a western, nineteenth-century, preoccupation with natural history:

> And what a science Natural History will be, when we are in our graves, when all the laws of change are thought one of the most important parts of Natural History.

In this quotation from an 1856 letter to J. D. Hooker (Darwin 1919, vol. 1, p. 439), Charles Darwin had in mind natural history as a synthesis of evolution (the laws of change) and ecology (the laws of the environment). Marston Bates, preeminent spokesman for the mid-twentieth century western worldview of natural history, expressed the idea this way:

> The diversity of life is extraordinary. There is said to be a million or so different kinds of living animals, and hundreds of thousands of kinds of plants. But we don't need to think of the world at large. It is amazing enough to stop and look at a forest or at a meadow—at the grass and trees and caterpillars and hawks and deer. How did all these different kinds of things come about; what forces governed their evolution; what forces maintain their numbers and determine their survival or extinction; what are their relations to each other and to the physical environment in which they live? These are the problems of natural history. (Bates 1950, p. 8)

And in the words of a paleontologist:

> It is the intertwined and interacting mechanisms of evolution and ecology, each of which is at the same time a product and a process,

that are responsible for life as we see it, and as it has been. (Valentine 1973, p. 58)

As the field of genomics blossoms, threatening to overwhelm us with vast amounts of information, and draws more and more research workers into tackling its fascinating questions, it is possible to forget that evolutionary biology begins with simple observations of organisms in nature, and seeks to explain and understand them. While there is so much to be learned, and now can be learned, by what is sometimes called "skin-in" biology, biology outside the skin can appear to be relatively unimportant, largely accomplished, ignorable. We subscribe to an alternative view (Grant 2000). The problem for evolutionary biologists is to explain the origin and maintenance of all living things, at all levels of organization, and in all of their amazing complexity and diversity. The biodiversity problem is as much a modern problem as it was in Darwin's day. The tools available to tackle the problem have changed so much, but the problem itself has not. Answers should be sought both inside and outside the skin and the cell, and that is what authors in this book attempt to do.

INFERENCES

Evolution is studied as a product of the past as well as a process of the present. The first three chapters address questions of identifying and dating fossils, characterizing the environments in which the organisms lived, and inferring the causes of evolution by linking organisms with their environments. The challenge for the authors is to come up with a convincing explanation of how the past gave rise to the present when only fragmentary information about the past is available, and alternative hypotheses are often exceedingly difficult to reject. Incrementally that understanding is being improved by discoveries of new fossils and of new patterns of variation in evolutionary success (chapters 1–3). The first has changed our views on particular evolutionary pathways taken, for example by reptiles in the origin of flight (chapter 3), by mammals in the origin of marine living (Gingerich 2003), and by fish in the origin of terrestrial living (Daeschler et al. 2006). The second has revealed strong patterns in the geographical origin, history, and fates of clades (chapter 2); as measured by the production of species, some clades consistently outperform others, some environments repeatedly foster proliferation of species more than others. At the same time the past is being illuminated through advances made in understanding living organisms, for example in the genetic control of development (chapters 5–8), and in the improved use of molecular genetic data to estimate phylogenetic relationships (chapter 4).

MECHANISMS

The chapters illustrate several ways in which inquiry begins with observations in nature, such as enigmatic fossils, albino fish in subterranean caves, nests of mice, or the taxonomic richness of lizards on islands. The investigation proceeds, guided by theory, to expose the underlying causal mechanisms. For many evolutionary questions, answers lie in understanding how hidden genotypes are translated into observed phenotypes; how genes are expressed in development, how one developmental pathway may constrain or facilitate elaboration of another, and how the environment drives or denies the direction of evolution of developmental programs. The full translation comprises a chain of connections between the genes that initiate the development of a trait, through the chemical and neurohumoral pathways that mediate its structural, physiological, or behavioral expression and integration with other traits, to the ecological interactions that test its performance and success. In short, it is a network of interactions among genes (Hinman and Davidson 2007) and between them and the environment. These are the concerns of several chapters, especially chapters 5–10.

Identifying biochemical pathways by which zygotes develop into adults tells us how the organic world is constructed and functions. By itself it is not enough. It does not tell us how organisms evolved, or why they evolved in the way they did and not in some other way. To answer questions of evolutionary change we need to understand germ-line mutation. First, and most important, what are the mutations that produce evolutionarily significant change? Second, are they in coding regions or in regulatory regions (near or far), when and where do they occur in development, are there many with small effects, just a few with large effects, or both, and how are their effects restricted to one developmental stage and one site when their potential expression is organism-wide (chapters 5–8, and 10)? Third, how do population size, structure, and the external environment affect their frequency and fate (chapters 13 and 16)? Add to mutation within the gene the other mechanisms of evolutionary change, such as recombination, chromosomal inversions and translocations, mobile elements, gene-duplication, gene-silencing, genomic imprinting, horizontal gene transfer, and introgressive hybridization, and the potential for evolutionary change seems enormous. Can this plethora of genetic mechanisms of change be reduced to synthetic order? It may be that more natural history of DNA and RNA is needed before a theory of evolutionary change can be built, one that make testable predictions about the where, when, and how of genetic mechanisms of change among the different levels of biological organization in the tree of life.

For understanding evolutionary change, it is just as important to determine the ecological mechanisms that govern fitness as it is to determine the genetic mechanisms that govern the development of a trait. In fact, the fusion of both in single studies is one of the exciting growing points in modern evolutionary biology. We see the empirical study of relative fitness in contemporary, evolving, populations as a key area for the understanding of microevolutionary processes in nature, and by extension for the understanding of broad patterns of evolution in the past (see also chapter 2). Yet it is an extremely challenging one, owing to the difficulties of measuring relative fitness and microevolution in nature, and identifying the causal environmental factors. Progress requires a good means of characterizing and quantifying environmental factors, experimentally tractable systems, relatively simple food web structure, and measurable variation in the evolving species under investigation, preferably at the single gene level. Chapter 16 provides an example of progress in a well-chosen system.

SYNTHESIS

It is well known that the New Synthesis of the 1940s integrated ecology, behavior, systematics, and population genetics. Today, specialization and the unprecedented rate of accumulation of new information threaten to fragment the synthesis. Yet in some ways evolutionary biology has never been more unitary. Paleontologists, for example, use information from molecular genetics to understand trait evolution, phylogenetic affinities, and the dates of taxon origin and extinction. Paleontology also connects with ecology (e.g., MacFadden 2000; Vermeij 2008). The common goal is to understand the influence of the environment on organisms. These are prime examples of cross-connections, both between disciplines and among the four sections of this book, further illustrated in this book by the use of molecular phylogenies by almost all authors. A general theme that emerges is an interaction between organisms and their environments that is mutualistic. In chapter 1 we learn that paleoenvironments that no longer exist were transformed in part by the activities of organisms, which in turn evolved in response to altered conditions. In this case the environment is physical and chemical, whereas in chapters 5, 11, and 12 it is biological as well as contemporary. The traits and fates of organisms thus depend upon responses to and from the environment. These chapters (and chapter 15) show that addition or deletion of a member of a set of interacting species alters the selective pressures on the remainder, which can lead to rapid change. Evolution of epistatic gene interactions can be thought of as a genomic equivalent. Inasmuch as some phenotypic traits are both evolutionarily more malleable and plastic than others, and the genetic architecture of traits varies, the question arises as to how factors that operate at the level of ecological communities exert an influence on the evolution of genetic architecture of individual organisms. Uniting ecology and genomics will produce a new evolutionary synthesis (again).

Major Challenges

We see two major challenges in the study of evolutionary biology that are not addressed in this book. These are the evolution of genomes and of microbes. Both are hidden from the unaided eyes of naturalists.

Genomes

Genomes vary enormously among taxonomic groups, in size, composition, and configuration; hence, they have undergone spectacular differentiation. How has this happened, and why? What is the functional and evolutionary significance of the differences? Why do structurally simple organisms such as amoebae have much larger genomes than more complex metazoans like birds and mammals? Why are genes packaged in four chromosomes in *Drosophila melanogaster* and in hundreds of chromosomes in butterflies? In most plants, mitochondria and chloroplasts are maternally inherited, but in pines, chloroplasts are paternally inherited and in Cupressaceae, both organelles are paternally inherited: why? Why do the chromosomes of female *D. melanogaster* recombine but the chromosomes of males do not? Why do mammals have much more repeated sequences of DNA than birds? Why are males the heterogametic sex in mammals but not in birds? These are just a few in a long list of questions about the structure of genomes and their variation among very different taxa. At first glance none of these phenomena has anything to do with external environment; they appear to be entirely internally driven, or perhaps they vary as a matter of chance. Is this correct? Can ecology be ignored?

As knowledge of genome structure increases, the list of questions will shrink, and then expand with new questions. And this is just structure: gene function varies across taxa too. A growing list of questions could be drawn up for gene regulation!

Microbes

It is said we live in a microbial world. Chapter 1 provides a needed perspective by reminding us that for most of the life of this planet there was nothing else but microbes. Thinking about the biological world has been transformed by recent discoveries of unexpected genetic diversity and biomass of microbial life in the oceans, and elsewhere. Yet there is very little understanding of how it evolved. We need to know how species multiply and diversify, but gaining that understanding with microbes is hampered by not knowing what a species is in the sense the term is used for macro (visible) organisms. The apparently widespread horizontal gene-transfer between bacteria, for example, prevents us from defining species by the usual genetic criteria. Recently discovered microbial richness thus widens the problem of explaining biodiversity, and simultaneously presents new challenges. To be able to enter the microbial world as a naturalist, and to define a species, recognize a population, and study its evolutionary dynamics, we need a new technology so that we can follow more precisely the fates of microbes in nature.

Their diversity is not the only reason why microbes deserve attention. The microbial world we live in also lives in us. Microbes reproduce, mutate, and evolve rapidly. They are symbionts, commensals, competitors, and parasites. At their most severe they kill frogs, oak trees, and us (e.g., HIV and SARS), and they may have caused, or at least contributed significantly to, extinctions without leaving any evidence of having done so. A problem for evolutionary biologists is to explain how microbes and macrobes coexist. Symbiosis is one answer; nevertheless, given their numbers, biomass, diversity, rapid evolution, and occasional hostility, why is the world not more over-run by microbes than it is?

Some Prospects

Several chapters in this book give pointers to new developments in the field of evolutionary biology. To elaborate: new developments arise from new techniques (e.g., polymerase chain reaction), new methods of analysis (coalescent theory), and concepts that are either new or old but much improved as a result of revision (group selection, sympatric speciation) or now seen to be relevant (canalization). We are at the beginning of a new, uncertain, and exciting phase in evolutionary biology, the genomics era. What will knowledge of whole genomes do to our understanding of evolution outside the laboratory, past and present? Are species that are studied intensively in the laboratory suitable models for related species in nature? What will anthropogenically caused climate change do to the environment as it impinges on the lives of organisms? Can this be turned to advantage, through the study of evolution as a process? As we gain more complete understanding of the genetic workings of individual organisms, we are losing more and more of the natural environments in which their populations evolved, so much so that there is now a rush to inventory our global biodiversity while we have it. Will the loss be stopped, and will those environments still intact be given the same extraordinarily detailed scrutiny as currently given to genomes? Will ge-nomics and eco-nomics be truly united as ecogenomics in the new evolutionary biology?

We conclude by raising some questions on three topics central to our research on speciation and adaptive radiation of Darwin's finches in the Galápagos archipelago (chapter 15). Restricted to neither finches nor archipelagos, they have wide ramifications in evolutionary biology.

Learning

For behavior to evolve, there must be genetic variation. This simple idea drives the search for evolutionarily significant genetic variation underlying behavioral variation. As chapter 10 makes clear, surprisingly few attempts have been made to study the genetic basis of naturally occurring behavior using modern molecular methods: behavior that is important in food finding, enemy avoidance, mate choice, parental care, etc. Our experience studying Darwin's finch evolution on the Galápagos islands tells us that learning to respond to the environment can be as important for finches as deploying inherited behavior (Grant and Grant 2008). For example, mate choice in these species is largely influenced by early experience. When it comes time for finches to breed, the morphological and song cues they learned early in life, through imprinting, typically on their parents, guide the choice of a mate. Learning thus affects a crucially important stage in speciation, the establishment of a pre-mating barrier to interbreeding with members of other populations. It also affects feeding ecology. Finches learn to feed on food items that can be handled efficiently, by copying others and by trial-and-error learning. Hence, early experience helps to determine fitness. When experience varies, and ability to learn varies, fitness varies. Therefore learning, and specifically the genetic variation underlying the propensity to learn, should be a major focus for evolutionary studies.

Learning ability varies greatly across the animal kingdom, and this raises many questions in evolution. In general language, just how widely applicable is the interaction of cultural and organic evolution? How does the interaction vary among taxa, and how does it work? What are the neurological and genetic rules that govern the evolution of an age-dependent propensity to acquire information from the environment and use it? Do the rules change gradually or in jumps? Where are the constraints that limit the evolutionary elaboration of learning abilities? When is learning selectively penalized? Generalizing yet further, phenotypic plasticity can be thought of as analogous to learning. Both phenomena involve responses to the environment according to information gained from the environment, actively or passively. Thus, learning is not the only solution to the problem of gaining information and responding in a way that enhances fitness. The evolution of information-processing according to the inherent properties of organisms provides a framework for uniting such disparate phenomena as the growth of seaweeds (chapter 9), where learning is surely not involved in adaptive responses to the environment, and the behavior of mice (chapter 10), where it just as surely is involved.

Reproductive Isolation

Intrinsic post-zygotic isolation can arise at any time after separate populations of a single species have become established, potentially early and rapidly, for example by meiotic drive of selfish elements (Phadnis and Orr 2009), or over the long term by genetic drift, even in the absence of selection (Muller 1940). For many organisms pre-zygotic isolation arises first, as a result of divergence in morphological and behavioral traits. Divergence is important. What are the genetic changes, step by step, that lead to the transformation of the key traits by which species differ, and how is variation within species thereby converted to variation between or among species? Differences between species are already being subjected to molecular genetic dissection in model systems; for example, the color and shape of *Mimulus* flowers (Bradshaw and Schemske 2003), physiological traits of *Helianthus* sunflowers (Rieseberg et al. 2003), skeletal traits of sticklebacks (chapter 6), and beak traits of Darwin's finches (Abzhanov et al. 2006). The next question, the key question, is how did these differences arise, and why? Stated in another way, differences are the result of a process, so what exactly was the process? Identification of the mutational steps and the time course involved is needed to answer the genetic component of the how question. The learning component, discussed above, may be just as important or more so. To answer the why question, a study of selection and fitness is needed (chapter 16). Given the prevalence of ecological determinants of divergence leading to reproductive isolation (Schluter 2009), knowledge of environments is as important as knowledge of mutations. The study of speciation at this time is both exciting and challenging because it unites genetics, ecology, and behavior.

Where is it heading? Looking to the future, we see transgenic manipulation as a tool in the study of speciation and further evolution (Grant 2000). Transgenic experiments are routinely carried out in the laboratory to investigate gene function. In the future, they will be used to alter genomes of existing species in order to reconstruct others that no longer exist, to test hypotheses about how evolution occurred, and to identify the most important genetic changes and the order in which they occurred. For example, species that share a common ancestor but are no longer capable of interbreeding could be altered genetically to reconstruct the intermediate stages in their divergence. There are obvious limits to success, in proportion to the genetic distance and time of origin of the taxa being compared. It may not be possible to choose between two or more alternative pathways of genetic change, and past environments that guided or drove the change will forever remain uncertain. Ambiguities of interpretation will not deter everyone, however, and even if definitive answers are out of reach, at least some alternatives may be excluded. The prospect is a better understanding of speciation as a process.

Introgressive Hybridization

Long-term study of Darwin's finch populations on the Galápagos Island of Daphne Major has revealed a porous barrier to interbreeding between resident finch species. They interbreed rarely; nevertheless, when feeding conditions are suitable for birds with intermediate beak sizes, hybrids survive well and backcross to the parental species. As a result a trickle of genes flows between species, elevating the standing additive genetic variation in each, altering their genetic correlation structures, and enhancing their potential for further evolution. We have speculated that the enhanced potential for evolutionary change could be an important factor in the early stages of adaptive radiations (Grant and Grant 1998), and have offered one example in which reproductive isolation was the outcome of hybrids or backcrosses interbreeding (Grant and Grant 2009; see plate 16 and fig. 17.1). We are not alone in suggesting hybridization in animals may increase their evidentiary

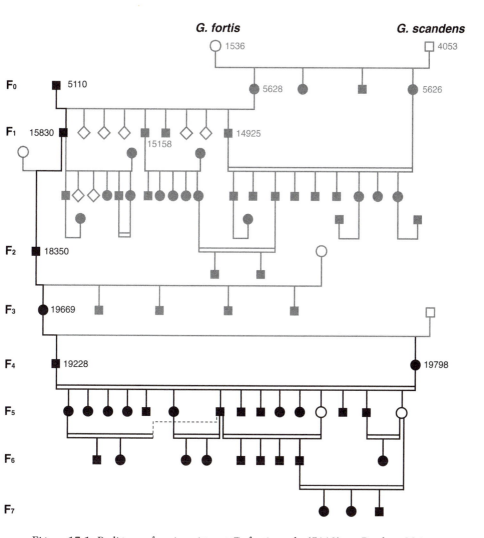

Figure 17.1. Pedigree of an immigrant *G. fortis* male (5110) on Daphne Major Island, Galápagos. The line of descent leads to an exclusively inbreeding (endogamous) group at generation F_4, reproductively isolated from residents by a behavioral barrier involving song as well as morphology, and highlighted in black. Males are indicated by squares, females by circles, and birds of unknown sex by diamonds. Individuals of unknown genotype are indicated by open symbols, and filled symbols refer to genotyped birds. Salient individuals in the pedigree are identified by their band numbers, e.g. the mate (5628) of the original immigrant (5110) is a *G. scandens* - *G. fortis* F_1 hybrid or backcross (4053) who bred with a *G. fortis* female (1536). Double horizontal lines connect close relatives that bred together. The frequency of inbreeding with close relatives is exceptionally high, even in the first few generations. Modified from Grant and Grant (2009) with better genetic resolution of the last two generations: see also plate 16 for an illustration of immigrants and residents.

potential; for example, Ole Seehausen (2004) has argued the same for cichlid fish (see also chapter 14).

The occurrence of rare introgressive hybridization, well known in plants, is being increasingly detected and reported in animals (Schwenk et al. 2008). Occurrence is not the same as significance or importance, however. The major question that needs to be addressed is whether introgressive hybridization has enabled evolutionary divergence to occur faster, or in new directions, than would be the case in its absence. Evidence for both has been in the literature for some time (Lewontin and Birch 1966; Svärdson 1970), but whether these are isolated cases of a generally insignificant and transient phenomenon or examples of an under-appreciated and important process in evolution remains to be determined by observation and experiment.

And so we finish the chapter where we started, drawing attention to the problem that unites evolutionary biologists, the problem of explaining the world's biological diversity.

REFERENCES

Abzhanov, A., W. P. Kuo, C. Hartmann, B. R. Grant, P. R. Grant, and C. J. Tabin. 2006. The calmodulin pathway and evolution of elongated beak morphology in Darwin's finches. *Nature* 442: 563–567.

Bates, M. 1950. *The Nature of Natural History*. New York: Scribner.

Bradshaw, H. D., and D. W. Schemske. 2003. Allelic substitution at a flower colour locus produces a pollinator shift in monkeyflowers. *Nature* 426: 176–178.

Daeschler, E. B., N. H. Shubin, and F. A. Jenkins, Jr. 2006. A Devonian tetrapod-like fish and the evolution of the tetrapod body plan. *Nature* 440: 757–763.

Darwin, C. R. 1859. *On The Origin of Species by Means of Natural Selection*. London: J. Murray.

Darwin, F., ed. 1919. *The Life and Letters of Charles Darwin. Including an Autobiographical Chapter. Volumes I and II*. New York and London: D. Appleton and Co. 1st edition 1887.

Gingerich, P. D. 2003. Land-to-sea transition of early whales: evolution of Eocene Archaeoceti (Cetacea) in relation to skeletal proportions and locomotion of living semiaquatic mammals. *Paleobiology* 29: 429–454.

Grant, B. R., and P. R. Grant. 1998. Hybridization and speciation in Darwin's finches: the role of sexual imprinting on a culturally transmitted trait. In

D. J. Howard and S. H. Berlocher, eds., *Endless Forms: Species and Speciation*, 404–422. New York: Oxford University Press.

Grant, P. R. 2000. What does it mean to be a naturalist at the end of the twentieth century? *Am. Nat.* 155: 1–12.

Grant, P. R., and B. R. Grant. 2008. *How and Why Species Multiply. The Radiation of Darwin's Finches*. Princeton, NJ: Princeton University Press.

———. 2009. The secondary contact phase of allopatric speciation in Darwin's finches. *Proc. Natl. Acad. Sci. USA* 106: 20141–20148.

Hinman, V. E., and E. H. Davidson. 2007. Evolutionary plasticity of developmental gene regulatory network architecture. *Proc. Natl. Acad. Sci. USA* 104: 19404–19409.

Lewontin, R. C., and L. C. Birch. 1966. Hybridization as a source of variation for adaptation to new environments. *Evolution* 20: 315–336.

MacFadden, B. T. 2000. Cenozoic mammalian herbivores from the Americas: reconstructing ancient diets and terrestrial communities. *Annu. Rev. Ecol. Syst.* 31: 33–59.

Muller, H. J. 1940. Bearings of the *Drosophila* work on systematics. In J. S. Huxley, ed., *The New Systematics*, 185–268. Oxford: Clarendon Press.

Phadnis, N., and H. A. Orr. 2009. A single gene causes both male sterility and segregation distortion. *Science* 323: 376–379.

Rieseberg, L. H., O. Raymond, D. M. Rosenthal, Z. Lai, K. Livingstone, T. Nakazato, J. L. Durphy, A. E. Schwartzbach, L. A. Donovan, and C. Lexer. 2003. Major ecological transitions in wild sunflowers facilitated by hybridization. *Science* 301: 1211–1216.

Schluter, D. 2009. Evidence for ecological speciation and its alternative. *Science* 323: 737–741.

Schwenk, K., N. Brede, and B. Streit. 2008. Introduction. Extent, processes and evolutionary impact of interspecific hybridization in animals. *Philos. Trans. R. Soc. B* 363: 2805–2811.

Seehausen, O. 2004. Hybridization and adaptive radiation. *Trends Ecol. Evol.* 19: 198–207.

Svärdson, G. 1970. Significance of introgression in coregonid evolution. In C. C. Lindsey and C. S. Woods, eds., *Biology of Coregonid Fishes*, 33–59. Winnipeg: University of Manitoba Press.

Valentine, J. W. 1973. *Evolutionary Paleoecology of the Marine Biosphere*. Englewood Cliffs, NJ: Prentice-Hall.

Vermeij, G. J. 2008. Escalation and its role in Jurassic biotic history. *Palaeogeogr., Palaeoclimatol., Palaeoecol.* 263: 3–8.

Chapter Eighteen

A Festival for Rosemary and Peter Grant

David B. Wake

The occasion of the formal retirement of Rosemary and Peter Grant from the Princeton University Faculty is an opportune time for a brief overview of the accomplishments of these remarkable scientists. I was honored to have been asked to introduce the formal program of oral presentations, which are represented by the chapters that constitute this Festschrift to honor Peter and Rosemary Grant. My comments here are essentially those I delivered in my brief introduction.

There are some parallels among the Grants and me. We were born in the same year (1936), although in different parts of the world (Peter in England, Rosemary in Scotland, me in South Dakota). Peter and I are also academic contemporaries, receiving our doctorates the same year (1964) and advancing through the academic ranks to full professorships at about the same time (1972–73). Peter did his graduate work at the University of British Columbia and while there met Rosemary, who had come to teach genetics, having received her BSc (Hons.) from Edinburgh University in 1960. Peter and I had the great good fortune to marry dynamic, strong, challenging women (in my case, Marvalee Hendricks Wake) who have been our intellectual peers.

While Peter followed what might be called an orthodox career path—a year of postdoctoral studies at Yale, then taking up a tenure-track job at McGill University—Rosemary followed a more indirect pathway in a pattern only too typical for women during the 1960s. She was delayed in her research career by establishing a household and giving birth to first one and then another daughter. Rosemary completed her PhD in 1985 working with Staffan Ulfstrand in Uppsala.

Rosemary's first publication, co-authored with Peter, Jamie Smith, and Ian and Lynette Abbott, an integrative treatment of Galápagos finch ecology, appeared in 1975, whereas Peter had an earlier start. His first publication, which I remember well because I read it eagerly when it appeared, was in 1964 and dealt with the avifauna of the Tres Marías islands. This work was coauthored with his major professor, Ian McTaggart Cowan (a student of Joseph Grinnell, first director of my own home institution, the Museum of Vertebrate Zoology). Between 1964 and 1975, Peter was author or coauthor of forty-eight publications by my count, the majority of them dealing with birds and islands, but also a number of papers dealing with vole ecology, and ecology more generally.

Since 1975, Peter and Rosemary have been frequent coauthors and nearly all of these publications deal with the Galápagos Islands and usually with finches. There are more than eighty coauthored publications, a number of them involving other collaborators as well. Most of these appeared in prominent journals or as chapters in well-regarded books. Their remarkable productivity, accomplished while spending extended periods of time in the field, is testament to the focus and dedication of this stunningly effective team.

I offer some perspectives on their wonderful careers.

Taking Darwin Seriously

Peter and Rosemary certainly were strongly influenced by the work of Charles Darwin, from his fieldwork in the Galápagos Islands and his observations on what are now known as Darwin's finches, to his later synthetic masterpieces, which continue to inspire young (as well as older, I might add) scientists attracted to evolutionary questions. Rosemary, Peter, and their daughters started serious fieldwork in the Galápagos in 1973, and they have spent literally years of most dedicated and focused field research studying Darwin's finches (see more below). This research has resulted in an impressive body of work, which has had immense and, I think it safe to say, lasting impact on evolutionary biology. Their many superb papers and several books are a striking tribute to what Charles Darwin began. They have indeed taken Darwin seriously, to the lasting benefit of the evolutionary biology community.

TAKING ISLANDS SERIOUSLY

Peter focused on islands from the very beginning of his research career. His earliest publications, on the Tres Marías Islands off the west coast of Mexico, attracted my attention because I was interested in the islands from a herpetological perspective. Islands long have been seen as having high value in studies of evolutionary biology, especially with respect to such major themes as species formation and historical biogeography and ecology. Early in his career, Peter conducted research on diverse islands—Tres Marías, Azores, Iceland, Newfoundland, Canary Islands, and others, before he and Rosemary determined to make the Galápagos their major and eventually sole research focus, following up in particular on the work of David Lack. And, what focus it has been! Not only did they revive interest in Darwin's finches, they also demonstrated in especially vivid ways the importance of selecting the right circumstances, especially evident in the archipelago they chose, to generate and test evolutionary hypotheses. Seeing the advantages that islands offered, they specialized in extreme habitats, which they studied year after year, eventually gaining a rich understanding of the significance of climatic fluctuations on the ecological and behavioral dynamics of finch biology. In the course of this work they made formerly obscure islands, e.g., Daphne Major, famous.

TAKING EVOLUTION SERIOUSLY

Rosemary and Peter have taken evolution seriously. The finch "system" has been used to study many of the central questions in evolutionary biology. Hypotheses generated in the course of their research have been tested and the results thoroughly integrated. A good example is the manner in which their studies of the relationship of body size to beak size in finches, taking into account phylogenetic data, historical data on movements, climate variability, behavior, and genetics, have been used to develop explicit scenarios that effectively communicate their views on process and pattern. The Grants are not content to deal with phenomenology, as shown by their recent collaborations with specialists in evolutionary developmental biology to explore in great depth the developmental genetics of finch beak morphology.

Taking the Public Seriously

I greatly admire Peter and Rosemary's gracefully written book, *How and Why Species Multiply* (Princeton University Press, 2008). This summary of their work on Darwin's finches is accessible to a wide audience, including students and general readers, but professionals also have much to gain from it. The topics in the book read like the outline of a course on adaptive radiation:

The biodiversity problem and Darwin's Finches.
Origins and history.
Modes of speciation.
Colonization of an island.
Natural selection, adaptation, and evolution.
Ecological interactions.
Reproductive isolation.
Hybridization.
Species and speciation.
Reconstructing the radiation of Darwin's finches.
Facilitators of adaptive radiations.
The life history of adaptive radiations.

Each chapter is a gem. Years of fieldwork are woven into the texture of every chapter, with development of a general message at the end of each. Rosemary and Peter are gifted communicators.

Being Taken Seriously

The world at large knows the work of Peter and Rosemary, thanks in particular to the compelling story of their lives and work written by Jonathan Weiner (*Beak of the Finch*, Random House, 1995), which won a Pulitzer Prize. It is just the most evident of the many manifestations of the seriousness with which the work of Rosemary and Peter is taken. While I will not review citation indices, the Grants are among the most widely cited of evolutionary biologists.

The last few years have witnessed an outpouring of awards and recognition for both Peter and Rosemary. Among the numerous awards that Peter has received are the following elections: Fellow, Royal Society of London; Foreign Associate, National Academy of Sciences, USA; Member, American Philosophical Society; Member, American Academy of Arts and Science; Fellow, Royal Society of Canada. Rosemary is an elected Fellow of the Royal Society of London; Member, American Academy of Arts and Science; Member, American Philosophical Society; Foreign Associate, National Academy of Sciences, USA; Foreign Fellow, Royal Society of Canada. Peter has been awarded honorary doctorates by Uppsala University, McGill University, University of Zürich, and the University of San Francisco in Quito, Ecuador, and Rosemary also has honorary doctorates from McGill University, University of Zürich, and the University of San Francisco in Quito. Peter won the Brewster Medal of the American Ornithologist Union.

It is appropriate that Rosemary and Peter share a number of prestigious awards, including the Darwin Medal of the Royal Society of London, the Balzan Prize, the E. O. Wilson Naturalist Award of the American Society of Naturalists, the Loye and Alden Miller Research Award of the Cooper Ornithological Society, the Joseph Leidy Medal of the Academy of Natural Sciences of Philadelphia, the Joseph Grinnell Medal from the Museum of Vertebrate Zoology in Berkeley (especially fitting given Peter's academic lineage), the AIBS Outstanding Scientist Award, the Darwin-Wallace Silver Medal from the Linnaean Society of London, and most recently the Kyoto Prize in Basic Sciences.

Rosemary and Peter Grant—A Complete Life

Rosemary and Peter have shared a life in science. They have enjoyed a successful marriage that has led to a loving family with children and grandchildren. Admiring students have become research collaborators and friends. Their academic legacy of successful mentoring is manifest in the students who are now influential, even senior, evolutionary biologists with established legacies of their own.

Peter and Rosemary are, above all, scholars. Their scholarship is founded in field research that has integrated ecological, behavioral, genetic, developmental, and morphological data in an evolutionary context.

Finally, as is evident in the book before you, Peter and Rosemary have merited the acclaim of peers in the scientific community.

While they have now retired in the formal sense of that word, there remain lessons to be learned from finches, and the Grants continue the quiet scholarship that so many of us admire. I wish them good health and happiness in the years ahead, and I await new lessons from my friends, Rosemary and Peter.

Index